Dedicated to our children and future generations

Photo by Adriana Powers, 2016.

Praise for The Permaculture Student 2

"The Permaculture Student 2 is well written and beautifully produced. Our children especially will thank Matt Powers for this terrific contribution to the permaculture literature."

—Larry Korn, author of <u>One-Straw Revolutionary</u>, and editor of Masanobu Fukuoka's <u>The One-Straw Revolution</u> and <u>Sowing Seeds in the Desert</u>

"Well researched, articulate and engaging, the Permaculture Student Two fills a key niche in the world of permaculture education. Matt Powers has reached his goal to convey the solutions and the science behind Permaculture to the standards and audience of a K-12 environment. Indeed, this book goes even further with its well organized curriculum, to gifting permaculture to the world at large."

—Erik Ohlsen, StoryScapes & Permaculture Skills Center

"There are so many places to start once you open the can of worms that permaculture is. It can lead you in many directions, but the Permaculture Student 2 is a great place to start to get an idea of what all the possibilities are. Matt Powers does a fantastic job of piecing together all the new and exciting developments underneath this big green umbrella!"

—Curtis Stone, <u>The Urban Farmer</u>

"Matt Powers has done a great service to the future by creating such a full-spectrum analysis of the solution-based system that is called permaculture. With clear and concise explanations, *The Permaculture Student 2* is a wonderful tool to empower people with a holistic set of tools to create a regenerative future."

—Hannah Apricot Eckberg, editor and co-founder of Permaculture Magazine North America

"The Permaculture Student 2, is easy to pick up, digest, and explore further. Thanks for your work Matt."

—Javan K. Bernakevitch, <u>AllPointsDesign.ca</u> & PermacultureBC

"Matt Powers is more than a Permaculture Teacher, he's an educator. We need many more like him to reach the multitudes who have never heard of permaculture. Matt is teaching graduate level permaculture ideas to entry level students in a way that's understandable. In the end it's applicable to everyone since we all need to refresh the basics again and again."

—Stefan Sobkowiak, of Miracle Farms & *The Permaculture Orchard*

"Diving deeper into more detail and covering extra subjects in more depth - Matt has out done himself with The Permaculture Student 2, meeting and exceeding expectations from The Permaculture Student 1."

– Danial Lawton, Permaculture Tools

The Permaculture Student 2

The Textbook

the 2nd Edition

by Matt Powers

Copyright © 2017-18 Matt Powers.
All Rights Reserved.
Written by Matt Powers.
Illustrations by Wayne Fleming, Alex McVey, and Brandon Carpenter as labeled.
Cover is a combination of Photography by Matt Powers with 1 image from Grant Schultz (bottom right) and one image from Geoff Lawton (bottom left).
Charts and Diagrams by Matt Powers unless labeled otherwise.
Photography by Matt Powers unless labeled otherwise—all photos are used with author permission.
Edited by Gabrielle Harris, Matt Powers,
and our team of peer reviewers/editors.
Formatted by Matt Powers.
Send all Inquiries to:
Matt Powers
PowersPermaculture123
28419 SE 67th St
Issaquah, WA 98027

Published and Distributed by PowersPermaculture123.
ISBN-10: 0-9977043-3-0
ISBN-13: 978-0-9977043-3-4
Printed through IngramSpark.

Please Note: Just as food that nourishes one person causes an allergic reaction in another, the same concept of situational complexity applies to soils, dams, medicine, mushrooms, and more. In Permaculture, complexity is embraced with the understanding that every situation and biome is unique. The information in this book represents research from sources listed–it is an educational and informational resource and does not represent any agreement, guarantee, or promise by any party associated with the creation or editing of this book. This book and its information and sources are not designed to diagnose or treat any medical condition. Consult a licensed physician for medical treatment and advice. The publisher, editors, and author are not responsible for any negative or unintended consequences from applying or misapplying any of the information in this book.

TABLE OF CONTENTS

I.	Introduction	1
II.	Principles & Responsibilities	4
III.	Patterns of Nature	13
IV.	Our Regenerative World	20
V.	Trees	40
VI.	Water	48
VII.	Soil	82
VIII.	Fungi	131
IX.	Earthworks & Earth Resources	150
X.	Permaculture Processes & Frameworks	167
XI.	Food Forests & Gardens	193
XII.	Tropical Climates	216
XIII.	Temperate Climates	227
XIV.	Arid Climates	241
XV.	Aquaculture	253
XVI.	Alternative Energy	276
XVII.	Urban Permaculture	294
XVIII.	Permatecture	306
XIX.	Invisible Structures: Commerce, Community, & Governance	313
XX.	Regenerative Agriculture	336
XXI.	Permaculture in Action	342
XXII.	The Permaculture Lens	390
XXIII.	Glossary	396
XXIV.	Index	399
XXV.	References	402
XXVI.	Editors/Peer Reviewers	408
XXVII.	Teacher's Rationale	409
XXVIII.	About the Author	410

I. Introduction

What is Permaculture?

In the simplest terms, permaculture is a way of seeing the world through nature's eyes, a lens through which to view the world based on the three **ethics**: Earth Care, People Care, and Care of the Future. This requires constant observation and adaptation as nature is constantly changing. Permaculture is action-oriented; it provides a framework for **ecological** and **sustainable** problem solving and design. It can be applied to anything because everything is sourced in nature, its patterns, and its cycles. The way cities are designed, food is grown, homes are heated, and communities interact can all be redesigned using the permaculture lens to be sustainable, ethical, and **regenerative**. Permaculture enhances, organizes, ethically frames, and solves.

For many who are concerned about the direction our world is headed, this book will come as a great comfort. It is a roadmap to a prosperous, sustainable, healthy, and ethical future where we will have enough for everyone because everyone in their communities and **biomes** can provide for themselves in abundance.

This book is an invitation to take action in your own life to create an abundance in the environment, for yourself and your community.

Design Ethics

Every decision has to find an ethical balance, the overlap between the three ethics where all three are being served. This is the basis for all of permaculture. It is a lens through which to analyze designs, methodologies, systems, choices, and more. When balanced, designs and decisions are always beneficial to the earth and all life, including people. The ethics don't just apply to gardening or agriculture but to all systems, economies, and social settings. The way we live, think, speak, and act can all reflect these ethics.

David Holmgren and Bill Mollison, the founders of the concept of permaculture, chose these three ethics because they were the core ethics of all the longest-lasting civilizations. Earth Care was of primary importance in these cultures because they recognized that all their

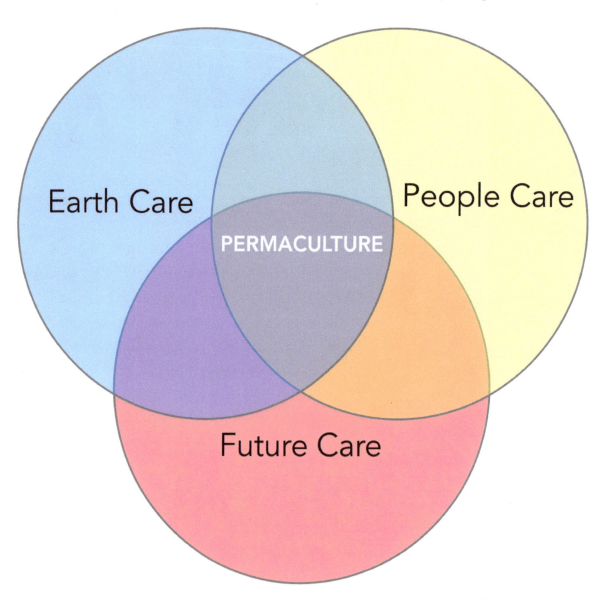

food, water, and shelter came from the earth and that they had a responsibility to behave within the boundaries of the cycles of their **bioregion**. Limits to consumption of non-renewable resources come as a natural result of Earth Care.

People Care is the both the recognition that we are all connected and rely upon each other as well as acts of **compassion** for others—in families, communities, and even strangers with food, shelter, water, clothing, kind words, and in other ways. Without People Care we cannot form safe and strong communities to raise families in.

Future Care has been referred to as Fair Share[1], Return of Surplus, and Setting Limits to Population and Consumption.[2] Caring for the future implies saving, sharing, donating, preserving, setting limits, reducing consumption, and returning both natural and man-made capital to support preserving both socioeconomic and environmental systems on into the future. On a pasture it can be as simple as spreading out the cow manure to sanitize the field and spread out the returned and changed **organic** matter. In our personal lives it could be volunteering at a local nursing home or donating to charity.

All people—past, present, and future—have relied, rely, and will rely upon the earth. It is critical that the same opportunities the earth has to offer everyone continue forward—not just for one but for all.

In our neighborhoods and communities, in our personal lives, in politics, behind closed doors, and even in our own minds, the ethics can guide us towards a better world. Every decision, interaction, and observation that flows through the ethics is refined by them. Use them liberally without fear.

The Permaculture Prime Directive

"The only ethical decision is to take responsibility for our own existence and that of our children. MAKE IT NOW"

—Bill Mollison, *Permaculture: A Designer's Manual*, 1989

---

[1] Lawton, Geoff. *The Geoff Lawton Online Permaculture Design Course*. 2014.
[2] Mollison, Bill. *Permaculture: A Designer's Manual*. 1989. p. 2.

II. Principles & Responsibilities

These **principles** and responsibilities primarily guide our interactions with nature, but they can also be used as lenses with the ethics, as a guide in other situations. There are innumerable specific principles to **microclimates** and bioregions, but these are a collection of universal principles and responsibilities—primarily sourced from _Permaculture: A Designer's Manual_ by Bill Mollison.

Principle of Observation

Observation and reflection are the greatest tools we have as humans. We can learn, adapt, and progress in our understanding limitlessly, but it can only be achieved through listening, observing, and pondering. We always run the risk of imposing solutions before we observe what the situation might actually be calling for. A lengthy observation is always better than a hasty action, especially when we are moving soil or building a home.

> **"When one observes, one learns"**
> –Hawaiian saying

Rule of Necessitous Use

We only use what we need, leaving nature alone whenever possible. We can already make sustainable living systems ethically on less land then we are currently utilizing. Despite drought, many regions are still getting enough rain—Brad Lancaster harvests enough rain in Tucson, Arizona to supply nearly all of his needs[3]. We are mining the planet's natural capital and destroying and disrupting its cycles, but if we only use what we need and in ways that serve the local ecology's needs as well, we will begin to reverse the trend.

Rules of Conservative Use

- **Reduce Waste and Pollution** - Everything is part of a cycle in process. All waste is a resource in the next step in a cycle or several cycles. It is our responsibility to cycle our waste safely and regeneratively; it helps to produce waste that can be cycled easily. Switching from plastic grocery bags to reusable natural fiber bags is a simple, easy step that dramatically changes one's impact when the natural fiber bags break down in the **compost** heap in a few weeks—whereas the plastic bags can persist for decades.

- **Restoration of Mineral and Nutrient Cycles** - Over the past century of industrial farming the perpetual tilling, expansion, and persistent use of salts to time-release fertilizers and biocides has left us with dirt: dead soils which are deficient in **soluble** nutrients and soil life. Through the return of **biodiversity** and the conservation of resources and ecologies, we can bring our soils back to life. We also need to turn around and mine the landfills we've created and begin to sort out all the resources we've mixed and to various degrees decomposed.

- **Careful Energy Accounting** - When someone buys solar panels, they often don't consider where the raw materials were harvested from, how they were harvested, how they were processed and transported, how those refined materials combined and were transported to you, and how much time and energy they will require to break down, including the panels themselves. Understanding the amount of energy it takes to complete these processes and to cycle the waste shows us the true cost and nature of that product or process.

- **Identify and Prevent Potential Long-term Negative Effects** - This may seem hard to do, but business plans in Japan traditionally looked 100 years into the future, so it's not unheard-of thinking. Generational thinking is needed and looking ahead for unforeseen problems is critical as we plan and implement our land restoration projects. We need to

---

[3] Davis, Tony. *Tucson's Rain-catching Revolution*. 2016.

look at the 50- and 100-year storms and see if our dams will be able to handle that amount of water safely. If we don't look ahead, we will find ourselves unprepared.

Principle of Cooperation

Cooperation is the name of the game in nature. It may be hard to see at times, but there is a constant exchange like a vibrant economy at work in our forests, fields, and soils. The wolves that pull down the sick or weak deer prevent disease from spreading, keep the herd healthier over time, and keep it close together and constantly moving, which is what the soil and plants prefer—and what is ideal for the deer's digestion. Everything works in cooperation: occupying niches, participating in cycles, balancing each other's populations, removing unwanted traits and trading inputs and outputs when seen from the perspective of energy, water, or the larger life cycles of that species or ecology.

> **"Cooperation, not competition, is the very basis of existing life systems and of future survival"**
> –Bill Mollison, <u>Permaculture: A Designer's Manual</u>, 1989

Permaculture Ethics in Landscape and Society

- **Care for Natural Systems** - The remaining pristine wilderness must be protected to prevent further habitat destruction and to preserve the natural systems that provide our air, soil, and water.
- **Rehabilitate Degraded or Damaged Ecosystems** - Most of the agricultural land used in the past 10,000 years is now infertile. Most of the primeval forests that dominated the landscapes of the globe are gone. In some areas we will be able to support and restore the struggling **ecosystems** while in other areas we may be reintroducing vegetation.
- **Create our own Beneficial Living Systems** - Using permaculture, we can create our own complex environments to support ourselves. Every homestead can be a beneficial microclimate of human habitation. When we take care of ourselves, we remove the pressure we are putting on the planet elsewhere.

Life Intervention Principle

Chaos creates the most opportunity to implement creative design simply because as designers, we can create order, and not in some linear, straight-lined, or square fashion, but in nature's **syntropic** concept of order which usually looks wild and disordered to the eye. From the **monoculture** fields to the chaotic ecologies coping with climate change and invasive species, they all provide an opportunity for order through a biological intervention whether it be through a human, a beaver, or termites—life is the catalyst in creating order out of chaos.

Principle of Return

As with the concept of whatever goes up must come down, this principle is about how every cycle and process in nature requires an input. Whatever we harvest, we must in turn sow. If we take, we must return.

Syntropy not Entropy

Entropy, the traditionally accepted idea that energy is always dissipating and that order is always headed towards disorder, is false in the sense that it overlooks how life systems work. Life is syntropic: it attempts to organize systems to trap and cycle energy in as many ways and as many times as possible while generating more and more life. All our systems and behaviors must be syntropic—creative designs that channel energy and generate more life.

Birch's Six Principles of Natural Systems[4]

- **Nothing in Nature Lasts Forever** - Though some trees can grow for thousands of years like olive trees that were alive during Roman times, all life reaches its peak in growth, maintains for a time, and then eventually declines into decomposition.
- **Natural Cycles Perpetuate All Life** - In the web of life, all things interact directly and indirectly through localized and larger natural cycles like the water cycle, the soil food web, and the seasons.
- **Extinction Occurs With Very High or Very Low Populations** - When vital resources become scarce in high population situations, extinctions can occur quickly like when large schools of fish get caught in shallow warm waters (that have low oxygen levels), and a mass die-off occurs. There is also a minimum genetic diversity needed to prevent extinction. With corn, you typically need a minimum of 100-200[5] plants to maintain enough genetic diversity to maintain that line and prevent inbreeding depression which makes the corn ears and kernels shrink. Some corns like Glass Gem and Painted Mountain

---

[4] Mollison, Bill. *Permaculture: A Designer's Manual*. 1989. p. 34.
[5] Deppe, Carol. *The Resilient Gardener*. 2010. p. 282

corn are genetically diverse enough that they show some resistance to inbreeding depression.

- **Every Species has Key Elements that it Depends on to Survive** - When farmers that use **biocides** focus on killing a certain plant or bug, they tend to cause a chain reaction that threatens critical pollinators and other vital cycles. Unintended consequences come into play when we don't see the full extent of relationships in an ecosystem. That is why we work with nature, observe nature, and learn from nature, because those key elements that intertwine in the ecosystems we live in, we depend upon as well.

- **Our Ability to Change the Earth Always Precedes our Ability to Foresee What the Consequences will be** - We make change faster than we can foresee results, so we must always observe, test, and plan carefully before we act. Always spend more time observing than in action.

- **All Life has Intrinsic Worth** - Everything living has a function even if we cannot readily perceive what it is. Observation and respect is needed when encountering or interacting with all life. With perhaps the exception of some humans, all biodiversity is trying to participate in the cycles of life and to create more life.

Mollisonian Permaculture Principles[6]

- **Work with Nature** - Instead of trying to control, pacify, "tame," or dominate nature, we need to work with nature and recognize the truth that we <u>are</u> nature, and our survival is dependent on our positive relationship with nature.

- **The Problem is the Solution** - Perhaps the widest-spread permaculture principle, the concept that our problems are really indicators of solutions is a powerful one. It puts a positive spin on negative situations while focusing on the systems surrounding the "problem." For example, excess manure from farm animals can become a serious problem if allowed to accumulate and then leach into groundwater, but it also can be a high quality soil amendment if properly composted. The toxic situation could be growing us healthy food instead.

- **The Least Changes for the Longest Term Effects** - To maximize our effect and minimize our efforts, we need to seek the longest term and most ethical solution that requires the least amount of input energy and maintenance. When we route our washing machine's **graywater** into our yard (while using safe non-toxic detergent), we save an immense amount of water and irrigate our gardens routinely.

---

[6] Mollison, Bill. *Permaculture: A Designer's Manual*. 1989. p. 15-16.

- ***Yields are Limitless*** - Only our imagination limits our ability to find more yields from a finite system. It is often described how Native Americans traditionally used the entire deer, and nothing went to waste. This abundance and efficiency perspective leads to viewing the world differently. Yields are also limitless in the sense that through stacking functions, they can diversify nearly endlessly even in a finite system. Total yields must be accounted for, not just individual yields.
- ***Everything Gardens*** - This can be easily observed in nature. The moles **aerate** the soils, redistribute seed, and fertilize it like tiny blind subterranean farmers. Birds spread seeds to the edges of forests and expand the system, as do deer and many other animals. All these animals fertilize the soils around the plants they eat from. Even non-living elements like the wind and water cycle contribute to the spread of more life. Everything works together in a symphony of gardening and food forestry (even if it's not always food humans can eat).

The Principle of Empowerment

The best natural systems are self-managed and need no human intervention. To relinquish power to the people, the land to the wildlife, to share responsibility, this empowers through **decentralization** and makes stronger, more resilient ecosystems, countries, companies, families, and neighborhoods.

Principle of Purity and Preservation

No methods can be used that will degrade, taint, or completely consume any resource. Because high-pressure hydraulic fracking for natural gas forces **carcinogenic** chemicals into the **water tables** permanently, it should be banned or boycotted when seen from this perspective. There is plenty of methane gas produced from our waste and our cattle's waste to provide all the gas we need. We need to preserve our limited stores of freshwater and keep them pure. This same principle can be applied to our soil, air, bodies, cities, forests, and more.

Principle of Choice

Dictating behavior or preventing natural behavior, both prevent choice, and, in natural systems, prevent systems from thriving. In schooling, student-centered learning, the practice of giving students choice in their studies, is well recognized for its power to boost engagement and achievement. The same concept carries in biological systems; the highest levels of vigor and **self-reliance** are found in wild systems.

Principle of Stability

Though more diversity is ideal, the **mutualistic** connections between elements in the ecosystem are the actual source of benefit. If we combined unrelated or redundant components in a system, we may have a high number of different living elements, but perhaps a low number of interactions or a high number of negative interactions. Often systems have a set of keystone species of plants and animals that all the rest rely upon. Everything interacts through the web of life and therefore holds intrinsic value, but certain associations are more beneficial than others, and we can select and combine them for greater stability.

Principle of Self Regulation

Self regulating systems are more stable and long lasting than systems maintained by inputs and managed by people. This only happens when elements are aligned to work together. In the wild, deer and wolf populations self-regulate through cyclical rises and falls in both populations in tandem. When one element is removed, like the wolves, it sends others into a state of imbalance.

Definition of System Yield

The concept of yield is not limited to what we harvest from a single crop; it is the total energy produced, recycled, reused, converted, trapped, and conserved over a period of time —taking into account the energy consumed by the processes. Only when we see a **holistic** picture of energy-in and energy-out can we see how efficient and effective a system is ecologically. For instance, a farmer's yield from his orchard goes beyond the apples harvested. If one farmer's rows are organized by ripening dates while another's are all mixed, their yields may differ significantly even if they sell the same amount of apples at the same price. The organized farmer saves in energy and gains an extra yield: more time. The freedom that their organization affords has its own psychological yields, such as greater contentment and correlated social yields as well like more family time or just richer relationships.

The Role of Life in Yields

All yields in nature come from biological processes. Without life, the sand, clay, and silt will never break down into soluble minerals and nutrients that plants and animals can absorb. The efficiency and diversity of the biology in a system determines how high yields will be.

Global Threats to Life Systems

- ***Soil Degradation*** - Soil loss through erosion, tilling, overgrazing, poor design, or natural means is happening at an alarming rate though exact numbers are notoriously difficult to calculate in natural systems, so there are only estimates. What is clear is that our agricultural and natural soils are all eroding and degrading observably and quickly.
- ***Deforestation*** - The complete removal of trees has a holistically negative impact on every ecosystem treated this way. Trees are long-term carbon traps. They turn water, sunlight, and carbon dioxide into carbohydrates and oxygen. They provide food, habitat, shade, water vapor, mulch, building materials, and more intangible yields like places to climb, play, explore, and observe. They make rain by condensing the air above them and creating ice **nuclei** with their dust and leaf litter. They are the lynchpins for human health and survival on this planet. Without forests we cannot clean our air and water naturally, and we can only do those things for a short time mechanically.
- ***Pollution*** - Toxins and pollutants that cannot be cycled easily or at all should be boycotted, refused, and rejected along with any systems that use them or produce them. The pollution we have currently can be remediated with living systems like **mycoremediation**, reed beds, algae, or microbiology, but the daily release and constant accumulation of pollution must end.
- ***Water Scarcity*** - Water is life as much or more than the soil is. Without them, we cannot sustain living systems on land. Water scarcity and water pollution are together making fresh water sources for people and living systems exponentially harder to find. **Desertification**, climate change, deforestation, watershed disruption, unsustainable drawing down of **aquifers**, and poor planning of human settlements are all combining to make fresh water less available. The water is still in the system but primarily located in the oceans and atmosphere. We need to restore our watersheds, bring back the forests, and recharge our aquifers. Only then will we see the groundwater return abundantly.
- ***Ocean Acidification*** - With carbon dioxide levels rising in the atmosphere, diffusion of carbon dioxide into ocean waters rises at similar rates causing **acidification** of ocean waters. This causes invertebrates to have weaker shells and coral to struggle to properly form their structures. It is affecting all levels of the food chain. With two thirds of the world's oxygen coming from phytoplankton, it is critical that these sensitive organisms maintain their populations.

- **Mass Extinction** - Related to climate change and the actions of humans, animals and plants are going extinct at an alarming rate all over the world with some estimates at 100 to 1000 times the pre-human natural extinction rates.[7]
- **Climate Change** - As the carbon from our soils and forests gets washed out, combusted, **oxidized**, or blown away, it acidifies our oceans and acts as a greenhouse gas in the atmosphere. Water is not being held on the land in forests or soils any longer, leading to desertification and water scarcity. The derailment of the carbon cycle has to end, and a carbon sequestration plan must be adopted to fight climate change realistically. Only then will the water and carbon cycles return to normalcy.
- **Human Disconnection** - Social, economic, and political conflicts, though they have long defined our history, are always caused by some form of disconnection from social groups and/or nature. It is this disconnection that allows for short-term thinking and ecologically and socially destructive practices in business, design, or society. It is also the same disconnection that allows for violence, sexism, racism, classism, and all forms of prejudice. Connection leads to an appreciation of what you are interacting with, and that gratitude leads to compassion and solution-thinking. It all starts with our relationship to nature, ourselves, and each other.

---

[7] Pimm, S.L., Russell, G.J., Gittleman, J.L., and Brooks, T.M. *The Future of Biodiversity*. 1995. p. 347.

III. Patterns of Nature

Memorable, repeatable, and found in our languages and cultures everywhere, patterns help us make sense of our world and ourselves. It is a collection of patterns that work together that form our ecosystems, our bodies, and our thoughts. We design in patterns to create beneficially interrelated systems. Patterning is good design.

Patterning is clearly seen in nature, but what isn't readily seen is perfection. Euclidian geometric perfection is not found in nature; nothing upon close examination is perfect. There is nothing perfectly flat, round, square, or straight in nature. We interpret nature with patterns, and all life replicates and thrives using patterns. If we recognize the patterns that exist, how they are used, when they are used, and where they are used, we can apply those patterns with relative assurance that they will behave predictably or "naturally." Our natural systems and patterns shape the world around us from changes in season, to the way snow melt travels down a slope, or to the way life reproduces; it's all pattern-based, consistent, and predictable.

How to Describe Patterns

Patterns are useful for designers and educators, but are infinite in form, so they don't fit easily into Euclidian geometric thinking. The simple pattern groupings we cover here are generalized abstractions of pattern expression but allow us to identify them in nature and use them in design. Patterning is vital to creating planting guilds and scaling up any design; it is the application of a pattern. Pattern literacy is our ability to read patterns and their interrelations. We need to know both to be good designers. Patterns are formed when two or more media interact. There are endless patterns, but here are some general ways of speaking about patterns.

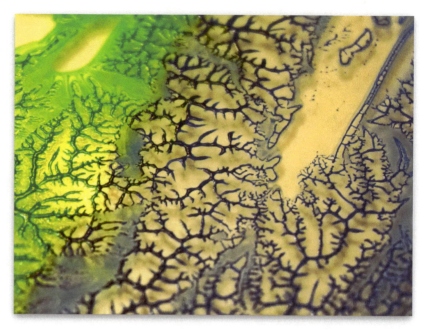

Dendritic patterns distribute, transport, and accumulate.

- **Shapes** - Their general outline
- **Branching** - The dendritic focusing and spreading of water, veins, shattered glass, or roots
- **Pulsing** - Repetitious patterning—waves, the winds, tides, and heartbeats
- **Scattering** - From wind or water, the drift and spread by pulsing stimuli like debris from a storm or clumps of trees in a windy region
- **Wave** - Created by pulsing, creates a flow between two media like waves in water or sand
- **Streamlines** - Created by forces over time like wind or water over rock, slowly shaping it with erosion, like a shoreline
- **Matrices** - Patterns that interconnect in a repeating pattern that spreads like honeycomb cells, cracks like netting in mud, or snake scales
- **Cloud Forms** - An explosion shape like tree crowns, puffy clouds, or mushroom forms like a

Cracking or netting in matrices conserves space and resources especially as they deplete.

A Torus

torus (many patterns are found inside the torus)
- **Spirals** - Found in sunflowers, galaxies, and whirlpools
- **Lobes** - Rounded edges of reefs or lichens[8]

Turkey Tail Mushrooms growing in a lobed pattern.

The Tree Pattern

When two media interact they often create a tree pattern: atomic bombs, mushrooms, bones, erosion, deltas, communication networks, and trees themselves are examples. Often the ends of these patterns are intensely intricate, dendritic, spherical, or spiral in expression. The chaos that occurs at the edge of this expression is usually where the most energy is manifesting. The fruit, the seed, the tender leaves, the initial intensity of an explosion, and mushrooms are all examples of the final expression of the tree pattern's behavior. Where the freshwater delta meets the saltwater edge, it is the most prolific even though it is the end of two different aquatic systems. Once we recognize a pattern, we can then begin to analyze its functions.

Edge Effect

Edge effect is pervasive. Once you begin seeing it, you can't stop. It's everywhere. It's in nature as much as it is in people systems. The idea is simple: where two different ecosystems meet, you have a multiplier effect of species and interactions. The two sets of species combine with a third set, the species limited to the edge ecosystem, the edge

---
[8] Mollison, Bill. *Permaculture: A Designer's Manual*. 1989. p. 72.

species. These can be animals, plants, fungi, bacteria, or even people. We often think of successful people as having broad appeal. They can interact with multiple groups but are of none exclusively; they are edge species.

Edges are also the most productive areas of nature. Estuaries, forest edges, along river sides, and coral reefs are all teeming with life greater than the two meeting ecosystems support independently. Increasing edges in a design increases habitat which increases the biodiversity and cycling of nutrients. It increases water catchment, accelerates the accumulation of organic matter, and generates higher yields.

By increasing the stress (or pressure), we can increase yields. Trees thicken their trunks in response to wind. The larger the windbreak, the more wind-borne nutrients, organic matter, seeds, and insects are deposited. The more biodiversity cycling through an area, the more organic matter accumulates and is cycled into **humus**.

We can physically increase the amount of edge in a system for higher yields as well. An undulating pattern increases edge interaction in a limited space where straight lines, being the shortest distance between two points, limit the edge effect. Human civilization often occupied these edge zones in nature: flood plains, river deltas, and coastal regions.

Even as as we increase yields, we are also boosting biodiversity by attracting new species through an increase in habitats using edges. We can further encourage such diversity by allowing for variance in the physical environment–everything, from the size of the holes of our fencing to the density of the trees or the types of trees we plant. We can build habitat diversity into our design interventions at every level using edges, from the shaping of the terrain with curved lines for smooth airflow, to creating deliberately jagged angles for greater disturbance, more shelter, and more significant airlift. These kinds of changes can lead to unanticipated yields that take time to observe, especially if they bring unique or rare kinds of yields.

Edge Cropping

We can use the edge effect to generate a yield rather easily. Through alternating large rows or lanes of a field growing annual crops and **perennials** or tree crops, we create a constant edge effect between the annual beds (that are disturbed regularly and often closer to the ground) and the perennial beds (which are not disturbed and grow tall and form a windbreak and create shade.) This system can be created on **contour**, in a **keyline** pattern, or it can be used to create even more edge by making the lines wavy. There is actually more room in a field of wavy rows than in straight rows.

By researching our plants and carefully matching them up with beneficial plants, we can prevent **allelopathic** plants from harming other plants. The juglone of a black walnut tree doesn't bother a mulberry tree though that is not the case for a peach tree. Luckily a mulberry tree can grow between them and prevent the allelopathy from reaching the peach tree through the **rhizosphere**. Beyond just providing a buffer, we can maximize synergy by pairing up plants that increase each other's yields and benefit the soil. Using **legumes** that fix atmospheric nitrogen in the soil, we can passively feed our fruit trees the organic matter and nitrogen they need to produce a regular and robust yield.

Alley cropping, a system that grows crops between rows of trees, is a form of edge cropping usually done on contour or in a keyline design; it is also known as **agroforestry**. It is fast becoming a recognized and proven way to retain soils, creating shade and mulch with plants on-site. **Silvopasture** is similar in that it alternates trees and pasture strips, using grazers and browsers to simultaneously prune the trees on the edge as they graze down the pasture strips. In both systems, trees can create too much shade over time, making it hard for annuals to grow or pasture to get enough light. This makes these areas perfect for growing mushrooms! Peter McCoy calls it "Fungalley Cropping" in Radical Mycology. The alternative is to prune to let in light.

Nature Strives for Balance

The drive for equality is as natural as the temperature between two media seeking equilibrium. In nature all differences strive for equilibrium. It could be the difference in **salinity**, temperature, moisture, or anything; when elements with such differences meet, equalizing begins. This principle of balance is key to design and planting guilds. Knowing this and managing it are two different things; every change we make will affect everything else in a system to varying degrees and over time. Every change creates greater complexity. If we have fertility in the soil, adequate water available throughout the growing season, and the right seeds, we can set the stage for nature to determine the balance in an area. Our role after that is to observe how nature is trying to balance the ecosystem and support those elements or cycles.

Spirals in Nature

We can observe spirals nearly everywhere, from the spinning turbulence caused by large rocks in a river, to the spirals of air caused by windbreaks that carry leaf litter into the sky and assist in forming rain. Spirals are found in galaxies, whirlpools, and in sap flows. Often these spirals look like a perfect Fibonacci sequence, but even these are imperfect copies of a general spiral pattern, or an adaptation thereof. We can use this predictable form to aerate, mix, maximize minimal space, focus accumulation, and increase biological interaction.

Applying Patterns

Throughout time, humans have used patterns in various ways to interact with each other and the natural world. We've crafted art, images, songs, and dances to house critical information about medicine, weather, history, humanity, nature, and more.

Where we don't readily see patterns, we create them to create meaning and understanding. For instance, we navigated the stars using constellations we invented to make the vast array readable. We also used the currents, the winds, water temperature, and more to understand the greater patterns at work in the oceans and weather. Not only does pattern recognition underpin maritime navigation, but it is also the foundation for all disciplines.

Pattern literacy education is necessary for our more modernized cultures to renew their understanding of patterns. All pre-modern cultures used patterns to maintain cultural knowledge. It is imperative we craft new songs, dances, and other forms of art that will store our understanding of our world's patterns and pass them down to the next generation in enjoyable, memorable ways.

Lastly, in our systems when we apply patterns, we must make sure we apply them to support and enhance the ecology; even though spirals and matrices are beautiful to look at, we must always make sure they have an environmental benefit.

Often when enough observation and reflection is made, patterns that are already present emerge. In this case, we as designers are aligning our designs to the natural patterns already at work on that site.

V. Our Regenerative World

Climate

Climatic factors determine everything we do or place on a site we are designing. They will influence every decision. If you mix up your climatic applications, your system will fail: a Cavendish banana plantation cannot grow in the cold temperate zones, for instance. This section will acquaint you with what you need to know for each general major climate zone and for some minor climate zones. There are also the local climatic and orientational factors that strongly influence design specifics on-site: the slope, watershed orientation, distance from large bodies of water, elevation, wind, sun path, and fire dangers. Beyond strictly environmental factors of climate, there is also the social climate and our own emotional, physical, and mental climate. These factors can sometimes influence what we do on a site as much or more than ecological factors.

Climatic factors are very complex and influenced by both regular and irregular factors like longer-term cycles of the sun, moon, and earth, atmospheric factors like volcanoes and pollution, and extra-terrestrial factors like meteors and solar flares. There is an accepted, general consensus in the scientific community that there is an observable rise in more violent and erratic weather patterns like flood, hurricane, and storms, as well as an unprecedented rise in global temperatures. It is best to plan for variable weather and temperature patterns by designing with diversity; climate change is already here and disrupting the patterns that we've come to rely upon. We can mitigate these extremes and begin to restore our atmospheric systems to normalcy through large-scale land restoration projects.

Understanding the historical highs and lows, the extremes over time of rain, wind, temperature, flood, migration, etc., is vital. This gives us a clear list of things to expect or to investigate for their absence. For example, the largest historical rain event is used to design all roads, dams, catchments, sills, and swales; we need to know how much rain the swales or dams should be able to hold. We can source libraries, elderly members of the community, town record halls, local historians, or even the internet. The information is usually available in your area through one of those channels.

Broad Climatic Zones

The broad zones below are meant to help guide us rather than restrict us. We may find our site is a mix of SubTropical and Mediterranean Temperate. Arid and desert zones are found throughout all climates as well. Looking at the climate map we can readily see that these zones are not evenly distributed, with the most obvious imbalance being the northern hemisphere containing most of the world's temperate climate landmass. We can also spot our area and its color and search out our **climate analogs** all over the world.

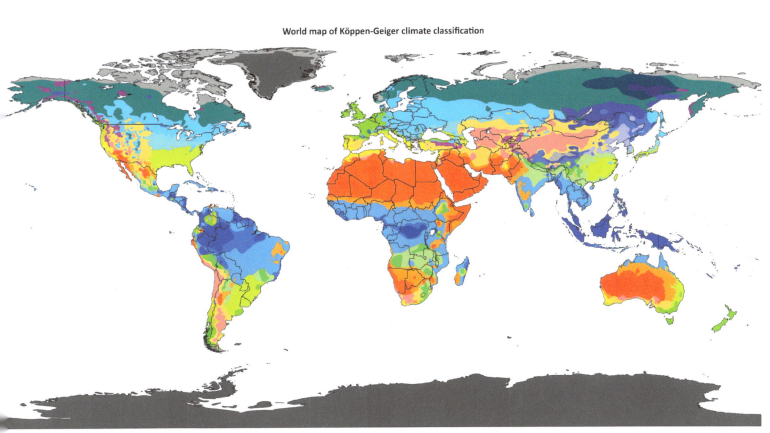

World map of Köppen-Geiger climate classification

Peel, M. C., Finlayson, B. L., and McMahon, T. A. Wikimedia, 2016

- **Tropics** - At and close to the equator with temps above 18°C/64°F year round
- **SubTropics** - Extends away from the equator to the Tropics of Cancer and Capricorn with temps never below 0°C/32°F
- **Temperate** (*Mediterranean-Warm-Cool-Cold*) - With temps below 0°C/32°F in winter and above 10°C/50°F in summer—this zone extends from the Tropics to the Arctic and Antarctic circles.
- **Polar** - Below 10°C/50°F year round, dominated by ice and snow
- **Arid** - 50cm/19.5" or less of rain a year
- **Desert** - 25cm/10" or less of rain a year[9]

Precipitation

Every climate has fluctuations and cycles that affect its precipitation; no average is reliably true in nature, only a generalized rule of thumb. We should always harvest rain even in areas with regular precipitation because of seasonal, man-made, or cyclical droughts. Researching drought intensity and duration can help us be prepared and design with these specifics in mind. We should look into the extremes of precipitation to understand how much rain we could get in a 50- or 100-year storm event; this will help us design our dams, **swales**, homes and gardens. Understanding the pressure on our dam walls is critical to protecting our homes, buildings, gardens, animals, and others from flood damage or worse. Consulting professionals when constructing dams, buildings, and anything that could be potentially hazardous is always wise.

The distribution of the precipitation over the course of the year is important to understand. In some areas, they get most of their rain in only a few days. Intensity of rains can also increase flooding, especially during the longer cycle storms like a 50- or 100-year storm. The intensity of precipitation influences how we design our roofs, roads, swales, dams, and more. Measuring the water that falls is a simple and easy way to calculate how much moisture an area receives, but it omits condensation's contribution which is often difficult to calculate. In this area, do your best and always accommodate for a margin of error. Build things a bit big, buy a larger storage tank for the rainwater, etc. Be prepared for excess precipitation especially as groundwater and ground ice continue to end up in the oceans and the skies.

Some areas are naturally drier. The constant upwelling of evaporation at the equator leads to rising columns of warm moisture in a spiraling cycle that generates predictable atmospheric patterns that are influenced by the shifting tilt of the planet; these patterns are called **Hadley Cells**. These atmospheric patterns tend to create areas of shifting droughts

---

[9] Ibid. p. 107.

and belts of **brittle** climates across the globe, 15-30° degrees from the equator.[10] Desert belts across the world—the Sahara, the American southwest, parts of Australia and Peru, the Middle East are all influenced by this effect.

In areas with only coastal fog, precipitation doesn't reach far enough into the land's interior as is the case with the Nimib desert.[11] The fog is caused by cold ocean currents meeting the shoreline and leads to a rich maritime ecology but a desert inland beyond the coastal fog's reach.

Orographic Effect

There are specific effects related to rises in elevation that are observable. Air flows cool condensing moisture as they rise over mountains and hills—this increases condensation which increases cloud formation and all forms of precipitation. In areas with a prevailing wind, storms tend to drop the majority of their moisture on the windy side and very little on the sheltered side because of the rise in altitude condensing the moisture in the airflows. This is called a **rain shadow** and is easily recognizable. The eastern side of the Sierra Nevada mountain range in California is visibly drier than the west side when seen from satellite or witnessed first hand (the **jet stream** flows west to east).

The Sierra Nevadas are a clear example of the rain shadow effect.

In addition, as you move up in altitude, it is the equivalent of moving up through progressively colder climate zones, even at the equator. The sun path and day length are still those of the equator, however, so transferring foods and seeds from a cold temperate zone to a mountainous tropical region (and vice-versa) doesn't work out perfectly. The seeds are all photoperiodic-specific to their original bioregion. The adaptation can be made but takes technique, patience, and work.

---

[10] Spackman, Neal. *10 Keys for Greening Any Desert*. 2016.
[11] Ibid.

Cyclonic Effect

When hot and cold media meet, especially in the atmosphere, they form cyclones of rising hot air and falling cold air. This can be seen in the form of a hurricane or tornado. This is also occurring in liquids that are stirred into a whirlpool–the water is being pulled up from the bottom as it is being pulled down from the top. The Archimedes Screw Pump is an excellent example of the cyclonic effect being applied to serve human needs.

Archimedes' screw. Public Domain. Wikimedia, 2007.

Convection Effect

Hot air rises from the desert or ocean into the upper atmosphere where it cools. Thinking this way illustrates quickly how we can use this effect to diffuse, concentrate, or channel hot or cold air by design.

Joel Salatin, the regenerative farmer, has portable chicken houses that have openings up top and down low to allow for hot air to rise and for cool air to be pulled in. This creates a consistent flow of air which cools and clears the air which both the chickens and the farm interns benefit from.

Dew

Dews are commonly caused by clear nights, rapid heat loss, and atmospheric moisture settling over coasts and hills. Dew coats plants and surfaces with moisture. As the sun rises, its warmth begins to evaporate the dew; a significant amount of moisture is soaked into the soil, consumed by organisms, and absorbed by plants. These processes are difficult to accurately measure as the functions are stacked, happening in real time, and outside of a laboratory.

We've yet to design machines as efficient, sophisticated, and reliable as anything found in natural systems and cycles.

Distillation

Water vapor rises from the ground into the atmosphere but can collect on all the leaf surfaces as it passes. If we dripline irrigate, even below the surface, we can rest assured that the vapor will distill on the plants above the surface. Having a closed canopy, having multiple layers in our food forest or agricultural system, and using shade, we can recapture moisture repeatedly as it cycles.

Fog

Often carrying more moisture than dew, fog can provide enough moisture to support life, even on islands without bodies of water or with low rainfall. Fog is a critical source of water that can be harvested carefully using trees and water catchment hardscapes like stone cisterns or, in more recent times, mesh fencing with gutters at their bases that drain into cisterns or water tanks.

Ales Krivec. unsplash.com. 2014.

- ***Radiation Ground Fog*** - In hollows or gullies where it is naturally shaded or cooler, fogs (and frosts) form
- ***Advection Fog*** - When warm air from the sea meets cold winds from inland, they form a thick fog that many a port city is often shrouded in
- ***Upslope Fog*** - Warm, humid airstreams rise and condense as the air cools (part of the orographic effect)

Condensation

When warm air cools, it drops the moisture it was holding onto surfaces. Water vapor can collect on surfaces from fog or even clear air when there is a cooling of the night air. This moisture is very difficult to calculate since it is vapor collecting on all exposed plant parts and, at the same time, being absorbed. In design, a common condensation trap is to arrange stones in a circle around a tree along the tree's dripline or strategically place small piles of stone to condense moisture. Often worms, attracted by this coolness, will begin working at the soil below these stones.

Moisture in the air collects on all the exposed plant surfaces especially in areas of fog.

Radiation

Solar Radiation

Solar radiation is the sun's energy which is mostly filtered by the atmosphere and found in excess primarily at the poles and high altitudes. The atmosphere filters the sun's energy perfectly for life systems to use; it is thinner higher up and thicker closer to sea level. From plants to animals, we all benefit from the stable temperature ranges the atmosphere and sun's energy together create. The sun's light is necessary for human health as well for the well-being of most ecosystems, plants, and animals. The tilting of our planet throughout the year creates a wax and wane to the solar radiation. This pulsing effect of more and less solar radiation is part of the most basic pattern for all life on earth. It directs the flowering of plants and the germination of seed. It powers our weather, seasons, and ocean systems. Solar radiation is the source of all energy on earth—even the earth's core is powered by the sun's energy.

Color

The darker the color, the more absorption of light and greater radiation release of heat while the lighter the color the more reflection and less absorption and later release of heat. We can observe this when we enter a black vs a white car that has been parked in the summer sun for several hours. If there is a black interior, it might even be too hot to touch. A home built in the desert that attracts heat with a dark coloring will store heat and give it off during the night while a home that is painted white will block heat and therefore not absorb and give off as much heat. Often greenhouses in cold temperate climates use black water barrels against the back wall of the greenhouse to store solar energy from the daytime and radiate it back out overnight.

Albedo Effect and Absorption

Albedo is the effect of light reflecting off natural surfaces and plants. This can be the winter snowpack atop Yosemite's majestic Half Dome or a white wall on a house. Light that is absorbed is turned into heat that is conducted and radiated outward. We can use this understanding to filter, reflect, absorb, focus, or transmit the solar energy we receive naturally.

Heat

- **Convection** - A low-grade heat that travels in liquids or gases such as air, wind, or water. The equivalent to this in the home would be an electric heater which blows hot air. It is an inefficient way of heating a home by focusing on heating the air instead of the objects themselves.
- **Conduction** - An intermediate-grade heat that travels solid-to-solid or fluid-to-fluid by contact. This can be the most effective way to transfer heat. This is the fundamental basis of the **rocket mass heater**; it conducts the heat through a mass.
- **Radiation** - High-grade heat that heats solids and liquids but affects air very little.

Thermosiphon

Applying heat to the bottom of a loop or chimney causes the air in the loop or chimney to move, allowing for the heat to rise using the convection principle. We can use this principle to draw out heat from an area as well. A **solar pump** or solar chimney is a tall black pipe that extends much higher than the house. It collects heat in the sun all day, drawing out heat from the house with the intensity of the heat at the top of the pipe. It creates a suction by forcing the air in the pipe to rise (similar to a rocket stove).

We can also fan hot air into cool sinks like a sunken path in a greenhouse. This air exchange is healthy for the plants as well. Glasshouses attached to homes can also serve as thermosiphons to warm a home in a cold climate (as seen in the diagram).

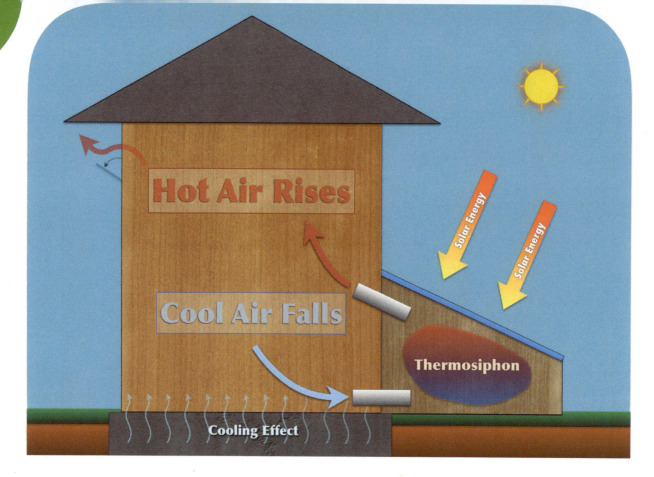

Heat Factors

- **Latitude and Season** - Distance from the equator and the time of year.
- **Slope Angle** - Facing the sun straight on (or perpendicularly) we absorb the most solar radiation; the greater the angle of the slope—the more indirect light, the less we absorb. Sun-facing slopes receive more solar radiation than slopes that face away from the sun.
- **Air Clarity** - The amount of dust, ice, vapor, or smoke in the air affects the capacity for heat to be absorbed or released. A cloud of hot volcanic ash can have a cooling effect by blocking the solar radiation.
- **Advection** - Winds bringing cooler or warmer air to areas of different temperature. Warm air that enters cooler regions creates condensation, as it does in forests.

The Thermal Belt

A strategy for many temperate climates and areas that are not too hot is to build home sites on sun facing slopes below hilltops and above the valleys and plains. Cold night air leaves quickly in direct early sun while the soil retains the sun's heat

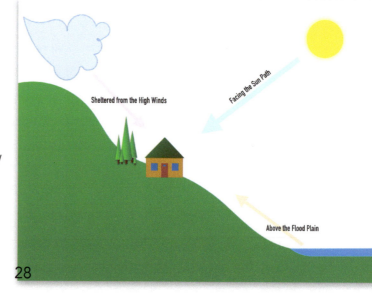

longer, extending warmth into the night. Frost and cold nighttime air lakes form in the valleys and gorges. Ridges and hilltops are windy. On hilltops water has to be pumped up to homes with windbreaks and extra insulation needed to keep houses warm. To save energy and time, plan to build in the thermal belt, or its equivalent, on your property.

Frost

When air rapidly cools over a still spot, frost can occur. To prevent frost, have steep hollows or steep-sided holes that are thinner than they are tall. The inverse creates frost: large open fields or holes with open sides that invite the wind. In design, holes can be dug with steep sides to prevent frost falling on sensitive plants, or clearings in forest can be narrow to prevent frost (half as wide as tall) and retain warmth.

Insulation

Insulation (earth, brick, stone, fiber, straw-bale, cob, or concrete) can prevent heat from being lost by blocking conduction into the ground or radiation into the air. The thickness of straw bale home walls provides more insulation than most conventional homes; this extra insulation lowers heating bills dramatically by greatly preventing conduction and radiation loss.

Heat Absorption by Plants

High biomass plants like trees can hold and radiate captured solar radiation throughout the cooler parts of the day and night . However, this effect varies depending on the season and the tree variety. Large, dark evergreens trap and exude heat while white birch trees reflect the heat. Trees with more leaves and branches absorb more heat than trees with fewer leaves. The opposite is also true: the larger the tree, the greater the shelter and shade it provides which is producing a cooling effect. It all depends on the tree, time of day, and the season.

Water, Earth, and Stone

Water, earth, and stone store heat well. We can capture heat from the sun in water, earth, and stone and use it to warm cold areas or focus the heat into a microclimate. We can also reflect and block solar radiation or allow light further into the home as well. We can even insulate the thermal masses to slow conduction and radiation loss.

Wind

Winds have regular patterns and pulses that vary according to their location on the globe, the season, and local effects such as topography. On slopes hot winds rise uphill during the day and cool winds fall downhill during the night. Using windbreaks and wind funnels of trees and/or berms, designers can block, lessen, or increase wind on a site thereby increasing or decreasing the amount warmth held in that site. In areas of intense winds, all design considerations come after initial windbreak design. Crops and animals are also affected by wind intensity. Winds can carry salt, sand, nutrients, pollutants, organisms, and other particles from one area to another which can have negative or positive effects on plant and animals systems. Many coastal, island, sub-tropical, and exposed sloped regions are subject to wind's cooling and erosive effects.

In some areas, hardy, wind-tolerant species arranged in a windbreak to shelter a garden from sand or salt are the only way to grow food. Windbreak is essential in these areas where heat- or light-related transpiration is already occurring. The addition of wind may well kill or stymie the growth of the plant. Windbreaks create still, protected microclimates in which more tender plants can thrive.

Tree Flagging

Though winds can be seasonally or daily bimodal, we can determine the prevailing wind by the shape of trees and specifically the branch crown in relation to the trunk placement (see picture). The wind literally pushes the tree crown away from the trunk to varying degrees which tells you the severity of the wind in that area. You can tell a lot about a site from the wind and water patterns that are already in place before a design even begins. When we are to thin an area, we should spend time examining the trees first. Identify the trees you wish to cut and make sure you aren't creating a wind tunnel or a frost pocket. Use the natural patterns to influence your design; whether it be the final design, thinning a forest, or building a house, look for indicators like tree flagging to tell you the story of that place.

Coastal winds can cause stark tree flagging. This is near Monterey, California.

Building Windbreak

- Research hardy, native species that are commonly found in local high wind areas including all layers from grasses to canopy trees.
- From this plant list choose the most beneficial **polyculture** to the system both from an ecological and human usage perspective.
- Use these trees and, if feasible, fencing, earthworks, or even stone to create turbulence and push the air upward while slowing the infiltrating air.
- In areas where establishment is difficult, layers of wind-hardy trees may be needed to create a fully protected inner zone. Some areas may need a salt-resistant layer first to withstand the desiccating sea breezes or desert sandstorms.
- In high wind areas, regular windbreaks within cropping systems may be needed to keep winds at bay.

Permeability Effects

While a full stop from a physical barrier causes intense turbulence in the air and frost pockets, allowing some of the winds to permeate through the windbreak is the most beneficial and influential arrangement possible. It pushes air upward, condensing the air as well as the moisture in that air which leads to increased precipitation. Warm winds entering a windbreak or forest cool and slow, allowing the water vapor they carried to fall and condense on the surfaces around them. When frozen precipitation falls, it can pile up and make windbreaks more effective. Permeability of windbreaks can also affect how much and what kinds of organic matter and nutrients accumulate as slowing winds drop whatever silt, seeds, sand, or organic matter they carry. Therefore, permeability affects how fast topsoils and organic matter content increase.

Hurricanes, Cyclones and Typhoons

Still air near the equator causes violent up-draughts of air over warm sections of ocean, creating the large, classic hurricane spiral which is drawn to land. The direction spirals turn in depends on the hemisphere (northern - counter-clockwise, southern - clockwise). Tornadoes are vortices of air that occur when hot and cold air meet, usually during a thunderstorm and can be very localized, dangerous, and brief. Tornadoes can happen inland and are not dependent on a temperature threshold. Cyclones occur in the Pacific ocean, and though less violent, focused, and intense, they are similar in behavior, destructive nature, and function to tornados (in certain areas intense Cyclones are reclassified as Typhoons).

Though not water-related, firestorms look and behave similarly to a cyclone. Dry winds and fire combine into strong vortices as intense heat pulls upward while dry cold air pulls downward to fill in the vacuum, creating a tight spiral of air. They are literally like tornadoes of fire.

A hurricane seen from space.

Landscape Effects

Continental Effects

The third most powerful effect on climate after temperature and rainfall is the effect caused by the distance a site is from the ocean or large body of water. Large bodies of water have a mitigating effect on the localized atmosphere, lessening the cold of winter and the heat of summer. The inverse is also true: the further you are away from bodies of water, the deeper the summer heat and winter cold. We observe this readily in the United States' Midwest.

Altitudinal Effects

Georg Nietsch. unsplash.com. 2015.

- As we rise in altitude, the relative climate changes even though our distance from the equator does not. Islands with high altitudes can cover a diverse range of climate-specific species, however it is not an exact analog to similar areas. At higher elevations, air is more rarified, the radiation higher, the precipitation higher, the temperature changes between day and night more drastic, and the air pressure lower than their lowland equivalents further from the equator.
- Snow cover can create a warm layer above the snow during the day in the sunlight by reflection. Inversely, it can create a very cold layer of air at night just above it, even though beneath the snow layer, a steady temperature of 32°F/0°C is maintained. Snow's high albedo reflects radiation efficiently which compounds the solar energy that plants, land, and animals encounter.
- Slopes away from the sun also have a pseudo-altitudinal or -latitudinal effect to varying degrees (like a deep crevasse that never sees the sun), but again, it is not quite a perfect

analog. The inverse is also true, wide valleys facing the solar path can trap heat, be drier, and generate strong winds (as warm air drafts rise). This creates a large microclimate that would mimic a less favorably structured valley closer to the equator. Cumulus clouds can be created this way in moist temperate climates. As you travel closer to the equator the clouds become more profuse until they remain clinging to the mountain tops and ridges.

Latitudinal Effects

Sub-polar latitudes have long day length during summer which compensates for their overall lack of sunlight during winter. Photosynthesis is curtailed by lack of light and heat during cold months. Certain sub-polar latitudes are fed by warm air and water currents that make for superb growing seasons creating their own large-scale microclimates (Norway, Scotland, Ireland, and Alaska are examples). This is the reason for Alaska's giant garden vegetables (and not Alaska's soil composition).

The equatorial tropics suffer from too much light, and this inhibits growth as much as a lack of light or heat does. Shade can raise yields significantly in these areas. Bright sunlight can inhibit growth above 77°F (25°C). When light intensity gets high enough, carbon dioxide must be provided to raise yields through use of animals in greenhouses, mushroom cultivation, or composting (all create CO_2). Using tropical plants in equatorial zones is ideal since they can handle low CO_2 levels and heavy light saturation. This is another reason why temperate farming techniques being applied in the tropics have been a historical failure and still are in many areas today.

Biomass accumulates seasonally in the high latitudes (each summer), and it accumulates continuously and inefficiently in the equatorial regions. This also accounts for the temperate climate's deep soils in comparison to the relatively shallow soils of the tropics. The riot of growth in the tropics belies the scant fertility in the soils.

The Carbon Cycle

Above all other elements, carbon is essential. It is the building block for all life, providing structure for plants and animals as well as fuel for reactions and food for all life. It is found in plants, animals, fossil fuels, the earth, the soils, microbiology, and in the air. Photosynthesis is literally synthesizing the sunlight, water, and air into usable energy for plant growth by forming carbohydrates (sugars) out of CO_2 and H_2O. Herbivores eat that **carbonaceous** plant material for their energy and structure while carnivores feed upon herbivores for their intake—omnivores eat both herbivores and plants for energy. The carbon source for all these forms of life comes from the carbon cycle.

> **"Life and all biological activity is composed of complex carbon-based molecules. Soil organic matter is all biological activity in the soil, plus the remains of all previous biological activity in the soil... Every single molecule in every bit of soil organic matter anywhere is made of carbon atoms"**
> —Alan Yeomans, *Priority One*, 2005

When grasses dry out to brittle straw, they've left behind nearly pure carbon; 97% of a corn plant's physical mass comes from atmospheric CO_2. When plants and animals decompose, they release carbon into the atmosphere and into the soil. The soil life releases and consumes carbon in all its interactions as well—forming soil structure and feeding multiple **trophic** levels. Animals, fungi, and plants (at night) take in oxygen and exhale carbon dioxide back into the atmosphere as part of **respiration**. Oxidation or burning of any carbonaceous material releases carbon dioxide—it is a process of oxidizing the carbon and changing it from a solid to a gas form. When soils are tilled, soil life is oxidized, mechanically destroyed, and dried out by being exposed to the air.

> **"Agriculture's most damning contribution to climate change is the release of carbon held in the soil, primarily from deforestation and land clearing"**
> —Eric Toensmeier, *The Carbon Farming Solution*, 2016.

Carbon sequestration is soil building. Since carbon is the primary component of soil organic matter, it is the best indicator of the current carbon levels in the soil. Carbon can be sequestered or trapped in plants, soils, and in soil life. Forests are large pools of sequestered atmospheric carbon—their structures are made out of carbon, and while they lightly exhale carbon dioxide at night they vigorously exhale oxygen all day, releasing 10 times the amount of oxygen they take in.[12] A third of worldwide fungal biomass is carbon (5Gt),[13] and **mycorrhizal** fungi are responsible for sequestering nearly a third of the carbon in soils, forming the soil structure ideal for growing plants.[14] Encouraging mycorrhizal fungal growth as a form of atmospheric carbon sequestration has not been adequately studied and easily comprises a greater potential than all vegetative pathways—which alone presents incredible potential.

---

[12] Ingham, Elaine. *Email*. 2016.
[13] McCoy, Peter. *Radical Mycology.* 2016. p. 38
[14] Wright, Sara F; Nichols, Kristine A. *Glomalin: Hiding place for a third of the world's stored soil carbon.* 2002.

IV

"A field of corn (Zea mays L.) captures about 400 times as much C (12.5 × 10-9 Pg of biomass composed of stover, roots, grains, etc., and containing 5 × 10-9 Pg C per hectare) as there is annual increase (12.5 × 10-12 Pg of C from an increment of 2 ppm of CO_2 per year) of man-made atmospheric CO_2 in the entire column of air above that field from ground to the upper reaches of the atmosphere (figure 2). However, most of the biomass C thus photosynthesized is respired back into the atmosphere. Nonetheless, the capacity of vegetation to photosynthesize atmospheric CO_2 can be used as a strategy to create a positive C budget (C input > C output) while also enhancing ecosystem services"

—Rattan Lal, "Managing Soils and Ecosystems for Mitigating Anthropogenic Carbon Emissions and Advancing Global Food Security," BioScience, 2010.

Professor Rattan Lal has said that we can sequester all the excess atmospheric carbon in global soils within 50 years, but with a combination of reforestation, wetland/riparian/

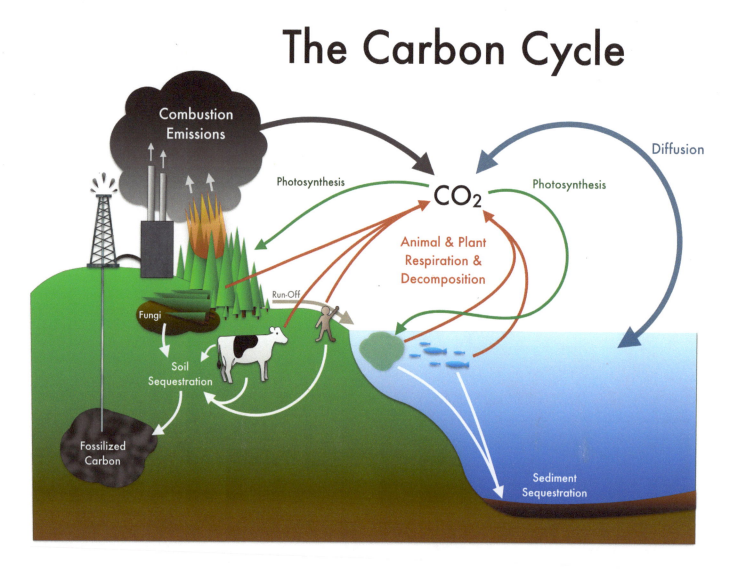

watershed restoration, and ocean restoration we can speed that up.[15] There have been claims that it can occur in as little as ten years time, and while that might be mathematically true, it hasn't proven true in testing and modeling for scale. (We can regrow forests and restore almost all the desertified land within that timeframe, but sequestering all the atmospheric carbon is different and likely will require time while the forests mature to fully achieve balance and mitigate the global water cycle). It is sometimes claimed that perennial grasses that are strategically grazed by high-density herds can sequester more carbon (build more soil) that a mature forest can—which may be true in some cases like Polyface farm—but in natural, untended systems, forests almost always sequester more carbon.

> *"Some types of grasslands have more soil organic carbon than some types of forest, but this is the exception, not the rule"*
> –Eric Toensmeier, <u>The Carbon Farming Solution</u>, 2016.

Algae and kelp forests can sequester carbon at least ten times the rate that perennial grasses and trees[16] can, but they need our help. The ocean waters have already absorbed half

---

[15] Lal, Rattan. *A Regenerative Future with Matt Powers*. Podcast. 2018.
[16] Olaizola, M., T. Bridges, S. Flores, L. Griswold, J. Morency, and T. Nakamura. *Microalgal Removal of CO_2 from Flue Gases: CO_2 Capture from a Coal Combustor*. 2016.

the carbon that has been released into the atmosphere,[17] and in addition, the US west coast kelp forests are over 90% reduced since the last El Nino. That is why there are concerns over acidification though the oceans are far from truly acidic—even slight acidification or ocean warming is enough to disrupt entire ecologies. The coral reefs are bleaching and shellfish are struggling to fully develop their shells. It is just like a shade plant being moved into full sun—it burns or, in this case, bleaches. As they bleach, they release more carbon.

Carbon levels in the atmosphere must be drawn down if we are to reverse climate change and all its connected imbalances. Carbon dioxide is the largest contributor to climate change though it is only one of several greenhouses gases (most notably: methane CH_4 and nitrous oxides NO_x). These gases act like greenhouse plastic covering the earth, trapping and focusing the energy of the sun. The soil is the best place to take back this excess carbon. Though we must also restore our forests, grasslands, oceans, and biodiversity, the soils are the foundation for all those systems; restoring our soils will sequester carbon as it supports and generates other carbon sequestration processes. While this is hopeful, we have to move quickly because we are exponentially releasing more carbon each passing year. It takes time for sequestration to occur in new forest systems though it accelerates after the first 5 - 10 years of establishment.[18] No-till agriculture and holistic management grazing reverse carbon-emitting practices while sequestering carbon in the soil, though this also takes time—the soils we have required millions of years to develop.

There is also a limit to how much soils can sequester, and it also should be noted: sequestration rates slow as they approach carbon capacity.[19] It will take time and an immense collaborative effort. Some studies estimate it will take 60-100 years to return atmospheric CO_2 rates to 1750[20], while others forecast a much quicker recovery given the many examples of rapid soil-building on small, intensive sites. Nonetheless, it is still a limited window of opportunity that will require unprecedented effort and requires farms to become atmospheric carbon sinks instead of sources. We have to phase out fossil fuel usage for these natural carbon sequestration methods to be effective; we have to start growing our fuels, fibers, foods, and resources, regeneratively and locally, while putting limits on consumption.

"Soil having an organic matter content of only 5%... holds over thirty-five times as much carbon as that causing Global Warming. That's why soil fertility concepts are so important and why increasing the fertility levels of our soils can fix Global

---

[17] Pickerell, John. *Oceans Found to Absorb Half of All Man-Made Carbon Dioxide*. 2004.
[18] Anderson, Jim, Beduhn, Rebecca, Current, Dean, Espeleta, Javier, Fissore, Cinzia, Gangeness, Bjorn, Harting, John, Hobbie, Sarah, Nater, Ed, and Reich, Peter. *Potential of Soils for Carbon Sequestration*. 2008. p. 20.
[19] Ibid.
[20] Lal, Rattan. *Managing Soils and Ecosystems for Mitigating Anthropogenic Carbon Emissions and Advancing Global Food Security*. 2010.

Warming... Prairie soils, such as those in the United States that once had up to 20% organic matter and have been gradually reduced to levels often below 5% can also be restored to their original richness. They can rebound rapidly back to those 20% levels. Recreating the richness of the soils of the American prairies could almost on its own normalize world carbon dioxide levels"

–Alan Yeomans, *Priority One*, 2005

The math is constantly changing as we release more CO_2, CH_4, and NO_x into the atmosphere, and while there is still debate over these processes and their conditions, it is clear that the soils are depleted of organic matter, the primeval forests are gone, and desertification continues to displace vegetation and contribute to the mass extinction of global biodiversity. The soil, forests, animals, and plants are the long-term storages of carbon for our planet. By dumping all the carbon out from those storages nearly all at once into the atmosphere, we've set ourselves up for a progressively hotter climate. We can reverse it by building soils, planting forests, reversing desertification, embracing carbon sequestering farming methods, rewilding our landscapes, and restoring and protecting the oceans and wetlands. It is entirely possible currently, but we must act quickly.

"To fix Global Warming we need to put that 0.081 inches (2.1mm) of carbon back into the 8.5% of the Earth we actively control. That quantity of carbon is the same as in a foot of soil with just 1.6% organic matter content. There is more than that in most desert soils"

–Alan Yeomans, *Priority One*, 2005

V. Trees

Trees are in a constant state of cycling the energies they interact with—from the air to the water to the organisms around them, they are in constant exchange. They take in and transpire moisture and gases. Their roots, branches, stems, and leaves release **exudates** that attract and feed microbiology. They create seeds, leaves, nectar, and fruit that are consumed by other organisms in trade for extending their genetic range. They create soil as their leaves, twigs, branches, roots, and, eventually, their trunks fall to the earth. With the help of the soil food web, the compounds that once comprised them are reconstituted in billions of other ways. Their seeds and seedlings progress due to a collection of interactions between animals, microbes, fungi, plants, and natural energies.

Trees interact with the wind, precipitation, radiation, and the atmosphere. Their interactions vary in degree and intensity constantly, making measuring these exact effects difficult—though we can observe and enhance them. Our human understanding of trees is limited because many do not grasp the vital role they play, nor the value that they bring. Without trees we would not have the favorable conditions for human life: clean water and air, topsoil, and moderate weather. Trees do not do it alone; they are dependent on ecosystemic interactions, but in many ways they are often the most visible point of convergence.

The biomass generated by trees (and all plants), the biological interactions, and natural energies and forces create soil. Above all plants, trees provide shade and shelter which promotes quick decomposition with less CO_2 loss than would occur in the direct sun. Trees also sequester carbon in their woody biomass. As mentioned in the previous chapter, trees release carbon dioxide and consume oxygen <u>at night,</u> but this amounts to only 10% of what they take in and release during the day. Though some perennial grasslands can sequester more carbon than some forests, most forests sequester more carbon, and grasslands are a step in succession towards a stable forest ecosystem as well. We cannot have one without realizing it is on a cycle of succession towards the other. Once a forest reaches its peak expression, it decomposes and returns to an early stage in the succession and begins its progress back to mature old growth forest again. We see this happening throughout forests in the form of open fields and glades. Large animals like elephants can clear areas and be a biological source of this kind of disturbance, as can invasive pests like processionary caterpillars on European pine forests and the Japanese bark beetle on North America's pine

forests. Some areas experience so much environmental stress that they cannot reach mature old growth forest—areas of erosion, drought, high winds, high altitudes, flood zones, etc.

The interactions between insects, fungi, birds, browsers, the soil food web, and other participants all contribute to the behavior and physical effects of the tree; it is difficult to separate the tree from all its interactions and related organisms. Trees represent a collection of organisms working together. Relying upon fungi, other microbes, and the physical processes available, trees are able to access the minerals they need from the soils around them, primarily in the top 6" (15cm) of soil.

Forests reduce erosion, increase groundwater recharge, reduce flooding, clean the air, increase precipitation, and increase stream flow. A thin forested strip at the base of a mountain slope or in an area that often floods can reduce flood damage and flood potential dramatically. Trees are also pumps—large trees like douglas fir pines or *eucalyptus globulus* will dry marshes and swampy areas and return the moisture into the atmosphere (and so not suitable for planting in arid regions).

In dry regions, regular windbreaks or forested sections increase precipitation by compressing air flows above them and amassing the moisture in a higher altitude which causes it to clump together and fall as rain, snow, or something in between. Trees also reduce salinity levels in the soil and water. This is a critical service for agricultural soils with high concentrations of salt—a common occurrence in more arid regions where constant evaporation leaves salts on the topsoil. Remediation of these soils is often only possible by planting trees.

Trees are fundamental to the constancy of our atmosphere and soil. Indeed, a world without trees would soon be uninhabitable. Most areas with a large human population, like cities, do not create enough oxygen to support their consumption simply because they lack the trees that once stood there. The water that once was on land and trapped in forests is in the oceans and atmosphere where it creates more erratic and violent weather patterns. Trees are the only way to practically remedy the situation.

Wind Effects

Some trees anchor themselves against the wind with their roots around rocks, others create a root mat, and still others change shape to preserve their structures. Trees thicken in reaction to winds, thus edge species tend to have thicker trunks. Wind can carry ice, sand, salt, dust, or silt. Forests net this diversity of wind-borne material and turn it into other forms of energy or nutrients for themselves and other lifeforms. As mentioned earlier, tree flagging has several effects. Older trees quickly reveal their experiences through the orientation of their trunk in relation to their crown. Trees also thicken in response to the wind's action, turning the wind's energy into wood mass and density.

Leaf Area

The more narrow the leaves, the less energy absorbed. Broader leaves mean more absorption. The more developed the hydrology of a system, the more broad-leaved plants can be in it. Trapping the most energy possible makes for the most stability possible.

Temperature Effects

- **Evaporation** - Almost half of all precipitation is returned through evapotranspiration (evaporation + transpiration). Evaporation has a cooling effect on the objects, plants, or animals losing the moisture. It is caused by heat, and while seemingly counter intuitive, we can observe this effect occur with our own bodies as we sweat to regulate our temperatures on a hot day or during exertion. Warm winds are cooled and humidified as they pass

through a forest or stand of trees. Though it should be noted that in hot humid tropical areas, trees can dehumidify the air! The forest works to moderate the temperature and humidity levels to be more conducive to the spread of life.
- **Condensation** - Cooling water vapor collects on surfaces releasing heat. Leaves are warmed by condensation and cooled by evaporation. These two tandem effects moderate the atmosphere around them and churn the water cycle in a way that is often hard to see. Trees and open bodies of water warm the air around them and also cool hot air.
- **Leaf color** - Green plants and leaves absorb nearly all the energy from the sun they receive while white or whitish plants block nearly all of the sun's energy. These are strategies that plants use to survive in areas of minimal and maximal sunlight.

Trees and Precipitation

Trees and forest edges cause a rise in the air flow which results in a spiraling down/up motion of air waves above the forest. This motion collects leaf litter, aerosols, and volatile organic compounds and draws them up into the air while creating a sheltered area beneath the tree canopy and past the windbreak for a distance many times the height of the trees. This leaf litter, dust, silt, plant aerosols, mushroom spores, volatile organic compounds, and other small bits of organic matter, together form cloud condensate nuclei (CCN) in the sky, which in turn helps form precipitation.

Less than half the air flow is absorbed into the stand of trees and dissipates as the depth and density increases. As the air slows, it deposits anything it is carrying—airborne dust is very minimal in the forest interior. More than half the air flow is forced over the trees. Pushing the air higher cools and compresses the flows—both increase precipitation. Tree evapotranspiration also contributes to the formation of precipitation by returning nearly half of all precipitation it receives.

All edge plant species are vital to a forest since they can weather the extremes of salt, wind, and sun that the more tender inner forest plants cannot. Trees are the greatest expression of the forest, the pinnacle of the succession both at the edges of and inside a forest. If the forest edges are removed, the inner forest suffers and can collapse. Protecting edges, expanding edges, and creating more edge is desirable and life-enhancing.

Trees shade snow from the sun, enormously increasing the snow's longevity and soil development beneath the snowpack—something especially important on high, cold or arid slopes. Unprotected snow evaporates into the atmosphere quickly while shaded snow slowly melts and is incorporated more into the soil and plants in the ecosystem. If we remove the trees, we remove the moisture they are sheltering and regularly gathering within themselves, the land, and undergrowth. Their removal dries up the rivers, streams, lakes, and ponds that we normally would see downhill from these initial catchment areas. It systematically causes drought, as can be witnessed in the Central Valley of California, where the removal and redistribution of the Sierra Nevada mountain range's annual water catchment has been happening for decades. This loss of habitat and watershed destruction is being observed around the world everywhere bad human settlement designs are implemented.

In a rainstorm, trees and bare soil are opposite in effect. Bare soil in any precipitation event experiences erosion, and soil is carried away; trees prevent erosion. When rain hits tree leaves and limbs, it explodes and turns most of its mass into airborne mist which combines with dust and pollen. This mist coats the tree's leaves and bark and gently falls down through the canopy. It is like a **compost tea** foliar mist for the undergrowth and is referred to as throughfall. There is often a runoff from the trunk that is similarly constituted.

The amount of interception is dependent on the crown thickness, density, leaf/bark type, season, intensity of rain, and evaporation after rain. Throughfall can be the majority of the water passed down to undergrowth or low canopy forests in tropical areas because the canopy is so efficient at intercepting the rainfall. Branch and leaf design and pattern all encourage water retention. Fungi, bacterial gels, root mats, mulch, and the actions of soil life and small animals retain more water on the surface and in the ground as well. All are supported by the presence of trees and forests in their ecosystems.

Emily Morter. unsplash.com. 2016.

Winter rains in areas with deciduous trees may see it go mostly un-intercepted, but evergreens and other trees that hold their leaves or needles throughout the year intercept rain consistently, so infiltration depends on the kinds of trees present in a forest.

The Conveyer Belt of Moisture

Coastal areas get inland humid cool air flows, especially in the evening. This moisture collects on all smooth surfaces and fine surfaces likes leaves and netting. Condensation on islands and coasts can account for almost all precipitation. A giant Til tree in the Canary Islands could condense enough water to support communities of people. Fogs and visible humidity condense the most moisture onto plants, but measurements are impossible to calculate accurately, despite the visibility and significance of the phenomena.

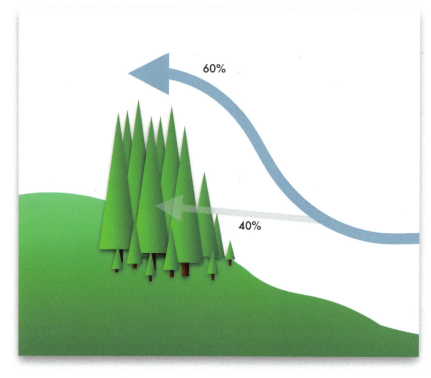

Sea evaporation of moisture is the primary source of water in ecosystems that are coastal, but as you travel further inland, tree evaporation takes over—but it is this conveyer belt of moisture from tree to tree and forest to forest that is the continuation of rain and moderated weather that starts with coastal moisture.[21] Remove coastal forests and you destroy the coastal forest hydrology and also dry out the inland forests. It all needs to be reconnected.

Forest with all the niches occupied and cycles that trap moisture in place also returns 75% of the moisture back into the atmosphere. In the Amazon rainforest, the incoming air's moisture is doubled by the natural action of the trees.

Forests also work like large invisible lakes of water that release a steady stream or river of clean, filtered water when the landscape itself is saturated. Soils can hold enormous amounts of water especially if held together by a polyculture of root systems and supported by earthworks or natural geography. Forests store and use water the most efficiently and are the most productive with what moisture they get, and they continue constantly to cycle

[21] Sheil, Douglas, and Murdiyarso, Daniel. *"How Forests Attract Rain: An Examination of a New Hypothesis."* 2009.

everything. No mechanization or technology surpasses the efficiency of trees, forest, watersheds, and natural cycles.

VI. Water

"The surface of the earth, human beings, and all creatures are roughly made of 70% water. Without water there would be no life. Water is the earth's blood"

–Sepp Holzer, <u>Desert or Paradise</u>, 2012.

On our planet there is precious little freshwater available for human use and consumption. What water abundances we've had, we've used, and we've cycled the majority of it out of our ecosystems. Industrialized countries are drilling down thousands of feet to pump freshwater up to the surface. These aquifers take thousands of years to develop yet we are consuming them in only a few short decades. Meanwhile we route rainwater, which is distilled freshwater, out of our cities into the ocean. We have both a design problem and a recognition problem. Water is precious to all life systems; we need an abundance of it to live productive and healthy lives, and we are squandering it. Wars over access to clean water are increasingly common in drier or desertifying regions around the globe. Global droughts and crop failures are also on the rise in tandem with decreasing freshwater sources.

For thousands of years, humans have primarily consumed fresh spring water; we are designed for it—well waters are too high in minerals and rainwater is too distilled, lacking in minerals. We can create earthworks that bring back the natural springs in our areas. We can catch all the rainwater that hits a city's or a home's hardscapes and use it for irrigation, cleaning, and more. We can intervene in the water cycle as nature does: slow, spread, soak, and store. We can bring back wetlands, marshes, deltas, and floodplains by regenerative design, restoration to historically and geologically indicated levels, and protecting the watersheds that feed them.

> **"Groundwater should not be considered a source of water, but a storage"**
> *–Neal Spackman, Sustainable Design Masterclass, 2016.*

Finally, we can protect the natural freshwater storages by avoiding disruptive activities and encouraging animal and plant habitat along the shores and within the lake. These bodies of water mitigate extremes in the local atmosphere, generate precipitation through evaporation, and provide key habitat, power, and nourishment. We can preserve, enrich, and enlarge our freshwater reserves with water-wise living standards in cities, towns, and homes. We can restrain ourselves from using industrial solutions that beget industrial-sized problems and restrict ourselves to natural solutions that beget products that feed other natural cycles.

Watersheds

Watersheds are land formations that direct and define where water will go. Ridges can separate different watersheds. As slope becomes more gentle, the dendritic patterns become more profuse, forming delta patterns. These forms occur to create and form larger catchments like headwaters or to disperse water, like an alluvial fan. The dendritic pattern is seen throughout nature in waterways, in circulatory systems, in neurons, and even in space.

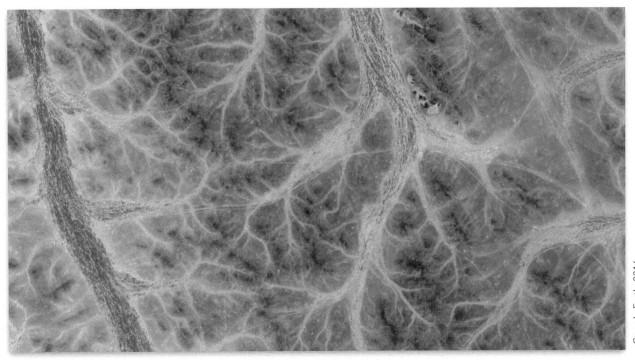

Google Earth, 2016

Regional Intervention in the Water Cycle

> "This [global] water cycle actually drives 95% of the heat dynamics of the blue planet, and so it's the engine room, it's the driver, it's the means through which this whole ecosystem on planet earth hangs together."
>
> - Walter Jehne, Sustainable Design Masterclass, 2018.

Cloud Seeding

Commercial cloud seeding involves releasing artificial ice nuclei into clouds to precipitate rain—usually from a plane. Though silver iodide is the commercial choice for cloud seeding, leaf litter is better cloud seeding material than silver iodide because it is natural, free of cost, non-toxic, and more effective. Cloud seeding with silver iodide can often overwhelm some areas with too much water while others get less with effects ranging over hundreds of miles. The unpredictable nature of cloud seeding with silver iodide from a plane makes it unreliable and can damage infrastructure and crops.

Natural cloud seeding is best. Rain is the natural result of a forest being in place—leaf litter carried by wind creates ice nuclei, while trees push air streams up where they compress and cool, causing rain. Rain dances, burning of some kinds of plants, and other indigenous rituals may have sent dust into the clouds to form ice nuclei effectively. Trees are more sophisticated, effective, and precise at cloud seeding and making precipitation than anything man-made.

Enhancing Orographic and Forest effects

If we plant trees that are superior at condensing water (like giant conifer trees along ridges that are aligned perpendicular to prevailing winds), we capture and release large amounts of moisture and compress the air flow, thus creating more regular and longer-lasting rain events. The removal of these kinds of forested areas will only cause and increase drought especially on steep slopes that face large bodies of water and ridges. Conservation and replanting of these areas is critical. If we just start with reforesting these areas, we will have the added precipitation, moisture, and windbreak that these areas provide to aid us in our efforts.

Soil Storages

Adding organic matter, **ripping** the soil, careful rotational grazing, compost tea'ing, and, most importantly, increasing the abundance of **mycorrhizal** fungi in the soil can all increase the soil's ability to store water. Runoff can be turned into abundance. If plants are in place, they will protect the soil's catchment of water. Soil can hold many times the amount of water held in open storages and streams.

Infiltration via Earthworks

Earthworks are cheap, relatively easy to install, and long-lasting: pitting, swales, level sills, gabions, check dams, *hugelkulturs*, and the like are all ways to get fresh water to deeper storage—they just vary in their ability to do so and in their function in a system. They can recharge aquifers, keep soils wet longer into the dry season, increase biodiversity, and feed springs. During the Great Depression in the United States, the Civilian Conservation Corps set out to restore damaged landscapes across America. Some of their work remains visible today as what remains of the three billion trees they planted and, most notably, the earthworks they made in Tucson, Arizona. These particular large earthworks harbor a thriving forest system that is surrounded by desert. It is a testament to the effectiveness of earthwork infiltration. By just digging the right kind of trench, you can create a microclimate and a moisture trap. It can be done anywhere.

Natural Water Storages

Swamps, marshes, wetlands of all kinds, and ponds can all be a result of moisture or precipitation levels being higher than evaporation or transpiration levels. These are life-rich areas that deserve protection as they generate life, store water long term, and drought-proof the land around them.

In US history, swamps and marshes were drained or pumped dry with trees like eucalyptus in order to open up land for agriculture and settlement, but we lost in the tradeoff: the wetlands can produce more value as wetlands. Fish farms and wetland plants can thrive alongside wildlife habitats thus protecting and enhancing both simultaneously. (This is dependent on keeping the water moving as stagnant water releases methane and other greenhouse gases and is unsuitable for a thriving ecology).

Biological Storages

Plants can store large amounts of water even in drier areas. Cacti in desert regions are critical sources of moisture. Soil and the soil food web can store large amounts of water. We can use plants, mulch, fungi, shade, and animals to help retain more water.

Tank Storages

Roof rainwater harvesting into tanks with a first flush diverter is a great way to get clean water (or water that is easy to clean). Runoff can often be laden with chemicals that need to be tied up in long carbon chains or consumed by fungi. Mycoremediation beds or **reed beds** that later are composted may help treat graywater both from the home and the first flush diverter.

Earthworks for Water Conservation and Storage

"Blue (Water) before Green (Photosynthesis and Trading) and Black (Carbon and Profitability)"
  –Darren J. Doherty, 2016.

We store water in the soil and biology as well as in dams and storage tanks. Earthworks recharge aquifers but above-ground tanks do not, typically, and hard sills or sealed diversions drains do not recharge aquifers at all. Using a combination of focusing, storing, and dispersing wisely, we can thrive with the water we get in our natural environment through precipitation and surface runoff. Small dams and earth tanks are superb for watering and storage during dry periods, which are globally increasing in frequency. Proper design of water harvesting and release is critical to avoiding ecosystemic collapse of local waterways and wetlands, to avoid damage to property and farm systems, and to having water stores last year-round.

Water storage strategies in drylands differ from those applied in more humid areas: open storages of water in drylands evaporate and therefore become salty and bad for people, plants, and animals. Water in drylands is stored in covered tanks or deep in the earth where the natural coolness prevents evaporation.

While dams are often seen as holding water inside them, they are also holding much more water in the surrounding landscape. Sepp Holzer will construct water retention sites that take years to fill because they are first filling up the landscape around them and the subterranean aquifers first. Only when they are saturated will the dam area itself fill.

For Sepp, dams and water retention sites can be anywhere there is natural deposition, where there are deep zones or depressions in the land. Clays and silts naturally collect in these depressions like large silt traps. Having both the materials for sealing the dam and a natural depression for catchment makes these areas ideal pond sites.

Many experts differ in their opinions as to what is ideal. Some recommend avoiding valley dams because we lose energy potential; others, like Sepp Holzer, argue that water is meant for the valleys and areas of deposition where clays naturally accumulate. Whatever you are designing, consult and work with experts when designing anything that could be destructive or dangerous. Restoration of natural wetlands is discussed later in this book in the Aquaculture chapter.

Common Dam Types

- **Gully or Embankment Dams** - A dam built across a depression, gully, or stream. This is the most common form of dam. **Spillways** and spillway pipes can be used. These dams can be used for irrigation or landscape water retention.
- **Hillside Dams** - These dams are built up against a hillside to catch runoff. They can provide gravity-fed irrigation to areas lower in the landscape. They usually have large walls and can require diversion drains or swales to magnify catchment—for those reasons, they can be costly to install.
- **Saddle Dams** - Sometimes the highest dam possible, it sits between two hilltops and has walls on both sides with spillways usually on both sides. It is ideal for storage, wildlife, and fish.
- **Ridgepoint Dams** - Wrapped around the end of the ridge or just below it, it forms a horseshoe or kidney bean shape and has one long outer wall and a spillway on one or both sides. This is great for storage and wildlife but of limited irrigation use.
- **Keypoint Dams** - "The keypoint of the valley is where the slope of that valley starts to flatten out and where deposition begins."[22] Keypoint dams can be fed by diversion drains in drier climates to focus water and swales in more humid climates to distribute the water, and sometimes keypoint dams don't even need anything more than the natural catchment already in place.
- **Contour Dams** - Ideal for flatlands (8% or less slope), these contour dams catch all the water that is running downslope for aquaculture and irrigation, and they even manage flooding in drier regions.

---

[22] Doherty, Darren J. *"Broad Acre Agroforestry Integration."* 2016.

- **Earth Tanks** - Commonly found in flatlands and made with machines, these tanks are made with raised earthen walls, usually steeply ramped, with water stored in their center. Earth is not removed from within the tank.
- **Ring Tanks** - Ring tank dams are like earth dams but they are pulling earth for the wall from the water storage area. It is like a sunken earth tank, but commonly only dug out in the ring to make the wall, leaving a perennial island in the center of the dam.
- **Excavated Tanks** - These dams are excavated downward into the ground. If the water table is high enough, these types of dams can fill up and maintain water level with the water table. If the tank is above the water table, seepage likely will occur.
- **Gabion Dams** - Wire baskets filled with rocks that stack to make a check dam—slowing the water but allowing it to flow through, leaving silt and rocks behind. This allows for water to collect and absorb and mitigates flooding events and their capacity to do harm. In arid zones where flooding is a real danger, we can mitigate or erase this threat with gabions and other earthworks, using them where appropriate. One criticism of gabions is that the metal wire baskets do not last, but there are also examples of gabions without wire baskets—these are built with giant boulders making up the main wall and then smaller boulders in succession behind them.
- **Barrier Dams** - These dams slow water as they cross stream beds but do not impede flow; these can take numerous forms. They can be used to induce meander or to slow water. They build enough pressure to provide for irrigation if located significantly higher than irrigated areas. These can also be used to generate energy or oxygenate the water to enrich the life in it. These dams can be made out of arranged stones, earth, logs, or even cement. More regenerative dam types and wetlands strategies are discussed in the Riparian Wetland Restoration section.

Petra's Nabatean Water-Harvesting Designs

Perhaps one of the most famous water harvesting systems in the world is also one of the most ancient. The city of Petra is carved from the sandstone walls of canyons and nestled in the middle of a Jordanian desert. This amazing city was only made habitable by the elaborate earthworks and water harvesting designs in place.

Above the city, the entire landscape is carved and arranged to encourage water to be slowed and diverted

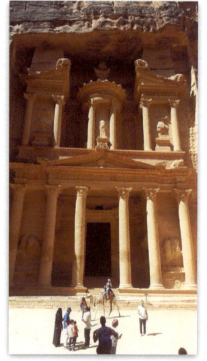

Petra's Nabatean water harvesting systems, if fully restored, could work again today. Historically, the Nabateans were very secretive about how they acquired their water in the middle of the desert. Now the secret is out! These systems can be replicated in designs everywhere regardless of the medium.

towards the catchment areas. The canyon walls have been carved as well to channel water smoothly. Ingeniously, the walls in many places have been made to prevent dripping and instead draw the water seamlessly down into troughs that run into silt traps before entering large underground cisterns for long term storage.

Natural Ponds and Pig Ponds

Ponds do occur naturally, and we can enhance these naturally occurring events by observing how the landscape is creating and maintaining them. Though we discuss how to "build" dams, we can naturally create them or enhance existing areas already in place.

Pigs at Krameterhof in Austria seal pond sites for Sepp Holzer.[23] Sepp goes in initially and creates a hole and lets it get wet. The pigs held in with an electric fence do the rest. They make a mud wallow, and in following their natural instincts, they seal the pond. Their bodies vibrant as they rub themselves in the mud, the pigs compact the subsoil layers below the mud until they are impermeable enough to hold a pond full of water. We can encourage pigs to wallow in areas of natural catchment to create small natural ponds without any excavation,

[23] Wheaton, Paul. *Can Pigs Build Ponds?* 2012.

building plan, excavator, shovel, or permit needed. Please note these ponds are not truly sealed like with an impermeable pond liner—they still allow water to slowly seep into the landscape.

Invisible Ponds

Though it may seem unethical to make pond building against the law, it is illegal in some places to put in ponds or requires an expensive permitting process. If this is your situation, you may want to make an invisible pond. You create them the same as you would any other pond site, but once complete you fill the bottom with large boulders followed by successively smaller boulders and stones until you've filled the dam with rock and stone. Fill the rock-filled pond up with water and see how much water storage there actually is—with a bit of practice you can create safe, stable pockets of water with the boulder arrangement and layering. If you designed it initially with a pipe and pump safely embedded within the stone arrangement, you can even use the water for irrigation or home purposes, once filtered.

Casing a Spring

Casing a spring begins when a spring is dug out, back to where the subsoil meets the substratum.[24] This is where the water filtering through the subsoils stops continuing downward and begins to travel more horizontally—this flow is what pops out of the ground and manifests as a "spring". After digging it back and keeping out all topsoil (and keeping all the soil types separate), lay a pipe slitted on top in the flow and see its rate. The flow rate determines the pipe size (always go with experienced professionals to learn and install). Covered by washed 1/4-1/2"(.5-1.25 cm) round gravel then surrounded by food-grade geotextile or clay, the pipe is now protected from sediment and contamination.[25] The water flows from there down to a spring box or another containment vessel and then onto irrigation or household use.

Building Dams

There are laws that govern pond design, installation, and usage in almost every town, city, state, and country. Permits are usually required. Designers and home owners are responsible for adhering to those laws and guidelines. Pond building, if done incorrectly, can fail and cause harm and even fatal

---

[24] Weiss, Zachary. *"Elemental Ecology."* 2016.
[25] Ibid.

harm. Work with experts, follow the laws, get your permits, and always plan for climate change to increase the frequency and strength of large precipitation events.

How to Approximate Dam Evaporation

"A rough figure for the evaporation loss can be obtained by taking two-thirds of the local annual evaporation and multiplying it by the top water surface area of the dam. For example, if the local annual evaporation is 1200 millimetres and the top surface area 5000 square metres (half an hectare) the volume of evaporation from the storage is approximately

$$\frac{2}{3} \times \frac{1200}{1000} \times 5000$$

which is equal to 4000 cubic metres or 4 megalitres"

Nelson, K.D. *Design and Construction of Small Earth Dams*. 1985.

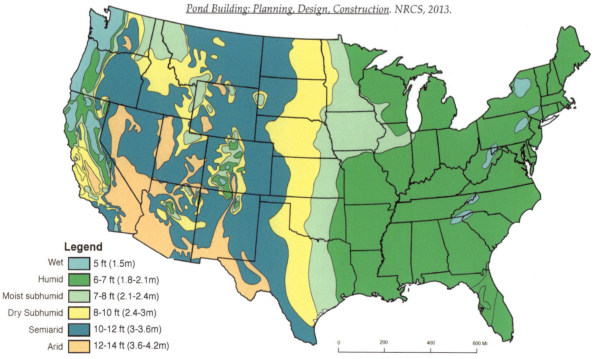

Recommended Minimum Depth for Ponds in the United States
Pond Building: Planning, Design, Construction. NRCS, 2013.

Legend
- Wet: 5 ft (1.5m)
- Humid: 6-7 ft (1.8-2.1m)
- Moist subhumid: 7-8 ft (2.1-2.4m)
- Dry Subhumid: 8-10 ft (2.4-3m)
- Semiarid: 10-12 ft (3-3.6m)
- Arid: 12-14 ft (3.6-4.2m)

U.S. Department of Agriculture, 2013.

Finding a suitable site

As Sepp Holzer suggests, any area of deposition can likely become a dam or water retention area.[26] Looking at a topographic map, we can spot areas like thumb prints or round oval openings in the contours, with pinch points. These whorls with open centers represent flat areas with easy dam wall sites—the closer the pinch point, the more inexpensive the dam. The economic efficiency of a dam depends on the ratio of its back-wall length to its dam-wall length, since dam wall building is the most costly item.

> **"A good site generally is one where a dam can be built across a narrow section of a valley, the side slopes are steep, and the slope of the valley floor permits a large area to be flooded. Such sites also minimize the area of shallow water. Avoid large areas of shallow water because of excessive evaporation and the growth of noxious aquatic plants"**
> –Pond: Planning, Design, Construction. NRCS, 2013.

Avoid letting livestock drink directly from the dam water—instead, divert water to troughs—in this way preventing the spread of parasites and disease. It will also keep your dam water cleaner and clearer. If runoff is the main source of water for the dam, make sure the watershed is well maintained to keep the water supply clean. A damaged watershed will bring sediments as it erodes and fill the pond with silt and clay. Prevent cattle from damaging the catchment and vegetation in the watershed. Silt traps can be used to keep water clear, but an eroding watershed will overpower any silt trap.

If dam sites are located in depressions or deposition areas, they will usually have the soil materials on location to make a dam or water retention site.

> **"The soil should contain a layer of material that is impervious and thick enough to prevent excessive seepage. Clays and silty clays are excellent for this purpose; sandy and gravelly clays are usually satisfactory"**
> –Pond: Planning, Design, Construction. NRCS, 2013.

Research your site and make sure that no electrical or sewage lines run through it and that no power lines run above the site. Always consult an expert before taking on any large project even if it is just for a site inspection or consultation.

[26] Holzer, Sepp. _Desert or Paradise._ 2012. p. 42.

VI

Calculating Rainwater Catchment in Metric
From Neal Spackman's *10 Keys to Greening Any Desert*:

"Every square meter of land receiving 1mm of water gets 1 liter of water"
Neal Spackman, Sustainable Design Masterclass, 2016.

1 acre = 4047m2

4" of rain = 102mm of rain

4047m2 x 102mm = 412,794 liters

(conversions here are rounded off to avoid decimal places)

In hectares: 1 hectare = 10,000m^2

Metric makes it EASY!

Calculating Annual Runoff from Brad Lancaster's
<u>*Rainwater Harvesting for Drylands and Beyond*</u> (2013):

Catchment area in square meters x Rainfall in Millimeters x Runoff Coefficient = Net Runoff in Liters

Catchment Area in square feet x Rainfall in feet x 7.48 gal/ft^3* x Runoff Coefficient = Net Runoff in Gallons

7.48 is needed to convert the answer into gallons.

Runoff Coefficient - This represents the amount of water that does not get caught or is evaporated. Loss can range from 5-20%. Hard surfaces like roofs and pavement would have a .95-.8 runoff coefficient range. 5% loss to runoff would mean 95% of the rainfall that fell on the roof stayed, flowed down the gutter, and was stored in the roof rainwater tank. Vegetated areas may be able to have zero runoff in some light rains.

100m^2 bare earth x 400mm of rain x Runoff Coefficient

600ft^2 bare earth x 1 ft of rain x 7.48/gal/ft^3 x .2 = 897.6 gallons of runoff a year

Keyway (Aquifuge)

In a zoned dam wall there is a dense core inside the dam wall consisting of hard-packed earth that is often called the core, the Keyway, or Aquifuge (water barrier). Homogenous dam walls do not have a core; they are made completely out of impervious materials. Some dams only have an impervious lining along the pond foundation and inside wall but no keyway. Others have a miniature keyway inside the wall called a diaphragm. Sometimes there is a natural impermeable layer in the ground that can be connected to—but not always.

Keyways and other impervious constructions are compacted and impermeable. Clay and silt are needed in high enough concentrations to guarantee impermeability—30-40% clay approximately: gravelly clay, clay loam, silty clay, etc. It has to withstand the dam's full pressure—even in a flood event. Dams requiring a keyway are created with heavy machinery like a backhoe or excavator that drives over the keyway repeatedly until the soil particles are tightly packed together such that water cannot penetrate. Water seepage will naturally occur in the areas that are more permeable and even in the impervious core though at a slower rate. It is key to maintain a dry top to the dam, or else this seepage could begin to erode the dam wall.

Based on Figure 29 from K.D. Nelson's <u>Design and Construction of Small Earth Dams</u>, 1991.

Sepp Holzer typically does not seal entire pond areas, just the aquifuge or core in the dam wall which he connects to an impermeable soil layer beneath the surface; he refers to these as water retention sites.[27] With only the dam wall being impervious, all the moisture the dam wall encounters gets backed up into the landscape behind it—only when the landscape is saturated will the pond fill. This process can take years. Sealing an entire pond cheats the surrounding landscape of moisture. Soils can soak up water until they are 90% saturated.

---

[27] Ibid. p. 43.

Unless you are on very specific soils that lack any impermeable layers, a pond or seasonal pond area can be made. Purchased clay or geotextile are an imperfect solution being imposed on the landscape, and, therefore, do not share the same functions as a natural pond.

Sealing a pond foundation is simple if you have the right materials available: you clear the area, fill in any holes with clay or other impervious material to seal them, disk or rototill the pre-moistened foundation area 16–18" deep (41–46cm), and then compress the soil by rolling over it with heavy machinery until it is compacted.

> **"Roll the loosened soil under optimum moisture conditions in a dense, tight layer with four to six passes of a sheep's foot roller in the same manner as for compacting earth embankments. Make the compacted seal no less than 12 inches [30.5cm] thick where less than 10 feet [3m] of water is to be impounded. Because seepage losses vary directly with the depth of water impounded over an area, increase the thickness of the compacted seal proportionately if the depth of water impounded exceeds 10 feet [3m] or more"**
> –Pond: Planning, Design, Construction. NRCS, 2013.

Spillways

When water catchment is at its peak or during a flood, the overflow needs to safely run off from the dam over an auxiliary spillway that spreads the water out evenly in a sheet to prevent erosion and, ideally, catches it in another catchment system below to further slow, soak, spread, and store it.

> **"Preferably spillways should be made from natural fresh cut undisturbed natural ground"**
> –Danial Lawton, PermacultureTools.com.au

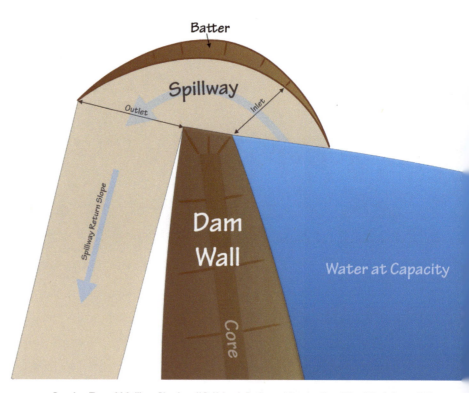

Based on Figure 36 Spillway Plan from K.D. Nelson's Design and Construction of Small Earth Dams, 1991.

A grassy, absorbent area is ideal for below a spillway. Spillways can also be channels that direct the water away from the area in case of emergency overflow (as in large hydroelectric dams).

Spillway sills can be compressed earth or even concrete or plastic hardscape to prevent erosion. The NRCS recommends small dam spillways to be designed for the large 25-year-frequency rain events, and large dam spillways to be built for 50-year storms. With climate change increasing, the large cyclical storm events are going to be occurring closer together and, perhaps, with greater strength than our historical precedents. To determine the size and scope of a spillway depends on precipitation, runoff, the watershed, and the fetch of the pond (or the distance from the dam wall to the farthest edge of the pond).

> **"No matter how well a dam has been built, it will probably be destroyed during the first severe storm if the capacity of the spillway is inadequate. The function of an auxiliary spillway is to pass excess storm runoff around the dam so that water in the pond does not rise high enough to damage the dam by overtopping. The spillways must also convey the water safely to the outlet channel below without damaging the downstream slope of the dam. The proper functioning of a pond depends on a correctly designed and installed spillway system"**
> –<u>Pond: Planning, Design, Construction</u>. NRCS, 2013.

Trickle Pipes (or Spillway Pipes)

A trickle pipe, or spillway pipe, is a pipe that acts as a release for small overflows that occur usually in the spring and fall. It helps maintain a set water level and keep the larger spillway sill dry. They redirect the water through pipes down past the dam wall and release it

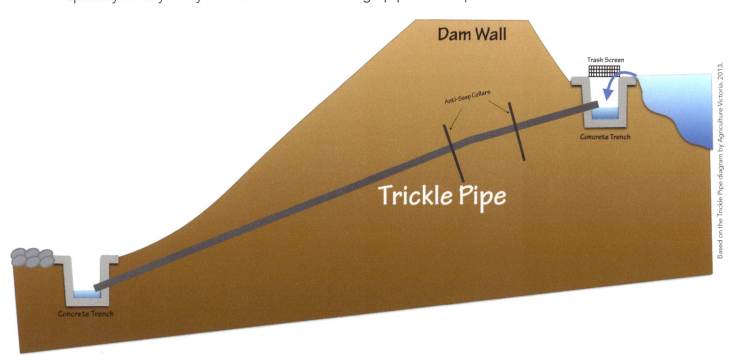

below where it naturally would drain to if there was no wall in place. Auxiliary spillways are for larger overflow events. The trickle pipe does not move nor can it be used to lower water levels as a valve release lower in the water profile would. Trickle pipes are permanent components of dam walls installed when the wall is built. They are built with anti-seep collars or rubber cut offs.

Backcut, Side Slopes, and Freeboard

The back of the dam or swale where the earth has been cut is the backcut. The backcut of a pond or swale should match the slope of the land below the dam wall. This helps slow the process of silt filling the pond over time and keeps the slope more stable.

Side slopes can vary in their design, depending on the concentrations of clay and silt in the soil—some can be steep while others must be gentle to avoid erosion. The freeboard is the distance between the height of the dam wall and the height of the water when it is flowing into the auxiliary spillway. Maintaining a dry freeboard is critical.

> **"If your pond is less than 660 feet [201m] long, provide a freeboard of no less than 1 foot. The minimum freeboard is 1.5 feet [.5m] for ponds between 660 [201m] and 1,320 feet [402m] long, and is 2 feet [.6m] for ponds up to a half mile [.8km] long. For longer ponds an engineer should determine the freeboard"**
> –<u>Pond: Planning, Design, Construction</u>. NRCS, 2013.

How big can it be? How deep? What shapes?

Look at the watershed—on paper and in person. Using topographic maps, calculate the square footage (meterage) of your site. Multiply that area by the annual precipitation to determine how much water you will receive, but make sure to calculate the runoff from that area as well (as best you can approximate). This gives you an idea of the forces at work before you intervene. Calculating dam evaporation rates and subtracting those losses from the runoff gains to potential dam sites will give a more accurate projection of the dam wall requirements, the dam depth requirements, and the expected overflows of average rain events as well as 25-, 50- and 150-year events. Always have an auxiliary spillway for large events and trickle tubes or spillway tubes for smaller events. The plastic or concrete tubes can handle regular usage, but the larger spillway sills of usually compressed earth need to stay dry between large events.

Farm dams need to be deep and large enough to mitigate evaporation and seepage losses while still providing irrigation and storage for human and livestock consumption. Arid climates need deeper ponds than more humid climate ponds to combat evaporation. Natural lakes and ponds tend to have deep zones that regulate water temperatures.

> **"Ponds that have a surface area of a quarter acre [1000m²] to several acres can be managed for good fish production. Ponds of less than 2 acres [8000m²] are popular because they are less difficult to manage than larger ones. A minimum depth of 8 feet over an area of approximately 1,000 square feet [93m²] is needed for best management"**
>
> –<u>Ponds: Planning, Design, Construction.</u> NRCS, 1997.

Dams or ponds should never have corners or straight lines. They should be curving and sinuous like natural rivers. lakes, and streams to encourage aeration and movement. Likewise, the dam itself should have deep and shallow areas to encourage water movement and provide critical habitat.

Two excellent resources for further study on building ponds are: the NRCS' <u>Pond: Planning, Design, Construction</u> and K.D. Nelson's <u>Design and Construction of Small Earth Dams</u>.

Sealing Leaky Dams

- **Gley -** a dense, oxygen-less, sticky soil layer made using an **anaerobic** fermentation process that can sometimes seal ponds with wet, green, sappy plant materials and/or fresh cattle manure beneath a layer of clay, cardboard, dirt, plastic, or anything that seals out the air. This can work in gentle sloping ponds. The slime that is created by this process can be a sealant. We can remove the covering cardboard or plastic once fermentation has occurred. For minor leaks in established ponds, adding green straw or hay to cause an algae bloom can have a similar effect to varying degrees.[28]
- **Water Fowl** - Ducks and other water birds can also be used to seal a leaky pond with their manure, similar to the gley method.[29]
- **Bentonite Clay** - A clay powder that is widely available though expensive. It can be tilled in and wetted to make a 5-7" (12-18cm) deep seal. There are several finer points to sealing that depend on your soil, so always consult with an expert when making dams. All clays tend to be expensive but can be useful to fix problem areas. Sometimes on soils that cannot hold any water only geotextile mats with bentonite clay sandwiched between them will work.

---

[28] Stone, Nathan. *Renovating Leaky Ponds*. 1999.
[29] Ibid.

> *"Sodium Bentonite is best, and a mix of both 5mm granules
> and powder is used for best results"*
> –Danial Lawton, PermacultureTools.com.au

Water Tanks

Tanks of plastic, metal, or concrete are expensive. Though surface water storages range in difficulty to install and cost, earth storages are the cheapest, most beneficial, and longest lasting. Tanks are best in urban areas and on flatlands while smaller dams are appropriate anywhere the landscape can support them. Dam walls get very expensive the taller they are, especially if they are holding a lot of water behind them; they also have more potential to have issues that can pose a danger to your land, infrastructure, animals, and family.

Earth Storage = Cheap Water Conservation,
Wildlife Habitat, and Drought Mitigation

Rainwater Tanks

Rainwater harvesting tanks are ubiquitous in Australia and noticeably absent in other areas of the world like the US, Europe, and India. Most areas get enough rain for all the family's needs if caught in the landscape and off the roof in a storage tank—this includes dry, arid regions like Arizona and Saudi Arabia though it can take time to establish these systems. Careful calculations are needed in terms of the rain coming down, the square footage or meterage of the roof catchment, the amount of storage space, and the frequency and amount of rain; even with attention to detail, calculating exactly the catchment and infiltration of an area is always impossible. We can only approximate as closely as possible by taking into account as many factors as possible. We can budget our usage based on our storage and water catchment capacities.

An urban rainwater demonstration tank system at the Avis home in Calgary, Canada. This system purposefully has mistakes to teach students: the IBC tank is too small and it should be covered to prevent algae and bacterial growth.

Using a first flush diverter (pictured), we can divert contaminated roof water that is collected when rain initially hits a dirty roof. Limestone, marble, or shell in a mesh bag can be suspended in the storage tank to mineralize or "harden" the water by making it more alkaline. Rainwater is naturally "soft", as is lake water, and is lacking in minerals. Giardia and other organisms that present health concerns grow in soft water as well. Keeping organic matter, insects, etc. out of your tank and tank in-flow is important: use gutters that block out material and add grates and screens wherever needed.

In many places in the US currently, there are laws against harvesting rain. Instead, they require all the residents to import water from a great distance at a great cost, have it treated with harmful chemicals, and then pay for it to be piped to their homes. Immense amounts of water are lost and degraded in a process which requires large amounts of energy. There's an easier way! We can all provide our own water in our local communities or even on-site.

Cisterns are wells that are designed to catch and hold rain and runoff for irrigation and home consumption. It can be as simple as a hand-dug well lined with brick or stone that is 6-7m deep and 3-4m wide with a smaller opening near the surface. These buried stone or earthen tanks are an ancient, reliable technology.

Swales

Swales are an earthwork designed to intercept, pacify, and soak water. They are level terraces of variable size with a soft, raised berm on their lower side that stops and absorbs the water captured by the flat area. It can be as simple as a foot path on contour. The un-compacted swale can flood feet deep in rain events yet soak it all in if designed in well-draining soils with level sill spillways.

Brandon Carpenter 2015.

To make a swale, soil is removed from the hill on contour and placed below that contour line forming the berm. As we dig down, the berm builds up. The path or road is perpendicular (at a 90° angle) to the flow of water—it's a full stop and completely flat. An enormous amount of water can be captured and soaked in with swales. Based on your research and planning, the swale can be designed to handle the largest rain or flood events in your area—but bear in mind that some areas already get enough moisture. Consecutive swales down a mountain side or hill can recharge dry springs and change an ecosystem fundamentally without irrigation. This occurs naturally often in a forest where trees fall at or nearly at perpendicular—they intercept and slow the natural pathway of water downhill, so in that area, it soaks in more deeply. We all can do this big or small—micro or macro. It will work to save the California Sierra Nevadas and other desertifying forests as well as urban areas in arid climates. It is a universal law that water is pacified when it is level, and it is also true that it soaks into unsealed earth easily. It is in essence a groundwater recharging catalyst.

A wide swale with a small berm with a furrow in it.

Swales are tree-planting systems at their core; the tree roots hold the berm in place. The more shade you have over the swale path and berm, the greater the water retention. It is therefore ideal to keep the width of the berm narrow enough to be under the canopy of the largest trees growing in the berm itself. They can also be ideal when designed into food forests, orchards, and perennial gardens with some annuals.

Swales can be used with grazing animals in the swale path. Just remember: any compaction in the soft berm area negatively effects the rhizosphere, or root area, of the plants living there—no grazing the berms. Grazing is ideal in large swale corridors with perennial grasses in the pathways (where the water is caught) with electric fencing lining the soft berms on either side, so that animals can graze the swale path and browse the edges of the berm without trampling the berm, but they must be moved through quickly to avoid compacting the soil and reducing water infiltration. Holistic management of cattle, which involves shifting livestock around different paddocks fenced with solar-powered electric fencing, can efficiently raise, not just cattle, but in turn also poultry, other dairy animals, food forest crops, and a myriad of other products. Tight control of animals in electric fencing imitates natural grazing patterns from when in the past predators were present and herd herbivores moved together very tightly as they grazed.

Swales can be smaller or closer together—especially on a hill where it is steep—dig them less deeply into the hill, and, conversely, the berm will be smaller. In order to fight erosion and capture as much of the water flowing over steeper slopes as possible, we need more aggressive trees, perennial ground covers, shrubs, or bushes to hold the soil together, stay in place year round, and spread vigorously (but not invasively), so they cover up the soil completely. We also need the plants to be synergistic—complementary in their inputs and outputs as in beneficial polycultures.

A swale creating a microclimate in the middle of the desert at the Al Baydha Project, Saudi Arabia.

To aid in absorption, the entire area can be ripped below the soft berm, along the swale path, or even the area where the berm gets made prior to cutting the swale. Ripping consists of running a straight furrow 1-2 ft (~.5m) deep into the earth—it doesn't overturn the layers, just rips a tear. This allows for water and root infiltration. Keeping compaction to a minimum is ideal even on the path—the more absorption the better. In sandy soils, wider shallow swale paths are ideal while in clay soils deeper and narrower swale paths are needed.

Seeding swales with annual legumes and planting nitrogen-fixing perennial seedlings or bare root stock and valuable tree seedlings or bare root stock (timber, fruit, nut, fiber, fuel, etc.) is a superb way to start a food forest. We can choose trees that will grow to shade the berms within three to five years or grow from seed, which takes longer. Their leaf drop, seed drop, and accumulation of organic matter like fine silts, pollen, insects, etc. through their action as windbreak will then provide the organic matter needed to feed the soil life to feed the trees and shade the entire swale end to end. This has a cumulative effect. Much in the way lenses in a telescope enlarge a distant object or the way they focus light in a laser, water can be focused in the soil with swales stacked above and below each other on a slope. The integrity and strength of each water-lens in the swale-stack depends on its shade, how wide the swale path is, how well water can infiltrate into the soil, and how much water is entering the site at any given time. It should be noted: swales must be able to safely overflow—spillways are vital.

Feeding swales with diversion drains will magnify the amount of water entering the swale area which can reduce or eliminate irrigation costs. In this way, swales can be used to catch road or any hardscape runoff and absorb it effectively into the landscape while feeding a productive ecosystem that provides an abundance of value to the community. Pitting the swale path is also an option (and can help with clay soils)—these pits can be filled with manure or compost or inoculated wood chips to feed the plants, support soil life, and aid in soakage. Paths should be wide enough to move materials along them.

At UC Davis in California, swales were used in a suburban housing community, and within a few years, water was penetrating 19 ft (6m) into the soil with swales self-shading around the same time.[30] In arid climates, swales can bring back forests, yield larger trees, and prevent soil degradation and salinification. In more humid climates, they yield larger trees, permanent **chinampas**, recharging of aquifers, renewal of natural springs, and a greater diversity of life on all trophic levels.

Swales are useful but may not always be appropriate (or allowed). Once called 'contour drains', swales have been historically used in the US and Australia in the wrong soil types, and instead of stopping erosion, these swales silted up and created more erosion.[31] That being said, in the right situation, swales are extremely effective.

In humid climates with heavy clay soils, trees planted on contour will be enough to mimic the effect of a swale—a swale in that situation would only make the trees grow rapidly, fruit lightly, and be prone to disease. Having too many springs appear in a pasture can also be a serious problem for some cattle who can suffer from hoof-related diseases if they are constantly wet. In other areas, it may not be allowed, and there are some simple tricks to make an invisible swale with either a keyline ripping technique or reducing the size of swales by filling them with wood chips or gravel like an invisible pond in concept. For those with pastures, ripping on contour is an ideal option to minimize the catchment and preserve the shape of the land.

Diversion banks and drains

Earthen diversion drains and banks differ from swales primarily in that they channel water towards a pond or swale to absorb or hold it, but within a few seasons in the tropics or humid climates, an earthen diversion drain's compaction is relieved by soil life and plant roots leading to it becoming absorbent, like a swale. Swales can even be used inside of large diversion drains to absorb water in the diversion drain beds themselves. All water has to continue safely in an overflow event and as much water as possible should always be soaked into the ground for storage—our earthworks will always be a pattern and a marriage between these two concepts.

Compacted earth, un-compacted earth, or impermeable layers: it depends on your needs, the local environment, and available resources. The amount of water coming into your

---

[30] Mollison, Bill. _Permaculture: A Designer's Manual_. 1989. p. 168.
[31] Yeomans, Alan. _Priority One._ 2005. p. 132.

area must be understood in order for an appropriate design to be made. If more water than the system can hold is coming in and if you don't have spillways, you could have a broken swale, plant damage, and unexpected and potentially dangerous flood paths. Using the largest rain event on record as a baseline, we can figure out what to expect when 50 or 100 year cycle storms hit. If we are prepared for these events and include spillways and large soakage areas, we can expect our earthworks to perform well in these times.

Clever additions can be made to the drains that act as portable or stationary spill gates to release water strategically, but good design initially is better than using more energy to intervene later on. The less we do, the more energy we save and can use in other areas of interest. That said, if these spill gates are used on a farm, two people can water many acres of land with minimal effort. Pipes through the diversion drain berms can also provide quick water release. Wildfire controls can be used with sheet irrigation (water released on a level sill) with automated gates set to infrared sensors to flood out forest fires early on and before any human response can reach that area.

Interceptor drains are sealed diversion drains that only direct water. They are sealed by ramming the earth, compacting it to the point that not even salt can seep into the soil. They can act and look like diversions drains or a swale but they are sealed. These are especially important in drylands where salting occurs. Extremely salty water must be diverted and cannot soak and seep into the soil.

Ethics of Earthworks and Water

All earthworks can be restorative in function as long as they are not being used for high water demand crops in drylands or some other function that removes that water from the natural cycle or bioregion. These earthworks are designed to slow, spread, store, and soak water into the local environment. Keeping a careful water budget is critical to ethical and responsible consumption. At the Al Baydha project in Saudi Arabia, they carefully measure rainfall catchment and soakage to make sure to never use more water in the dripline system than the site soaks in. They have a food forest emerging out of soil that any **agronomist** would say is non-arable with an average of 3" (7.5cm) of rain a year, some years getting no rain at all. Earthworks can work what seem like miracles but are simply re-aligning natural cycles for greater abundance, resilience, and water retention. It all can start with a shovel, a positive attitude, and a good design.

Graywater and Blackwater Conservation Starts Before Usage

Though it seems rather simple, if we focus on using less water in general in our homes, we will have less graywater and blackwater to filter. Graywater is wastewater lacking fecal matter or contamination; blackwater is from toilets and, sometimes in states like California, includes the kitchen sink water. We can lessen the amounts of graywater and blackwater we generate by clever reusages and minimizations. Sink graywater can provide the toilet flushing water, but it must be a low flushing toilet to keep pathogen numbers low. New hand washing sinks that attach to the tops of the toilet are being introduced—you wash your hands with the water refilling the top of the tank, so your hand washing water will flush the toilet for the next person using it. Dry urinals and compost toilets are also gaining popularity as they save enormous amounts of water. There are numerous inventions and designs out now that can save immense amounts of water like the tippy-tap, initially created in the 1980s in Zimbabwe, where a bottle of water is suspended above hand-washing areas and manually tipped with a food pedal to release small amounts of water (see image).32 This concept of releasing water in only a trickle with a foot pedal can easily be adopted into mainstream designs all over the world.

Graywater and Blackwater Treatment Systems

Laundry detergents, shampoos, and all soaps used in our homes need to be able to be safe to add to our graywater, so we can route that waste water into the garden or food forest. Detergents and soaps should lack salts, boron, and chlorine bleach—consulting databases that analyze and rate household products like those of the Environmental Working Group (EWG.org) can be helpful. The ingredient names can be confusing and knowing how products compare to each other can be impossible without a laboratory and scientific know-how; having both, the EWG also explains their ratings systems clearly in their online database.

Reed beds and mycoremediation wood chip beds can also be used to purify water before it waters fruit trees. Separation and diversion of urine is ideal as it can go immediately into the garden especially if watered down. Urine is a perfectly balanced ratio of 1:1, carbon

---
32 Teutsch, Betsy. *100 under $100: One Hundred Tools for Empowering Global Women*. 2015. p. 56.

to nitrogen, so plants and soils love it. It also can be reserved to treat leaf fungus or mold in higher concentrations or to feed the plants in lower concentrations (raising protein levels in rice specifically).

Graywater can maintain a large, vibrant garden or food forest if managed properly. Blackwater can be composted and used in the garden as well. Keeping these two waste streams separate and as uncomplicated as possible is ideal and saves time and energy.

Because we are eating out of the garden, certain things should never be allowed into that waste stream. Prescription drugs, industrial chemicals, and non-compostable trash must be kept out of the graywater and blackwater waste streams. As with most short-term solutions like plastic "To-Go" cups, they tend to stick around for the long term in a municipal waste center or a landfill. If we are responsible with our choices, and use what we already have in these forms wisely and carefully, we can avoid these kinds of long-term problems from the start. This means starting out with the expectation that what we have (and what we will acquire) will be used for as long as possible and recycled into other uses—also for as long as possible—to keep these substances out of the waste stream.

The Purification of Polluted Waters

Healthy forest ecosystems, protected watersheds and ridges, and an abundance of life at all trophic levels—these are the only long-term protections possible for maintaining a regular and pure source of water for any large population. Water scarcity and impurity can both be remedied with these natural systems. Banning of all biocides, industrial contamination, and human contamination from waterways is critical. All human waste must be handled at its source: we must stop exporting our waste—all of it. We are gathering all the nutrients, trace minerals, and exotic materials we can find in the world and dumping them into our landfills, oceans, and atmosphere. We need to reverse the process. Luckily, we can.

Common Issues

- **Turbidity** - The water is not quite clear; it has suspended silts in it.
- **Bacterial or Organic Pollution** - E.coli, fecal matter, decaying organic matter, etc.
- **Metallic pollutants** - Mercury, cadmium, lead, etc.
- **Biocides** - Pesticides, herbicides, insecticides, and fungicides. If persistent, these can prevent healthy gardens, soil, and life from developing.
- **Fertilizer Runoff** - The runoff from **petrochemically**-derived nitrogen, phosphate, and potassium fertilizers is creating massive toxic algae blooms (**eutrophication**) and dead zones like the dead zone in the Gulf of Mexico just below the Mississippi delta, where Iowa's and several other states' runoff and soils all deposit.
- **Acids** - Increasing acidity is causing metals to become soluble and to bind with plant proteins and animal fats, which eventually leads to human ingestion. Cooking with acidic water only magnifies the issue and adds more metals to the foods being prepared especially if cooked in something that the acids can break down, such as aluminum. As carbon dioxide levels increase in the atmosphere, a correlating effect occurs in the ocean; as carbon levels rise there, it becomes more acidic. In fact, today we have ocean waters that are so acidic that organisms like clams and shellfish are having difficulty developing their shells in some areas. Some coral are not developing at all, and coral reefs like the Great Barrier Reef are bleaching and receding.

Common Water Treatments

- **Aeration** - Adding oxygen to water purifies it in several ways. It promotes **aerobic** bacterial growth which inhibits **anaerobic** bacterial growth (which is negative for plant and animal health). Aeration can be created with a check dam, a water fall, a pump on a solar panel, or even with phytoplankton.

- **Settling** - Pacifying water allows silts or undesirable elements to settle out. Upper levels of water can be removed while lower water and the muddy bottom can be dredged if need be and processed separately.
- **Skimming and Sieving** - This removes large pieces of organic matter
- **Filtration** - Filtration can be through biological means with pond life cycling it, or it can be via the physical action of passing through natural filters like sand, unglazed clay, or activated charcoal. Sand bed filtration methods are widely used—a steady trickle or drip into a large sand bed will allow for unclean water to be filtered by bacterial life in the sand. Soil works in a similar and superior way because it has more soil life and plants as well to help. Carbon is vital for removing nitrogen. This is why we balance manure (N) with straw (C). This process also ties up toxins into long carbon chains. This is essentially what is happening with sand and charcoal filters.
- **Coagulation or Flocculation** - Chemical or natural additives are used to bond with the undesirable elements which then settle or float to be dredged or skimmed off. Adding crushed marble, limestone, or shells causes particles to settle through flocculation.
- **Biological Removal** - Plants absorb the toxins and then are removed to be composted. The toxins are then ideally trapped in the soil though some persistent toxins need yet more remediation. Soil can be further improved by being mixed with wood chips and **inoculated** with vigorous mushrooms like oyster or king stropharia to further **bioremediate** the soil.
- **pH Adjustment** - By adding calcium or sulfur compounds to change the pH. As plants and pond life do a better job of maintaining pH, any chemical adjustment we make will be temporary as the natural ecology will reset after it processes the chemical amendment. Using a careful selection of plants and pond life to achieve and maintain a particular pH is a more long-term solution.

Drinking Water and Ponds

Purity of drinking water is of critical importance though, collectively, we use and abuse our drinking water in wasteful ways that can be avoided. The water in ponds is alive with positive organisms in a way that chlorinated municipal water is not. The same is true of every moving body of water—every spring has its own unique blend of bacteria and fungi in the water. Our gut biomes are full of life and need perpetual nourishment and living waters. Stagnant water always begins to lose its vitality and require chemical treatment or filtration and aeration.

Using water lilies or water hyacinth, we can exclude sunlight from hitting the pond surface and promoting algae or bacterial growth. Stabilizing plants along the water's edge like reeds or bamboo lower turbidity.

Trickling, dropping, or dripping water initially into loose pebble beds and then into sand beds for bacterial absorption of contaminants, followed finally by passing through a sand column from below can prepare water for biological filtering.

Water can be treated by mollusks—mussels filter out bacteria and turn it into waste that forms a phosphorus-rich mud base below them. They also serve as the "canary in the coal mine" since they die at 5.5 pH or lower and are sensitive to biocides. Using living indicators can often be our best method for preventing and detecting contamination. Following this, the water can be processed by aquatic plants like watercress which can then be composted or fed to animals. Water then can be fed through an activated carbon filter like charcoal or burnt oats, rice, or wheat husks. Using gravity, this system can create clean water without energy inputs.

Sepp Holzer prefers to let the landscape itself filter the water and balance the mineral content. He does this by routing a sub-surface flow to his front yard's bubbler, a small bubbling fountain that keeps the water aerated. Sepp also has an idea for keeping water fresh with a Ring Water Feeder, featured in <u>Desert or Paradise</u> (2013), where water is continuously cycled through a house system in a circle with the input at a higher level than the end of the ring, so pressure is maintained. The pressurized water is then returned to the start of the ring. There is an overflow valve for high water-volume time periods and to release pressure. This keeps the water aerobic and safe for human consumption without chlorine. It also is very similar to ancient roman plumbing where waters were constantly moving to create pressure and remain aerobic.

Sewage Treatment Using Natural Processes

In small municipalities sewage can be treated initially with freshly burnt lime (kiln-baked seashells and limestone) as it settles initially, then following that, it is fed through sand beds or trickle towers to remove bacterial contaminants and ammonia then fed into large lagoons to allow a natural wild-food cycle to begin to take in the excess nutrients. Biochar can likely substitute for the lime. In larger municipalities especially in cities, they cannot rely on these methods alone to clean and filter the waste. There are too many combined and complicated waste streams in these larger systems for water to get as clean as it would in a natural system. All too often, the sewage treatment policy is often one seeking water that is "clean enough", not perfectly clean. These large municipal or privatized sewage treatment systems are flawed in many ways while the small scale, more natural systems are more efficient and effective.

What is Sewage?

Raw sewage is primarily comprised of human waste (humanure) which equates to a combination of nutrients, heavy metals, carbon compounds, nitrogen compounds, and other trace elements. In addition, there are now substances present that were never intended to be part of the sewage treatment system like prescription drugs and industrial chemicals which are difficult and often costly to filter out. It is critical that we are strict with what goes down our toilets and, looking to the future, what we put in our mouths. The higher the quality of the sewage, the more useful it is and the quicker it breaks down into safe compostable material. A balanced, mostly plant-based diet yields the best sewage in terms of compostability and balanced C:N ratios (heavy meat eaters tend to have humanure that is high in N). This is well documented in historical accounts of "night soil" systems in both Japan and China.

Stages of Treatment

- Anaerobic or Methane-Producing Ponds
- Facultative Ponds (both anaerobic and aerobic)
- Aerobic Ponds

During anaerobic bacterial action, heavy metals are absorbed and made inert. Sulphur-loving bacteria dominate but few algae can; this is what gives anaerobic ferments that rotten egg smell. The anaerobic settling ponds are especially critical for removing heavy metals especially. The resultant sludge also releases methane which can be a great source of fuel if used in a closed loop system. This is a smelly but critical stage of the process. It is similar in smell and function to a swamp or marsh.

Facultative action happens with phytoplankton: algae blooms, which are aerobic, and anaerobic bacterial life work together in one setting with some bacteria thriving in both conditions. Sulphur-preferring bacteria persist initially in the sludge but are not present when passing onto the next stage. It is similar to the edge or release areas of a swamp or marsh.

Oxygen-rich environments have lots of life, and it's all aerobic. These can be both producing oxygen and requiring oxygenation through physical or biological means. Zooplankton can thrive at this stage and support larger organisms as well. Some of the remaining metals are taken up by zooplankton. Algae (and rushes) can transpire hydrogen and can break down herbicides, pesticides, and other halogenated hydrocarbons. They can process and release into the atmosphere mercury, phosphates, chlorines, and more volatile chemicals. At this point the concentrations of all undesirables are within safe levels.

Since methane is produced throughout the process, it can be used to power the machines aerating and agitating the anaerobic and facultative sealed ponds, or it can be pumped from below to provide aeration and disturbance as well. Often pools are made shallow or with glass tops to magnify heat and speed up the process while trapping methane buildup.

The final stage is only a refinement of the aerobic environment with higher order plants like water hyacinth, rushes, or reeds cleaning the water. It should be noted that certain plants are useful in pulling out certain contaminants. Water testing and diversity are critical to knowing what the system needs to further purify the water—your water might be higher in copper or lead specifically. Careful selection of the proper plants for your area and spectrum of issues is needed. Depending on the kind of waste water, the treatment system can range from a living biofilter area with gravel and sand filters for a natural swimming pool to settling ponds with stages of bacterial digestion separated out for the most effective processing of sewage. Most municipal sewage treatment plants are doing similar things but chemically or mechanically which is costly to the environment and in terms of the energy input. We can keep things uncomplicated, small, and beneficial with home, neighborhood, and small, local systems. The smaller the amount of waste to process, the simpler and faster the processing of raw sewage can be.

These areas can be sheltered and lined with plants and trees that tolerate and even thrive on sewage's dense nutritional profile. Comfrey can grow in the presence of raw fecal matter extremely well. It can be planted around animal pens to catch any overflow, prevent it from spreading, and provide superb mulch ready to compost beneath plants or later to be used as animal feed. Willows have been documented growing healthily around outhouses as well. More research is needed, especially into mycoremedial options, but it is clear that we have amazing natural systems that are capable of handling diverse waste streams and translating them into more balanced and useful outputs.

Water leaving these treatment areas can be used to raise cattle, grow food, or water yards and landscapes. To a large extent, by reusing our wastewater, we can replace the freshwater we are currently wasting.

Compost Toilets

There are many different kinds of compost toilets currently available. The basic concept is that human manure (humanure) is composted by mixing it with a carbon-rich, high-surface area material such as wood chips, straw, hay, seeds, or sawdust.[33] These toilets can compost the humanure directly below the toilet: drop, tumble, and aerate, or they can be removed to an external composting area. Most simply, one can add one's deposit of humanure to a bucket and add carbon-rich, high-surface area material. Once full, the bucket is taken to an outdoor composting area. From this simple method or from a system with a high degree of sophistication, including automated, electric toilets, humanure moves through a specific composting process, conserving and returning its valuable nutrients and biological communities back to the soil food web cycles.

Composting of humanure can either be **thermophilic** (hot) and quick or cold (mouldering) and over a long period of time. Thermophilic composting creates a sterile product quickly while **mouldering** compost might allow parasitic eggs and pathogens to persist and would require thorough examination before that kind of compost would be determined safe. It is important for anyone looking to start working with composting toilets that they respect both the hazardous nature of fresh humanure as well as the management required. Carbon to nitrogen ratios must be maintained, and the mixture must be well-mixed, and often stirred, to guarantee even composting and a safe end-product. Several municipalities are considering bringing compost toilets into their building codes; the demand is high, and the science is proven. Gord and Ann Baird of British Columbia are people who used and proved compost toilets worked, and now they are helping draft the British Columbia building codes for them.[34] In Sonoma county, popular demand is fueling compost toilet research.[35]

"The toilets of the future will also be collection devices rather than waste disposal devices. The collected organic material will be hauled away from homes and composted under the responsibility of municipal authorities, perhaps under contract with a private sector composting facility. Currently, other recyclable materials such as bottles and cans are collected from homes by municipalities; in some areas organic food materials are also collected and composted at centralized composting facilities. The day will come when those collected organic materials will include toilet materials"
—Joseph Jenkins, The Humanure Handbook, 2005.

---

[33] Jenkins, Joseph. *The Humanure Handbook.* 1999. p. 122.
[34] Baird, A., Baird, G., Hill, G., Hoeppner, E., Payne, M., Seymour, M. *Manual of Composting Toilet and Greywater Practice.* 2016.
[35] OAEC.org. *Compost Toilet Research Project.* 2016.

There are more and more commercially available compost toilets each year. There are also designs for septic tank systems that compost rather than grow in toxicity and toilets that separate urine from humanure–these designs could also collect biogas. Innovations like the Tiger Toilet, a latrine using tiger worms to digest humanure while aerobically composting the urine, are providing affordable solutions that are small enough for a family home.[36]

Natural Swimming Pools

Natural swimming pools are pools or ponds that use biofilters to clean the water. There are numerous do-it-yourself models online as well as professional services for natural swimming pool installations. Hot tubs are being filled with plants and pebble beds to filter the water for the main pool area. Aeration is needed, but can be attained in several ways–especially if converting a standard, in-the-ground pool. Without altering the hardscape at all, a bio-filtration area can be directly next to the swimming area at a low cost, cleaning the water with plants before it is released back into the pool. The pool water can stack functions as a place to raise fish and store water for irrigation, human consumption, and fire fighting.

Frank Golbeck of Golden Coast Mead's natural swimming pool in Southern California. It is a converted standard pool with a liner, fish, plants, external bio-filtration system, and pumps.

---

[36] HydrateLife.org. *Eco-Latrine of the future: Tiger Toilets*. 2012.

The plant filtration station for Frank Golbeck of Golden Coast Mead's natural swimming pool, designed and installed by Eddy Garcia of Living Earth Systems. Vetiver grass, onions, strawberries, herbs, and more feed on the fish manure in the water, cleaning it in the process.

Water to Power

Humans have a long history of working ingeniously with water power. The Archimedes screw drill, water wheels of all kinds, dam turbines, **hydraulic ram pumps**, **trompes**, and even giant, stream-side washing machines are examples. We use gravity, water pressure, and the constant flow to creatively address our needs.

Designer's Water Checklist

- Find where water enters into the site and assess its quality and quantity.
- Develop a plan to clean and store that water
- List site goals to identify areas of usage for that water
- Use gravity to save energy and stack functions
- Develop a list of plants that don't require irrigation to shade and populate the landscape to promote more absorption and retention.
- Determine the directions of water-flow on the land - graywater, blackwater, roofwater, road runoff, swale overflow, etc.
- Plan for worst case scenarios of flooding and precipitation
- Plant trees on swales and berms around earthworks to soak up excess water, stabilize the earthworks, and to desalinate the water and soil if needed.
- Slow, Spread, Soak, and Store Water Wherever Possible

VII. Soil

edited by Dr. Elaine Ingham

"A teaspoon of productive soil generally contains between 100 million and 1 billion bacteria"

—*Soil and Water Conservation Society, 2000.*

Soil is the living skin of the earth. It is the most diverse and life-rich medium on this planet, and all things spring from and return to it. Soil is created through many physical, biological, and chemical processes. Clay, sand, and silt are commonly seen as the primary components of soil, but organic matter and the soil food web are as vital, if not more critical, for growing food and supporting life than the ratios of the clay to sand to silt—though understanding it all is critical.

Oxygen dissolves in rain and oxidizes iron in the rocks and soil. Frozen water and plants expand in the cracks of rocks to force rocks apart. Glaciers grind against rocks, turning them into fine mineral powders. Fungi are decomposing rocks with acids and organic matter with enzymes. Waves on the beach are slamming microscopic pieces of coral, crystal, and shell against each other creating ever finer sands, perpetually.

Any soil from anywhere on earth can grow plants. The nutrients may not be soluble, but the action of soil life can make them soluble. For decades we've been measuring only water-soluble forms of nitrogen (N), phosphorous (P), and potassium (K) without considering how fungi, soil life, and plants work on rocks and soil elements. It was once thought that soil needed to be a near-even mix of clay, sand, and silt to be perfect loam, but Dr. Elaine Ingham has proven otherwise in the field with commercial growers on large acreage in challenging climates like in Nevada with soils of pH 11. Soluble nutrients in the soil are misguiding indicators—all soils have all the micronutrients, trace elements, and macronutrients available in non-soluble form. The mechanism to release these nutrients is the soil biology.

Top soils, the refined, most life-rich top layer of soil, are being lost at a horrific rate through modern agricultural practices. Those same farmers could be building those soils each year through natural cycles. Herbivores and perennial prairie grasses worked together (along with their entire ecosystems) over hundreds of thousands of years to create the largest carbon sink and deepest soils in the United States, only to be washed down the Mississippi river to the detriment of all life downstream. Experts haggle over how many years of topsoil we have left to support agriculture, but they almost all overlook how quickly things can change. The places where soil is being generated naturally—grasslands, lakes, ponds, pastures, and forests—are dwindling; even most commercial no-till farms require organic matter or compost from offsite. We can, instead, rebuild the soils the same way they originally were generated: with herbivores and perennial grassland and savanna polycultures.

Soil Erosion

Whether through soil structure collapse or wind or water erosion, the loss of soils, especially topsoils, threatens the ecology dependent upon it. Though soil is created through erosion, this is not the same thing in this instance: it is the loss of soils rather than the formation of them. Poor agricultural practices like tillage, biocides, and synthetic fertilizers (which are salts) accelerate the natural processes of erosion. Rain and snowmelt can also erode soils without plants or mulch covering them. Earthworks, trees, mycorrhizal fungi, grasses, mulch, and organic matter in the soil are keys to stopping erosion.

Soil food web

Our soils are maintained, built, and nourished through a series of cycles involving many levels of a soil food web. From the smallest bacteria to the earthworm to the tree to the mushroom sprouting, they are all connected and interacting. Through consumption, reproduction, predation, waste, and movement through the soil, these soil-dwelling organisms cycle the elements of the soil, air, and water, changing compounds from one form to another, allowing for new opportunities for other trophic levels to participate at each stage

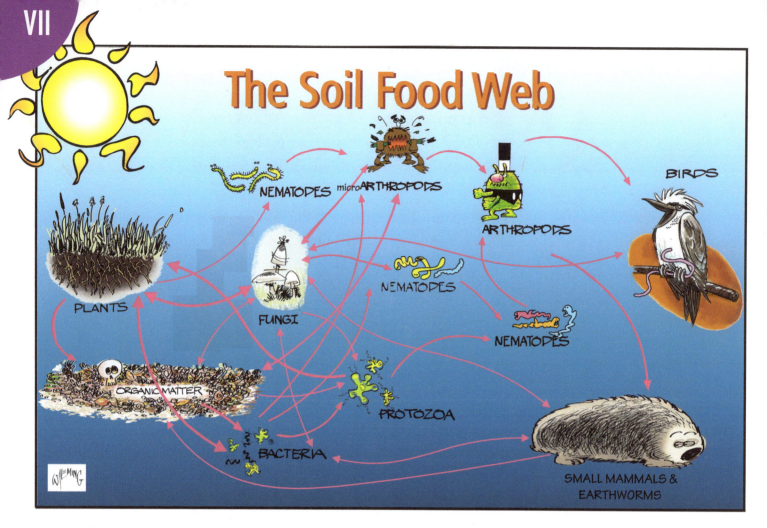

of decomposition and growth. The soil food web is critical to clean air, water, soil, and, thus inherently, clean food and bodies. These organisms process, trap, and transform toxins, biocides, and excess nutrients as well as build humus and soil structure. Though we rarely look up close, most of the life in a forest or field is below the surface and in the soil. Only

> "The Soil Food Web Performs The Following Functions:
> 1. Cycles nutrients in rocks, sand, silt, clay, organic matter ('total' nutrient pool) into plant-available forms, i.e., soluble nutrients that plants can take up.
> 2. Retains nutrients in soil so leaching losses and erosion do not occur.
> 3. Protects all surfaces of plants from disease, pests, and parasite attack. If a plant is dying, it is because the proper biology was not present to prevent an attack.
> 4. Decomposes toxic chemicals that would otherwise kill plants; converts wastes into plant-beneficial materials.
> 5. And, possibly the most important, builds soil structure. It converts dead plant residues into well-structured, well-aggregated organic matter allowing the free movement of atmospheric gases, water, and roots through the soil."
>
> —Dr. Elaine Ingham, Phd., *Permaculture Magazine North America*, 2016.

when we use microscopes can we truly be aware of how much life and potential there is in just a gram of soil.

Plants using photosynthesis produce exudates (mostly sugars and small amounts of protein and carbohydrates) to attract desirable sets of bacteria and fungi—each root hair can put out a different set of exudates. Exudates can exude from all plant surfaces but primarily the roots. Plants feed on the nutrients released by the nematodes, microarthropods, and protozoa feeding on the bacteria and fungi attracted by the exudates—in fact, for plants to be healthy, all trophic layers need to be active and in balance. Plants also participate in intimate exchanges with fungi and bacteria where they discreetly and directly exchange nutrients. Plants are creating the setting for the foods they need to be available with the kinds of exudates they put out. Dr. Elaine Ingham compares the exudates to cakes and cookies that fungi and bacteria feed on which attracts higher level soil life and begins a whole series of desirable cycles that feed the plants exactly how they need to be fed (not force fed through water soluble fertilizers). In essence, plants feed themselves when soil life is rich and diverse. The soil food web's activities provide enough nitrogen and carbon to support annual gardens on their own, independent of any fertilization. Plants need to be free to develop their own associations for the soil food web to be strong enough to accomplish this feat; chemical fertilizers prevent the soil food web from functioning properly. Compost teas, in contrast, set the stage for plants to choose their own associations by providing a concentrated solution of soil microbiology. This allows the plant to pick and choose which aspects of the brew to proliferate.

In terms of categories, there are at least five trophic levels to the soil food web. The 1st trophic level includes plants and organic matter which provide all the building blocks and food for the next trophic levels. The second trophic level is populated by decomposers, mutualists, and less desirables like pathogens, parasites, and root-feeders: root feeding nematodes, fungi, and bacteria. The third trophic level is one of grazers, predators, and shredders: protozoa, fungal and bacterial feeder nematodes, and shredder arthropods. The fourth trophic level is comprised of higher level predator varieties of nematodes and arthropods. The fifth trophic level is occupied by small, animal predators like birds, moles, and voles. All the trophic levels have waste that cycles back to the start of the soil food web's trophic cycle; it all becomes food for plants, fungi, and bacteria again, and all are dependent on the energy from the sun feeding the initial trophic layer. Complexity supports and enriches the soil food web while increasing the number of times energy, nutrients, and water are cycled in an ecosystem.

Bacteria

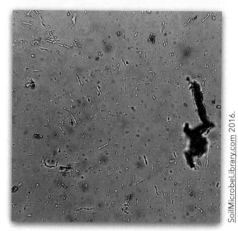

"Rods of Bacillus sp" Bacteria

Bacteria are single-celled organisms that feed on simple compounds. Soil bacteria usually feed on plant exudates or residues and can be found focused around the roots' exudates, decomposing organic matter in high concentrations, and miles deep into the surface of the earth where greater diversity exists though overall populations are lower for the most part: "Bacteria are found in high concentrations in the deepest oil and natural gas wells ever drilled."[37] Some bacteria are photosynthesizing (creating sugars with sunlight and CO_2), but almost all soil bacteria are decomposers with the exception of those bacteria getting their sustenance from plants roots—like **rhizobia**. Bacteria and fungi are both critical to the breakdown of organic matter and the building of soils, but bacteria are faster at returning to devastated areas than fungi or plants, traveling on the wind and on surfaces of all kinds. They arrive and establish the soil environment necessary for fungi and pioneer species plants to move in though there have also been studies showing fungi may be critical for bacteria to establish initially in arid areas.[38] They embody the highest amount of nutrients and provide along with fungi the foundation of the soil food web.

Bacteria create glues that hold **micro-aggregates** together. These aggregates hold water and nutrients. Soil fungi's hyphal strands as well as fungal glues connect micro-aggregates to make macro-aggregates; creating soil structure and interstitial spaces for water and air.

Bacteria primarily fall into four groups: decomposers, mutualists, pathogens, and lithotrophs or chemoautotrophs - though there are phototrophic (light eating) bacteria as well like cyanobacteria. Most bacteria are decomposers that feed on simple carbon compounds from plant exudates and residues. Mutualists, like nitrogen-fixing bacteria that live in root nodules, form a mutually beneficial relationship with another organism. Pathogens are disease-causing bacteria. The final group, lithotrophs and chemoautotrophs, feeds on specific forms of "nitrogen, sulfur, iron, or hydrogen instead of carbon compounds."[39]

Fungi

Translators of nutrients, water, and minerals in the soil, fungi also populate our bodies, the air, all plants, and the water, and they span in their expression from one-celled yeasts, to

---

[37] Ingham, Elaine. *Email*. 2016.
[38] Worrich, Anja, Stryhanyuk, Hryhoriy, Musat, Niculina, König, Sara, Banitz, Thomas, Centler, Florian, Frank, Karin, Thullner, Martin, Harms, Hauke, Richnow, Hans-Hermann, Miltner, Anja, Kästner, Matthias, & Wick, Lukas Y. *Mycelium-mediated transfer of water and nutrients stimulates bacterial activity in dry and oligotrophic environments*. 2017.
[39] Ingham, Elaine R., Moldenke, Andrew R., and Edwards, Clive A. *Soil Biology Primer*. 2000. p. 18.

long chains of cells called **hyphae** that form fungal bodies called **mycelium**, to lichens when partnering with algae, and to the large structures known as mushrooms. Fungi are amazing!

Decomposers of wood lignin, fibrous materials, cellulose, soil humus, and other complex carbon compounds, they eat what bacteria do not. Since these complex carbon compounds don't exist until later in the succession of growth, bacterial action usually dominates at first. Though fungi in different forms are always ever-present at all stages of succession, fungal dominance can only come later in succession when forests develop with mature, acidic soils. Fungal-dominant soils are found predominantly in old-growth forests.

The highway of mycelium in one cubic centimeter of undisturbed soils spans kilometers, where it acts as a communication network and delivery service for plants. Plants can respond and adapt to environmental changes quickly through this communication network, and also transport nutrients between each other, all thanks to these fungal partners.

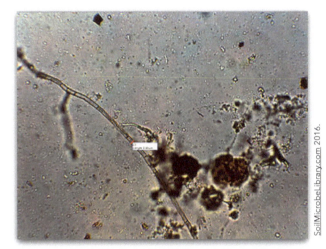

Hypha strand at 200x magnification

Fungi can also transport nitrogen from other areas of the soil to assist in breaking down complex compounds like the lignin in wood. This is why many food growers have concerns about using wood chip mulch in their gardens or farms: they fear the wood will temporarily lock up all the nitrogen in the soil as the wood breaks down. But in a mature no-till system, soil mycelium can bring in extra nitrogen, enabling soil life to access nitrogen from the inorganic sources, such as soil minerals. Green plants are primarily broken down by bacteria because nitrogen is still present in high enough levels—they don't need to bring in nitrogen to aid in decomposition (as we do in composting).

Fungi come in many groups, but for soil and plant concerns four are of primary interest: saprophytic fungi (decomposers), mycorrhizal fungi (the mutualists), **endosymbionts** which are found ubiquitously inside of all plants, and parasitic and pathogenic fungi. These groups complement bacterial roles in the soil and provide all trophic layers of the soil food web with critical components for their own cycles. More on fungi in the next chapter.

The Nitrogen Cycle

Weeds thrive in alkaline soils where only nitrate (NO_3^-) is available. This happens when there is perpetual disturbance as with tillage-based agriculture. This is why modern

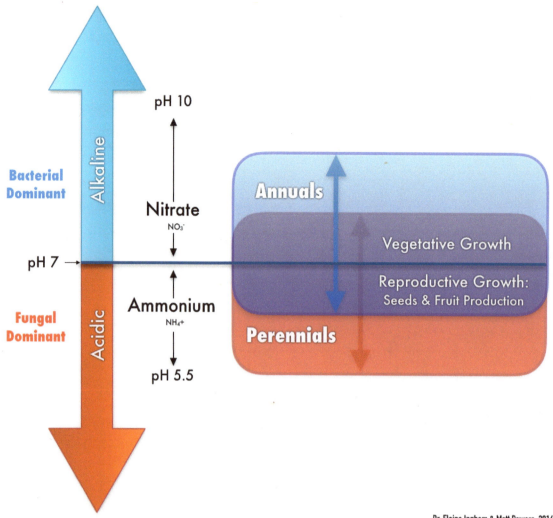

Dr. Elaine Ingham & Matt Powers, 2016.

agriculture always seems to fighting off an invasion of weeds. Primarily it is because they destroy the fungal hyphae and set the stage for bacteria to dominate. Adding salt-based fertilizers and biocides makes things worse—they serve as bacterial foods once they begin to break down, further favoring bacteria over fungi. NH_4^+ (ammonium) is turned into NO_3^- (nitrate) by these bacteria as they create the glues for soil structure after tilling. NO_3^- can also be released back into the atmosphere as N_2 by denitrifying bacteria, and it can be leached out of the soil through watering or precipitation events.

Most plants need a balance of fungal and bacterial elements, ammonium and nitrate, but they can't get that if the soil is bacterial-dominant and alkaline. They also can't get access to both forms of nitrogen if the soil is a uniform pH or all-acidic. Diverse pH zones are needed for plants to obtain the form of nitrogen they need, when they need it, and in the amount they need. Nitrates feed vegetative growth while ammonium feeds seed and fruit growth.

Remember: above pH 7, the form of nitrogen in the soil is nitrate, but below pH 7, it is ammonium, so soils need a range to give garden plants a full spectrum of choice. They will specialize around each root hair, so we need to give them a wide selection of microbiology to source.

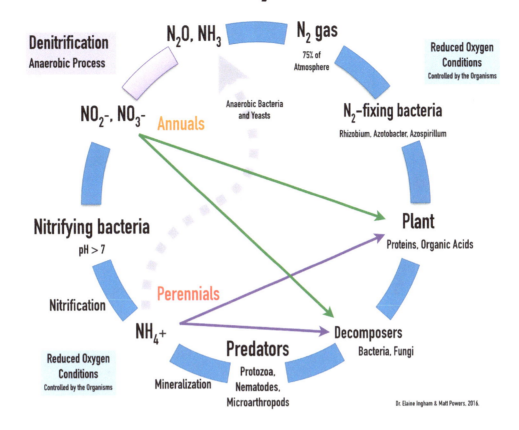
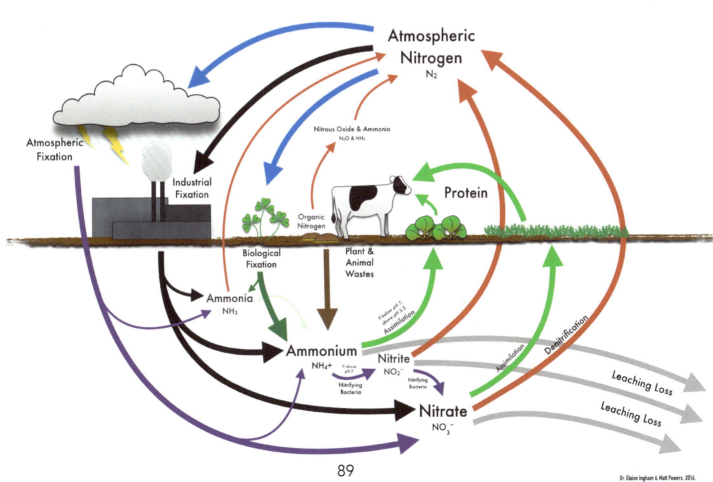

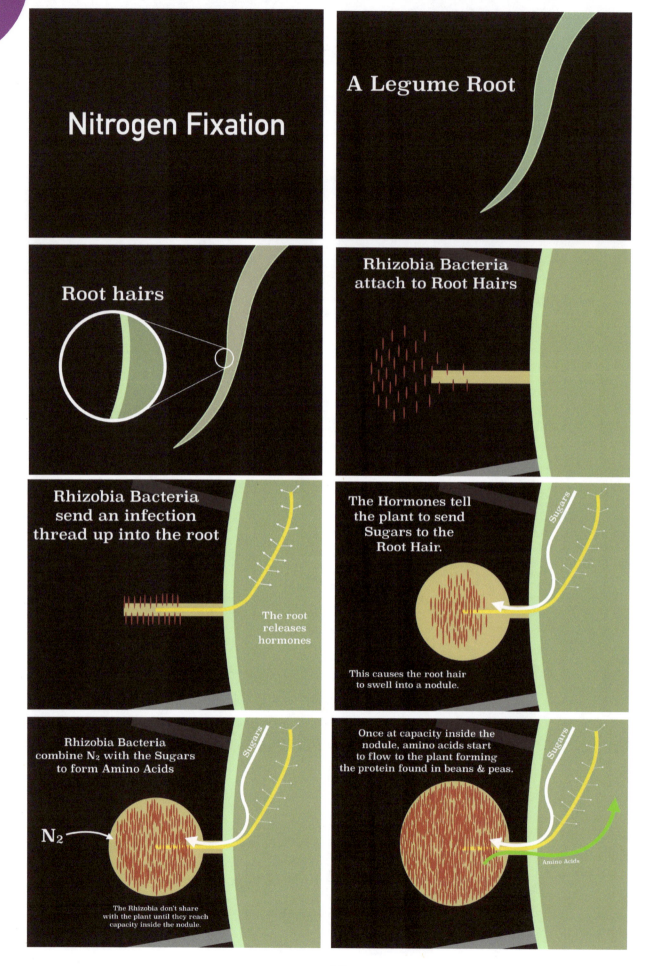

Protozoa

Protozoa are single-celled organisms, including ciliates, amoebas, and flagellates. Just like other soil organisms, protozoa move in the water films on the surfaces of soil particles and in aggregates. Protozoa are five to 100 times larger than bacteria. Flagellates, amoebas, and ciliates consume bacteria for the most part. Large protozoa will eat smaller protozoa, some amoebas have an ability to eat through the cells walls of fungal hyphae, and thus consume the internal contents of fungal strands. Protozoa are eaten by predatory nematodes, microarthropods, earthworms, and larval stages of many insects.

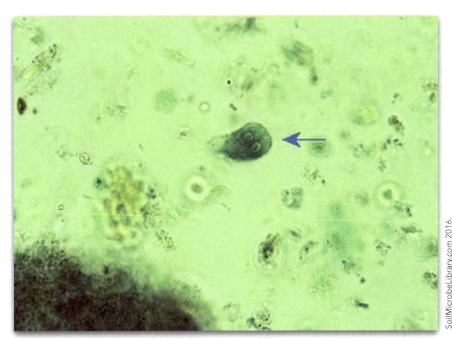

A Flagellate, Protozoa

Since the concentrations of nutrients inside both bacteria and fungi are much greater than protozoa require to maintain their own bodies, the excess nutrients (N, P, S, K, Fe, etc.) are released as soluble forms that plants can take up. The majority of flagellates and amoebas require strictly aerobic conditions in order to thrive and reproduce. Ciliates grow better at reduced concentrations of oxygen because their enzymes can continue to operate at lower oxygen levels than flagellates and amoebas. When oxygen is low, and competition from other protozoa is reduced, ciliate populations can reach high levels. Thus, ciliates are considered to be indicators of low-oxygen (anaerobic) conditions. When oxygen drops below 4mg of oxygen per liter, however, no protozoa can survive, and they will encyst, or go dormant, in order to survive these conditions.

Protozoa, microarthropods, and nematodes act as shredders of organic matter, allowing for more surface area to be accessible to bacteria and fungi to break the organic matter down further. This action feeds the fungi and bacteria and increases their populations. Protozoa and nematodes consume fungi and bacteria and release the nutrients stored in the consumed fungi and bacteria in their waste which plants feed upon.

Nematodes

Nematodes are non-segmented and, usually, microscopic worms. For a long time, nematodes were given a bad name, but it was primarily because no one knew what their roles were exactly. It was thought that all nematodes were root feeders while the truth is only some of them are. There are also bacterial- and fungal- feeder nematodes as well as predator nematodes which feed on protozoa and other nematodes. There are even omnivore nematodes.

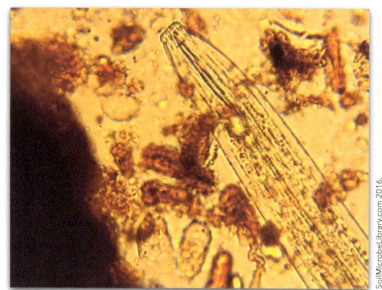

A Fungal-Feeding Nematode

Nematodes regulate the populations of other soil food web participants, mineralize the soils, provide nutrients for the soil food web, and consume pathogenic organisms. The presence of undesirable nematodes indicates soil conditions that are anaerobic; aerobic compost teas, ripping, or some other aerating action is needed.

Microarthropods & Arthropods

Microarthropods Arthropods are invertebrates with leg joints that prefer the surface or top 3" (7.5cm) of soil to live in: insects, sowbugs, scorpions, millipedes, beetles, ants, crustaceans, swobs, arachnids, and others. They can range from visible and quite large to microscopic. Microarthropods mostly shred materials as protozoa and nematodes do but are not consumed by predator nematodes, only by larger, predator arthropods and animals. They can be classified in four groups: shredders, herbivores, predators, and fungal feeders, though most eat other organisms and fungi. Their waste creates soil structure and enhances microbial activities while they regulate populations within the soil food web and keep disease-causing organism populations low—though it should be noted that several herbivore varieties are considered pests because they feed on plants.

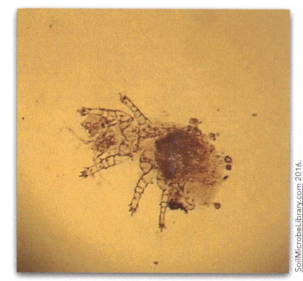

A Mite, Arthropod

> "Like earthworms, microarthropods encase their fecal pellets with a layer of mucus, thus helping to build macro-aggregates. Arthropods of all size 'rearrange' the positions of micro- and macro-aggregates, helping to create large, medium, and small size pores within the soil. The largest arthropods in soil build channels, just like earthworms, and thus are very important in burrowing air tunnels deep into the soil"
>
> –Dr. Elaine Ingham, 2016.

Earthworms

Earthworms are found in most soils in many places all over the world—where they are not found, **enchytraeids** perform the same function in the soil food web. Earthworms are small animals that gently till the soil beneath the surface efficiently and aggressively without damaging or oxidizing it. They eat bacteria, fungi, protozoa, nematodes, microarthropods or any living organisms they can crush when the soil gets inside the earthworm, and excrete a worm casting that is an ideal garden soil amendment or bacterial-dominant compost tea source. These worm castings along with the tunneling action create soil structure that is ideal for plant roots, water infiltration, and other soil organisms to utilize. Worms redistribute organic matter as well.

Earthworms fall into three categories: *epigeic* (in surface soil and litter), *endogeic* (in the upper soil), and *anecic* (deep-burrowing). These worms all perform the same function in their habitats. Whether it's animal waste, kitchen scraps, yard waste, or mulch, worms process their food quickly into rich, worm casting-strewn soils. An abundance of worms indicates healthy soils and can be also used as a feed source. They are high in protein, and they are often consumed by other components of the soil food web and can be raised for poultry or fish feed (as can many other invertebrates).

> "[Worms] can turn over the top six inches [15cm] of soil in ten to twenty years"
>
> –Soil and Water Conservation Society, 2000.

Plant Roots

Roots are critically important in that they hold soils together, provide exudates for the soil life interactions, loosen soils, prevent erosion, provide food for decomposers and root feeders, fix atmospheric gases, sequester carbon, and support a plant that is adding organic matter to the topsoil. Most of the life in the soil is focused around the roots of plants. Bacteria and fungi focus on feeding off the root exudates, while nematodes and protozoa mostly feed off the bacteria and fungi. Almost all of this happens in the first 6" (15cm) of soil. Fungal

Alex McVey 2016

associations with roots are common, numerous, and can change—single plants can associate with many different types of fungi at the same time. Rotating crops more often creates a greater diversity and greater overall amount of food for the soil life by introducing a diversity of plant roots over a short period of time. No-till cultivation combined with crop rotation creates a diversity of decomposing plant roots as well—giving roots and the soil food web a variety of foods to choose from.

> *"Trees, on the other hand, [their] roots can go down 150 feet. The world of landscaping states that roots of trees and shrubs only go down a few feet, and then they go sideways... The roots of all our plants should be going down at least as far as they grow above ground."*
> –Dr. Elaine Ingham, "Lecture notes, Work in Beauty workshop", 2015.

Perennial and native grasses have been recorded with roots of over a dozen feet deep (2-3m). These plant roots provide deep storage and soakage of water, a powerful energy storage system for dormant periods, and a fast carbon sequestration pathway as the entire plant can die down and generate soil each year while the plant roots continue to develop and improve. Many studies have assumed the deepest roots are getting harder-to-reach deeper nutrients, but there are more available nutrients in the top 6-8" (15-20cm) of soil than anywhere else because that is where the soil food web does most of its work—though soil food web members are found up to 16 miles (26km) deep into the earth.[40]

Roots respond creatively and sometimes unpredictably to their conditions. Robert Kourik's superb book, _Understanding Roots_, contains many root drawings that demonstrate plant root formation, granting the reader valuable insights:

- The above-ground form of a plant does not reveal or indicate what its root structure is.
- Roots do not grow deeper in drier times to reach water, instead their root systems shrink accordingly.
- Roots extend well beyond their dripline.
- New, feeder roots grow past the dripline, so mulch starting at the dripline and move outward from there.
- Most roots are found in the top 6" (15cm) of soil with almost all focus being on the top 6-18" (15-46cm) of soil, so don't mix that top 6" (15cm) of soil with any of the subsoils!
- Less than 2% of trees have a taproot, and they only form a taproot when they are grown from seed.
- Roots also redistribute water by pumping it up from the soil depths to soils closer to the surface which in turn provides moisture to other plants and the soil food web itself, benefiting all.

---

[40] Ingham, Elaine. _Lecture Notes, Work in Beauty workshop, Gallup, NM -- November 7, 2015_. 2015.

Soil Activity

Soil life is most active during the growing season. In some climates, winter may be the time of maximum moisture and growth with summer being a dormant and dry period. In other areas, summer is overly-wet and winter provides a respite that allows growth to return.

MacroNutrients

- **Phosphorus** (P) - Sources: bird manure, bat guano, unlocked by fungi, gathered by roots, or naturally occurring in kainite, sedimentary rock, and igneous rock. We can create bird habitat or start keeping birds to add their manure to our composts to guarantee this element in our soils, or we can increase the soil food web activity.
- **Potassium** (K) - Sources: potash is generated through composting, especially vermicomposting. Worm castings have up to 11 times the potash levels normally found in garden soils. Kitchen scraps and yard waste can easily become potash.
- **Nitrogen** (N) - Sources: manure, n-fixing plants, electrical storms, rain, and bacteria. Balancing our soil pH with organic matter and soil life keeps nitrogen cycling and avoids negative nitrogen imbalances (like nitrates leaching into waterways). Avoiding tillage also keeps nitrogen in forms healthy to people and useful to plants.

Calcium (Ca), Magnesium (Mg), and Sulfur (S) are also considered Macronutrients. None of these macronutrients need come from an industrial, chemically-derived source. They all can be generated on a homestead at little-to-no cost and used to amend the soil with time and labor as the only investments and local biology as the only input. Micronutrients can be released via the action of soil life, as can primary nutrients, independent of any soil additives we generate on the farm. Compost teas can save enormous amounts of time and energy in this way.

Soil types

Soils have been categorized and analyzed from many different perspectives and by many different cultures. Color, taste, clay/silt/sand ratios, nitrogen content, drainage or water capacity, texture, structure, and more have all been used to classify soil. Soils are a focal point in many cultures for good cause. Every climate and even every site has soils with unique characteristics. All are improved with raised organic matter levels and soil life.

Brittleness Scale

Developed by Alan Savory, the soil brittleness scale spans from 1-10, from the most humid climates to the most arid. Brittleness is determined subjectively by observing the soil and plants, but it is guided by how fast plants break down. In brittle, arid environments, plants and animals take a long time to decompose while in humid or non-brittle climates, they decompose quickly. The scale also helps determine if herbivores can be used in desertified land recovery—their removal in humid, non-brittle climates allows for quick recovery while it does not in the brittle climates. Brittle areas don't have rain most of the year, and they might even get rain only once a year in a true desert. Brittle environments must be carefully managed with holistically managed herbivores—lightly, intensely, and quickly—followed by periods of rest. Non-brittle environments can be restored with nothing more than a few seasons of rest. Here is an easy test to determine the brittleness of any area: does a bare patch of earth heal quickly or slowly on this site? If it degrades or spreads further, you are definitely in a strongly brittle climate.

pH

pH is a scale that measures the concentration of H+ hydrogen ions in a solution and indicates how acidic or basic a liquid is. We can suspend soil in water and test its pH as well or even test our urine or saliva as doctors do during routine physical examinations. pH is an important indicator: if ground or rain water is too acidic, heavy metals become soluble, and then plants can absorb them—though it should be noted that rainwater rarely has acidity this high. The more acidic the water, the more heavy metals can be made soluble and absorbed, so industrial sites and areas of high pollution should be closely monitored. That being said, pH is not as accurate at measuring soils themselves.

While water is easy to capture, mix evenly, and measure, soil is spread out and so

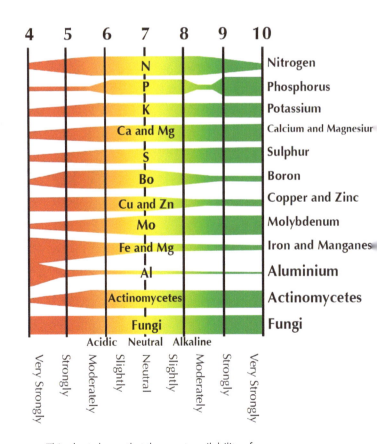

This chart shows that the most availability of desirable nutrients is found in the 6-8 pH range.

diverse that pH in undisturbed soils can change every inch, millimeter, and micrometer of soil in all directions. This is because plants affect the pH of the soil around them using their root associations. Plants release exudates that attract specific populations of soil life, and their waste changes the pH of the area around the root hair; the exudates put out can change entirely from root hair to root hair favoring fungal or bacterial processes. Therefore, using soil pH to determine what to add to the soil is an inexact science; we'll never get an accurate reading, and it's easier and less costly to just have the plants calibrate the pH themselves using compost teas. In natural systems, both anaerobic and aerobic sites appear and disappear regularly as part of the natural cycles of the ecosystem. In undisturbed soils (untilled) we have layers of sediment and organic matter that have built over time with only biological "turn-over" which means that each layer is unique and has served as habitat for billions of bacteria and fungi for hundreds and even thousands of years, so it would make sense in the face of that complexity that each root hair would respond uniquely and autonomously to each unique site. That being said, pH is important.

> **"pH is determined by whether aerobic, beneficial fungi or aerobic bacteria are predominant. When protozoa, nematodes, or microarthropods eat bacteria or fungi, ammonium is released. If aerobic fungi are predominant in the soil, the organic acids released by fungal metabolism will maintain soil pH between 5.5 and 7.0, thus keeping ammonium in that form. However, when bacteria are predominant, the mucigel or glues that bacteria produce to hold themselves on the surfaces of the root, the mineral particles or organic matter will shift soil pH into alkaline ranges, thus allowing nitrifying bacteria to express their DNA to convert ammonium (NH_4) to nitrite and then nitrate. Plants will typically control the predominance of bacteria or fungi by releasing foods that will select for bacterial or fungal growth, and thus controlling pH"**
> –Dr. Elaine Ingham, 2016.

If we are tilling our soils, we are cutting up all the fungi hyphae strands and favoring a bacterial-dominant soil environment, and the soil environment turns alkaline, produces nitrates, and then favors weed growth—which can be seen as pushing the soil and plant succession backwards away from fungal, forest soils. We can "see" pH to a degree in these settings.

Organic Matter is Key

If we raise the amount of thoroughly composted organic matter in the soil, or humus, we can see dramatic changes happen very quickly because bacteria, fungi, herbivores, and plants all rely and thrive upon humus-rich soils. High enough levels block and trap toxins. The nutritional density of our food and the diversity and concentration of living enzymes in our food is directly correlated with the humus content of soil. It is shocking that soil agronomists didn't take this into account long ago.

While some talk of a minimum 10-20% organic matter in the soil being needed, natural soils range from nearly pure organic matter (such as peat moss) down to nearly pure minerals found at the base of a glacier grinding over rock. In soils high in organic matter, soil biology traps heavy metals in insoluble forms. These soils also hold more water due to organic matter's spongey nature and the soil structure formed by the soil life feeding on the organic matter. This organic matter can be grown on-site and does not need to be imported in most instances.

On brittle, delicate, or shallow soils, it is critical to use plants, fungi, and animals to build and protect the soil. Animals are critical in these systems, but they can easily be overused, and they too can add to the degradation of the land. We need to observe and actively manage our soils, plants, and animals to see if we need more diversity, more aeration, more compost, fewer cattle, or to move the cattle twice a day instead of once—it all depends.

Soil Structure

When soils have a crumb structure, it allows for interstitial spaces for gels, water, life, and air to move easily through the soil. Only soil life can give proper soil structure ideal for plant life. Tillage destroys soil structure. Soil life need these tender highways to transport water and gases for their exchanges with plants and each other.

"When soil is tilled, fungi, protozoa, nematodes, microarthropods, larger arthropods, spiders, etc are killed or flee from that destruction. All that dead biomass is a source of food for bacteria, most of which are not killed by tillage. But the food and the bacteria are mixed, giving the bacteria a massive amount of food that they are now near. And bacterial growth increase[s] by a massive amount. The CO_2 released is from that use of organic matter by bacteria. But, adequate structure to prevent erosion requires aggregates built by fungi, but the fungi have been harmed, and their aggregate broken up, the structure of the soil destroyed. In addition, [the area] where the plow share, or tillage blades rested against the soil as the tiller was pulled through the soil was compacted. Above that point the soil may have been fluffed, but the plow depth is compacted. Water moving through the soil cannot move from the soil with fluffed texture into that compacted soil, and the water backs up at that layer, puddling, preventing oxygen from moving into the soil. The compacted layer goes anaerobic, with all the unpleasant consequences.

> *Additionally, since the water backs up, it will end up eventually going sideways and down the hill, taking soil and organic matter with it"*
> –Dr. Elaine Ingham, 2016.

Soil Water

The water that soils hold is full of life and nutrients. It's like a compost tea but not hyper-aerated–in areas where there is poor drainage, soil water can easily become anaerobic, but once that area dries out the remaining soil moisture can return to an aerobic or mostly aerobic state. Soils can hold large amounts of water and prevent evaporation if they are well vegetated or covered with mulch. Soil water, colloids, and air or gaseous spaces are vital to soil structure and fertility. Waterlogged soils can be assisted with earthworks such as swales or even *chinampas*, to raise the soils above the water table, or soil conditioning, like deep ripping or amending with sand, is also possible. Poor quality water, like salty water, or extreme atmospheric pressures also inhibit water availability. Ideally the water in the soils should be slightly acidic so as to make the soil elements more readily available to the plants and soil life. This is essentially the original compost tea that nature creates through a symphony of biological processes.

Soil Gases

Similar to grasses, plants, and trees, the soil acts as a pair of lungs for the atmosphere; soils take in air and moisture and release recycled air and moisture. Oxygen and nitrogen are commonly present, but in a diverse soil food web each area is a unique microcosm of diverse gaseous exchanges, much like a street corner in NYC is entirely unique but understandable. The moon's tidal push and pull works on the soils of the wetlands as well, releasing gases and injecting air, alternately. Changes in air pressure can also cause this effect. Animal-made holes and burrows also serve as air pumps for soil gas and air exchanges.

Carbon Sequestration

When organic matter becomes humus, it forms long-chained carbon compounds that sequester carbon in the soil which prevents the carbon from entering the atmosphere. Holistically managed herbivores, reforestation, no-till agriculture, mulching, growing our own firewood and timber, covering up bare soils, and using less wood in general, all these combined will help. However, primarily shifting our agricultural land into carbon farming and ranching methods while replanting, properly managing, and protecting the wilderness, watersheds, riparian areas, coasts, and wetlands, in order to sequester carbon on all fronts, is the only solution that will reduce the atmospheric and oceanic carbon levels to pre-industrial levels.

> "If we increase soil carbon by 1.6% in a foot of soil on the world's agricultural soils then we'll download 100 parts per million of carbon dioxide out of the atmosphere"
>
> –Darren J. Doherty, Permaculture Voices: PV3, 2016.

Fungal:Bacterial Ratio and Succession

Fungal to bacterial ratios change as soils become more acidic or, inversely, more alkaline. The ratio of fungi to bacteria concentrations in the soil on productive agriculture sites is usually 1:1, while forests can range from 5:1 in deciduous forests to 1000:1 in coniferous forests.[41] The closer to old growth forests we get, the more fungal-dominant the soils become. The closer to the beach or desert we get, the more bacterial-dominant the soils become.

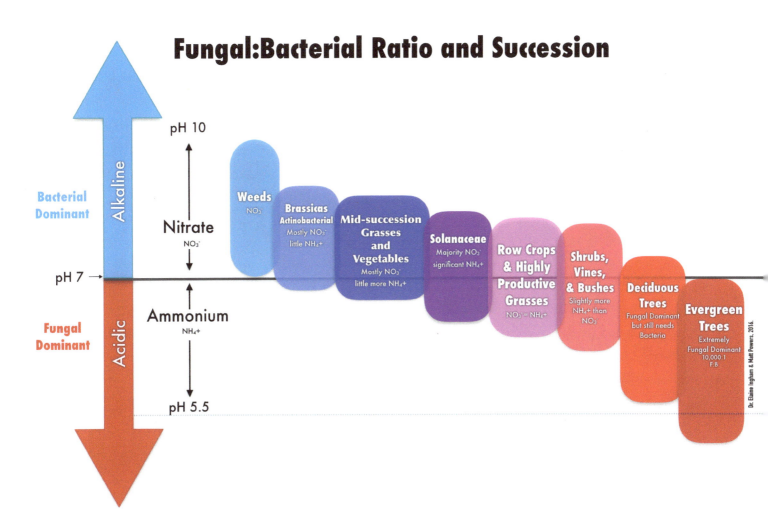

Annuals prefer more bacterial-dominant soils, though mycorrhizal fungi are still present ubiquitously, while perennials and trees primarily prefer more fungal-dominant soils. Low tillage operations tend to lead to more fungal-dominant soils as disturbance is lessened and soil life is allowed to work undisturbed. Annuals thrive on disturbance which destroys fungal hyphae and soil structure. Understanding this critical difference in annual and perennial crops will help you decide what kinds of ingredients to add to your compost teas, what composts to add where, and what plants to associate together in your polycultures.

Concreted or Cemented Soils

Though there are many types of difficult soils to work with, it should be noted that their remediation techniques are universal: ripping, earthworks, mulching, **green manure**, legumes, compost teas, strategic planting, and herbivores. Earthworks in naturally concreted or cemented soils may need heavy machinery or even explosives to create areas for mulch to collect and decompose. Allowing the soil life to work on the severely compacted soils will change the composition quickest. Ripping and injecting compost tea will allow the same process to happen but beginning in the ripped cut. If the ripping is done on-contour at the top or bottom of catchment, with parallel or on-contour rips between them, the effect will magnify and the soils will change even more quickly. Planting out the soils with aggressive green manure crops like cowpeas, field peas, or vetch will change the soil biology and add organic matter to the soil. Adding in herbivores to process the green manure crops on site, and then introducing chickens to spread the manure out and scratch it into the soil will lessen labor and increase soil biology and organic matter. Allowing the area to rest and then repeating the process when it is ready is a sure way to develop that soil into a desirable growing medium.

Hydrophobic Soils

Some soils repel water and are difficult to wet and keep wetted (sometimes caused by fungi). This can be remedied with organic matter as food for the biological reactions, clay for water retention, ripping to open up access to the subsurface, and earthworks to trap organic matter and water. Clay pans that form below excessively tilled fields can be treated in a similar manner as well though sand can be added if easily acquired or on site. Bringing in materials to the site should be avoided wherever possible to lower energy input and make the land itself self-sustaining. Soils are primarily repaired quickest and in the most desirable ways through the actions of the soil food web.

Ethical Interactions with Soil

No-Till or Low Tillage Annual and Perennial Production

Exposing soil life to the air exposes it to oxidation, killing the biota, and while it does make for a short-term windfall as their passing provides food for plants, the long term effect is loss of topsoil, soil life, and nutrient diversity. That being said, annuals need disturbance to thrive. Light or shallow tillage (1-2"/3-5cm) to allow for annual plantings can be ethically managed easily in cool and cold temperate climates, but the closer to the equator you get, the more difficult temperate-climate-style annual production becomes. The more we till, the more nutrients leach from the soils and into the local waters over time. We also change the nutrient composition when we till, making temporary blooms of excess nutrients, which over time deplete the soil of nutrients. For example, tilling turns ammonium into nitrate by favoring bacteria and favoring pH conditions which leads to weeds (which are reparative species of plants). When tillage does not occur, the polyculture of roots, that are continuously decomposing, provide nutrition to both plants and the soil food web perpetually.

Organic no- or low-till operations need to source their organic matter, their mulch, or their compost from their own farm if they are to be sustainable or regenerative. For what nutrients and organic matter the farmers take out of the soil and export off site, they have to return to the soil. This is an immutable equation and must be included in the holistic management of any production operation that requires soil disturbance.

In natural ecosystems, organisms consume organic matter, transform that matter, and pass on the transformed organic matter through their waste to the soil, other organisms, and the atmosphere. When we produce large amounts of annual vegetables on a piece of land and export it, all the water, nutrients, and carbon contained in those foods leave with them, and that piece of land now has a standing deficit that must be addressed. The future may well have farms collecting composting materials as they drop off CSA boxes of produce or even using local, fully composted, and pathogen-free humanure to replace exported organic matter, or the future may just have families composting their waste at home after growing mushrooms on them first, gaining a secondary crop before turning it into a garden soil amendment.

While no tillage is vital to preserve the soil, to

James Powers, 2016.

sometimes cleanse the soil of layers of settled toxins, we may need to turn or rip the soil and then add compost teas and organic matter. It should also be noted that before 10-11,000 years ago, we had Mastodons in North America and their elephantine cousins roaming the earth creating large, heavily-manured disturbances for regular successional reboots, forming a fully functional ecosystem where annuals and perennials were in a constant state of coming and going. Disturbance is natural, to an extent—though it requires rest to recover—it is different from modern agriculture's usage of constant disturbance. Currently, we have so much carbon in the air that we need to take it back into the soil as soon as possible, so we should avoid soil disturbance as much as possible.

Aeration

Adding air to compacted soils can be vital—even though air oxidizes soil life, it also makes soil life more aerobic. Using a pitchfork or a broad fork (as pictured), we can plunge their tines deep into the soil and then gently rock back and forth to aerate our soils in the garden without turning over any soil layers. Over-aerating will collapse the soil structure just like over-tilling would. Earthworms, moles, gophers, and much of the soil life also provide excellent aeration. Often in rain events or snow-melt events, the aeration provided by these organisms becomes a pathway for water infiltration and soakage.

Yeomans Plow or SubSoil Ripping

Using a straight sub-soil ripper attachment on a plow avoids turning the soil; instead the ripper creates a 0.5 meter-1meter deep rip in the land. A "shoe" at the bottom of the shank breaks up compaction in the sub-soil, which promotes subtle water catchment in pasture and can also be used to control wandering root systems such as honey locust in orchard rows. Ideally done on contour or near contour for the strongest effect, this allows deep water infiltration, deep compost tea application, and/or tree planting all on contour or in a keyline pattern. The trees or plants with aggressive root systems will break up the soils and allow for more water infiltration as they begin on the process of building up the organic matter. Compost tea applicators and seeders can be attached to the ripping setup as well—stacking functions!

Using Less Land

Because we can stack functions and multiply yields in a single space and keep it local or home-based, we can use less land. We can also increase our soil-building cycles to increase our yields on less land. If we increase the nutritional density in our foods, then our bodies will not need as much food and medicine which leads to us using even less land and less money. The more land we can design to be regenerative the better, but we need to focus on the land already in production and not expand any further. We need to return what land we can back into wild habitat.

This does not mean producing less, in fact the end result will be that we produce more. Jean-Martin Fortier in Quebec is making $100,000 a year on a 1.4 acre (5000m^2) farm. We all can use less land to be more productive, but we have to be smart in how we do it. Joel Salatin's farm is 550 acres (222.5 hectares) and is considered a hobby farm size, yet his farm is serving thousands of families and his practices are building soil and sequestering carbon annually and measurably. If we use less land, we can be better stewards of it, better managers, and better designers. Large sprawling monocultural farms cannot harmonize with nature; they're too big and monolithic. Only smaller, adapted, and specialized farms can accomplish this.

Building Soil

Through grazing, manure, mulch, compost tea, soil life, trees, and other perennials, we build soil. We've lost so much soil so fast that it is critical that we create it faster than it has been lost. We need massive composting, mulching, and fungi operations to tackle this issue. We also need to leverage our technology and machinery to do this, so that we can do it in a shorter time than it has taken us to create the damage. If we have a plan and widespread adoption, we can see real changes in topsoil generation and carbon sequestration in only a few years.

VII

Conservation of Soils in the Wild

Bare soils are the biggest issue in both open, untended land and in agriculture. All bare soils erode. Native forests, prairie, wetlands, and the whole diversity of natural habitats need to be brought back—but with climate change in mind—so they can be successful and resilient. Preventative thinning, large-scale restoration earthworks, and burns will be needed to fight back megafires; also needed will be a massive effort to replant the burned-out, at-risk, and degraded areas, followed by an active silvicultural practice to manage the forests. Large holistically managed grazing operations need to be used as well since the original grazers have already been removed. If we want the carbon in the air to return to the soil, natural cycles are the fastest way to make that occur.

Hugelkultur
translation: Mound+Gardening

Earthworks

Earthworks like *hugelkulturs*, terraces, and swales build soils. Swales and terraces pacify water in rain events which allows the water to drop whatever it was carrying. This leads to a rich flood plain of soil accumulating over time that can be left in place on a terrace or harvested to maintain the level and regularity of the swale and to be used in a garden or on the soft berm below.

Hugelkulturs build soil quickly by imitating deadfall in a forest. Associated often with Austria's Sepp Holzer, *hugelkulturs* are mound-cultures in translation: you mound dirt on top of a wood pile which speeds up decomposition of the wood and feeds and heats the plants in the mound. This is also a great way to compost without losing by-products into the atmosphere, especially if it is too hot. Composting does release gases, but *hugelkulturs* work like a contained compost where the plants, soil, and soil life intercept the released forms of nitrogen and carbon. Water is also held extremely well by the punky, fungi-inoculated wood inside the mounds. These kinds of earthworks could be used to build back the soil in the forests of the Western United States and many other degraded landscapes worldwide. They can be used as firebreaks and areas to replant the new polycultures. Read more in the Earthworks and Earth Resources Chapter.

Building Material

Natural building with soil can take many forms and is one of the longest lasting and energy efficient forms of building we have: mud-brick, rammed earth, earth bag, adobe, cob, natural plaster, earth-sheltered, earth-banked, and more! All these techniques leverage the unique power of earth. It also can be molded into anything: lines do not have to be straight. In addition, earth is easy to acquire and can be excellent insulation. Using earthen building materials can save energy, time, and money.

A cob wall under construction in Peru. The darker colored cob is freshly added and still drying.

Legumes and Soils

Most but not all legumes form a symbiotic relationship with nitrogen-fixing bacteria (Rhizobia) that "fix" atmospheric nitrogen inside a root nodule and upon decomposition, release their accumulated nitrogen. (There are also fungi-like nitrogen-fixing bacteria called Frankia as well.) It should be noted that nodules do not form on roots in soils that are too acidic (under 5 pH) or in areas with high nitrogen levels. Initially, the plant sends out exudates to attract the exact type of rhizobia bacteria needed. Rhizobia bacteria then attach to a root hair and send an infection thread through the root hair up into the main root (much like AM fungi). The bacteria then send hormones to the plant that cause the plant to send sugars in large quantities down to the root hair where the bacteria feed on the sugars and generate amino acids, which cause the root hair to swell and form the root nodule.

Root Nodules that house the nitrogen-fixing bacteria on Autumn Olive tree roots (above) and Siberian Pea Shrub roots (below).

Inside the root nodule the bacteria form a colony that in the center is anaerobic. Inside where there is no oxygen, the bacteria fix atmospheric nitrogen (N_2) by turning the sugars and the N_2 into amino acids (the building blocks of proteins). The plant only sees a return on its investment once the nodule is full to capacity with amino acids. From then on, the plant and the bacteria are storing amino acids with nitrogen bound up in them. Only when they decompose or are consumed and digested do they release nitrogen back into the soil. This is why legumes are chopped and dropped or tilled in and why manure is a source of N. Otherwise they very slowly release the nodules which break down and feed neighboring plants and the soil food web.

Some non-legumes such as Western Redbud and Buckwheat also develop these relationships and fix nitrogen. It should be noted that in nature all plants and animals return some form of nitrogen as waste or through decomposition to the soil and air.

In the tropics, it may be easier to build soils by chop-and-drop mulching or by plowing in than by composting because compost cycles so quickly that it can often yield little and release the vital nutrients as gases. Fast growing green manure legumes like cowpeas are ideal for developing soils in the tropics.

Phytoremediation and Mycoremediation

This is the removal or degradation of toxins or pollutants though phyto(plant) or myco(fungi) remediation. Rice plants remove arsenic from water as do many aquatic plants. Currently there is an epidemic of arsenic in commercially available rice, demonstrating that there is no consideration given to the possibility of either the water being contaminated with arsenic or the management practices making soluble arsenic available (anaerobic soils, synthetic fertilizers, etc.)

If we are to use fungi or plants to remediate the soil, we have to compost the byproducts of those processes and not eat them because they retain some of the original toxin—or a degraded form of that toxin—in their biomass.

Insects and Worms

Insects like the black soldier fly and earthworms like the red wiggler compost worm can consume compost heaps incredibly fast. Insects and worms create compost and reproduce quickly; they can also be a feed or supplemental feed for poultry, reptiles, and fish. There are even insect fodder trees that can be grown to attract insects for farm animal consumption. Super worms and meal worms have also both been documented to digest styrofoam and to transform it in their digestion back into organic matter, spurring the work of Living Earth System's Eddy Garcia to turn all old surfboards into trees. We are just scratching the surface of what insects, worms, and all the members of the soil food web can do!

Inoculants

Inoculating seeds or transplants with beneficial bacteria or fungi gives plants and soils a head-start on working together proficiently. Inoculants can be purchased and sometimes made at home. N-fixing bacterial inoculants can create a perceptible difference in clover plantings because if bacterial associations are not present in the soil food web, or not yet sufficiently accumulated, it will take time to develop these bacterial associations. Sometimes local, native populations of soil microbes are more appropriate inoculants for plantings, and these can be made using fungal or bacterial-dominant compost teas. Lima beans, for instance, originated in Lima, Peru where the bradyrhizobium bacteria preferrred by lima beans is already present in the soils. Be sure to match your inoculant to your legume type for effective applications.

> **"Some common legume species followed by the Rhizobia species which can infect them are: Alfalfa and sweetclover = R. meliloti; True clovers = R. trifolii; Peas and vetch (true) = R. leguminosarum; Soybean = R. japonicum; Birdsfoot trefoil = R. loti; and Crownvetch = R. spp."**
> –The Pennsylvania State University, 2016.

Windbreak

Though it may not seem obvious, windbreaks build soil by slowing the air such that airborne particles, silt, and seeds drop. Windbreaks always also foster unique habitats, which increase catchment and cycling of organic matter.

Heirloom and Landrace Varieties

Heirloom seeds are annual garden varieties that grow true-to-seed; they have a stable predictable genetic expression. Saving seeds from these varieties means reliable results year after year. Landraces are collections of genetics within a certain type—a landrace of cowpeas would have all sorts of colors and sizes of cowpea seed and pod. All heirlooms were selected from landraces or bred from wild sources. Older varieties and landraces of annuals and perennials all tend to need less fertilizer to grow a high yield. As more attention is focused on heirlooms, rediscovering lost plants, and breeding for new ones, the vigor and efficiency of heirloom plants in relation to their interactions within the soil is becoming more widely recognized as superior to any "modern" seed—commercial hybrid or **GMO**. As we face more drastic changes in the climate, landraces are also becoming an area of greater focus since they will provide the genetics we need to breed new drought-tolerant and hardy annuals and perennials for the future.

Reduction of Compaction

Preventing grazing, foot traffic, over-tilling, cropping, or driving in an area may be temporarily or permanently vital if we are to protect soil structure and prevent compaction, especially in an area of compaction recovery where our designs are just taking root. If the compaction in agricultural soils alone was reversed, the amount of extra moisture and soil life retained would be immense.

Ponds or Ditches

These areas generate soil quickly because they capture silt (filling them up slowly), retain most of what they catch, and have plants, animals, and their waste break down in the water as well. Ditches or greywater treatment areas, that become overgrown with reeds or bamboo, routinely need to be cleared, so they can continue to function; small ponds work the same way. The sludge is anaerobic, but given some time exposed to the air, it becomes aerobic. Forming *chinampas* in a waterlogged area goes through the same process. It leads to very rich soils and high yields because those sludges are so concentrated with nutrients.

Restoring Healthy Nutrient Cycling

Soils have increasingly been depleted of soluble nutrients through tillage, repetitive cropping without resting or replenishing the soil, and biocide usage. As these factors simplify soil life and soil structure, even more leaching of vital nutrients occurs. Our foods lack the historically documented nutrient densities that our ancestors or even grandparents were accustomed to; many feel there is a correlation between this and the overall health decline. Luckily this can be reversed in several ways.

Though we can spot specific mineral or nutrient deficiencies and treat those specifically directly in the soil, we aren't addressing the overall situation. There are many books complete with pictures of plants exhibiting deficiencies for identification and natural treatment, for example crushed eggshells for bottom-end blight in tomatoes, and these can work well temporarily, but generalized treatments, like compost, are more holistically beneficial and longer lasting.

We can have an in-depth soil analysis conducted through services like Soil Foodweb Laboratories, staffed by students trained and certified through Dr. Elaine Ingham's programs. When we know what our soils contain, we know what will be too much or too little. We can also use the plants themselves as indicators. Once herbivores pass through, followed by chickens for a season, you will be able to see a difference in the pasture or orchard soil, native plant population (weeds), and the valuable trees or plants.

We can also speed up processes that increase nutrient levels in the soils by chopping and dropping green manure or mulch plants and using compost teas to break down non-soluble compounds in the soil. All soils everywhere have the necessary components to make all nutrients available. Everything is present everywhere for plant life, though sunlight- and season-dependent. The nutrients are all locked up in non-soluble forms: the clay, silt, and sand. We have to use the soil life and organic matter to unlock these mineral stores if we want sustainable and stable mineral levels in the soils.

Health and Soil

The health of all life is tied up in the processes and diversity of soil life. When we consume nutrient-dense foods, we don't need to consume as much food, our digestion becomes more efficient, and we absorb more nutrition with less energy expended. Dr. Elaine Ingham has said that we don't need to take daily multivitamin pills if our soils are healthy because our foods will be high in beneficial nutrients—minerals, phytonutrients, vitamins, fatty acids, amino acids, and carbohydrates.

We are What We Eat

> *"Oil is everywhere in [industrialized] farming - chemical fertilizers and pesticides, intense mechanization, unnecessary packaging, global distribution - and so, increasingly, we are eating oil"*
> –Vandana Shiva, *SoilAssociation.org*

Biocide-free, GMO-free, and toxin-free fresh foods are the safest and healthiest choices in the grocery store. Processed foods contain preservatives and artificial additives; even certified organic foods that are processed are not as healthy as organic fresh foods. As the practice of medicine has evolved in the West world so has cancer, autism, mental disorders, and chronic disease, which seems illogical until you examine what we are consuming nutritionally.

Deficient soils mean deficient food which means deficient bodies and minds. Humans arose on this planet living on foods grown from rich soils—without petrochemicals. Healthy soil, specifically humus, is the primary ingredient of good food, not oil, and when foods are grown on petroleum products, they are made out of oil. Studies on ubiquitously-used plastic hoop houses and plastic row covers reveal that they leach **phthalates** into the vegetables grown under those conditions which means that people are consuming phthalates in their garden vegetables if using these plastics.[42][43] Phthalates can act as endocrine disrupters, they

---

[42] Fu, X., and Du, Q. *Uptake of Di-(2-ethylhexyl) Phthalate of Vegetables from Plastic Film Greenhouses*. 2011.
[43] Fu, X.W., and Xia, H.L. *Uptake of di-(2-ethylhexyl)phthalate from plastic mulch film by vegetable plants*. 2009.

GMOs and Epigenetics

Genetically modified organisms are a fundamentally dangerous and unscientific way of creating new plants and animals. Genetically modified organisms are combinations of plant and animal genes which would never have occurred in nature and cannot occur on a farm using traditional hybridization methods. When GMOs transmit their genes through breeding with natural species in the wild, other farmers' plants, or in our guts, they spread transgenic genes, a mixture of natural and unnatural.

Published in Discover Magazine in 2006, *DNA is not Destiny* derailed the mechanical paradigm of genetic modification's basis—DNA was more complex than was previously thought. Each sequence of nucleotides is actually a bundle of phenotypic expressions that are dependent on an **epigenetic** switch that is on the side of the sequence. That epigenetic switch is controlled by the behaviors, foods eaten, and experiences of the organism, the organisms' ancestors' behaviors and experiences, and even the organism's own perception. If we are depressed for an extended period of time or if we drink alcohol to excess for an extended period of time, it will affect our genetic expression and begin a game of probability where health problems will become more and more likely the longer we continue in that pattern. In other words, epigenetics prove that the nature vs nurture argument is not either/or, but both, spanning 3-4 generations.

Worker bees switch from maintaining the hives to gathering pollen during their life cycle, yet it is all bundled on one gene seemingly. The dance they do with their feet, the skills for finding and selecting flowers, and the navigation skills for returning to and fro all day—it is a complex set of skills, but all found in one sequence. GMOs are based on the idea that genes are far simpler in their function and can be exchanged sequence for sequence without syntactical errors or complications between different types and species. GMOs are reckless, unscientific, and unethical. The practice of patenting life and controlling seed should become illegal once again and, in the interim, boycotted. Life should not be patentable.

> *"Many human genes also code for multiple functions as well. It is false to believe that you can inset a gene for 'one' trait. There will be multiple effects, some of which may be unknown"*
> Jeanne Wallace, PhD, CNC

A final note about epigenetics is quite fascinating. Since the environment informs adaptation to genes through epigenetic switches, how we perceive our environment is the ultimate control in how our bodies adapt to our environment. A positive outlook and attitude therefore has epigenetic effect and can even act as a buffer to environmental stress, genetically speaking.

A revision of
You Are What You Eat:
We are what we eat, drink, breathe, experience, and believe.

can damage developing reproductive systems, and they can damage developed and developing vital organs.

> **"Plants grown in living soil and ecologically diverse environs develop high levels of phytochemicals aka phytonutrients that help protect them from pathogens and pests. These same phytonutrients–thousands of which have been discovered to date, like lycopene, quercetin, resveratrol […]–promote human health. In fact, they are capable of changing gene expression, turning off the expression of genes related to disease"**
> –Jeanne Wallace, Phd, CNC

This is likely why concentrating plants through juicing or distilling essential oils can have medicinal value. Living within a bioregion also adapts us to that area's pathogens through the local food we eat over time.

Reading the Landscape: Soil

Soils

- **Depth** - Soil depth determines tree height, speed of soil water evaporation, and the amount of nutrients available for plants. Stunted trees can indicate shallow or rocky soils as they do in shallow soils over bedrock.
- **Water Reserves** - By observing where plants congregate, especially more water-dependent varieties, we can identify underground springs, clay pockets in sandy, arid areas, and areas that would be ideal for planting trees and shrubs. Our dams don't have to be in the keypoint if a spring is manifesting in another area; the keypoint might not even be on your property. We can put them anywhere there is an area of deposition.
- **pH** - Though not an exact indicator, a general understanding of the pH preferences of plants will give one the ability to recognize alkaline or acidic soil conditions by simply observing a site. Sorrel and oxalis often indicate compacted or acidic conditions in pasturelands while snails indicate more alkaline conditions. Dandelions and other deep-rooted plants indicate soil compaction while thistles and other fire plants return phosphorous to the topsoil after a fire. Plants and polycultures all have stories to tell if we are observant and put in the time. Over-fertilizing the soils tends to make them alkaline and thus unfriendly to perennials looking for more acidic soils–this often brings a flush of weeds determined to balance things out. It is always best to let plants determine their pH by focusing on soil life and organic matter. If you do get weeds, chop and drop them in place

before seed heads develop (and remove the root structure only if you must as it will harm the soil structure to remove it, and it will usually be better suited as a mulch plant in place).
- **Mineral Status** - Reading plant health via the conditions of leaf, stem, fruit, and root will let us know what the site needs in terms of minerals that can be released from the non-soluble soil components with soil life. There are many books and online visual guides to help determine the mineral deficiency from leaf or stem appearance.

Site

- **Fire** - Persistent fire removes phosphorus from the topsoil and kills off soil colloids, preventing proper soil structure and soil life interaction. Fire also favors specific sets of plants that require fire to germinate properly. When frequencies of fires increase beyond the tolerable life cycles of these plants, however, they cannot compete and begin to die back leaving perennial, resinous fire weeds in their place.
- **Frost** - Frostlines often determine which plants stay where as they inhibit their spread and growth. Frost-intolerant plants will quickly indicate natural frost lines especially if our garden tomatoes cross over it. The same principle holds for frost pockets.
- **Drainage** - Mosses, anaerobic soils, standing water, water-loving ground cover plants, or sundews indicate poor drainage which would require ripping for drainage, water pump trees, and perhaps earthworks like berms to keep plant roots above the water table.
- **Mineral Deposits and Rock Type** - Plants often grow in association with particular mineral deposits in the soil as in honeysuckles indicate silver and gold, or rue or violets indicate zinc. If we know what plants indicate what minerals in the soils, we can also know what plants would dislike or prefer the same soil conditions. Animals can also indicate site toxins, pollutants, or heavy concentrations of minerals; these toxins can be detected in their waste or in their bodies. Heavy metals get trapped in fats much as they they can get trapped in soil colloids.
- **Overgrazing and Compaction** - If grazing operations aren't holistically managed with rotational grazing, over time this oversight will only lead to compacted soils and increasingly lower the quality and quantity of plants for the animals to graze upon. Weeds will appear as quality drops lower, until pasture is barely fit for grazing. Daily paddock shifting with recommended herd densities can reverse the effects of overgrazing and compaction.
- **Animal Effects** - Animals and insects can indicate soil structure and content as well as provide extra drainage and soakage areas for rain events. Look for tracks and signs of animal activity—the kinds of animals present will tell you a lot about the surrounding

ecology. A rise in populations of ants, moles, and ground squirrels during a drought prepares the soil to accept the rains and soak them in deeply when they eventually return.

Building Soil

Mycorrhizal Fungi

By increasing mycorrhizal fungi concentrations in the soil, we can build soil incredibly fast. Dr. Christine Jones, an Australian soil scientist, describes the key soil-building elements to be: "plant roots, liquid carbon [her word for root exudate sugars], [and] mycorrhizal fungi. Many scientists have confused themselves – and the general public – by assuming soil carbon sequestration occurs as a result of the decomposition of organic matter such as crop residues. In so doing, they have overlooked the major pathway for the restoration of topsoil."[44] With the ability of mycorrhizal fungi to boost photosynthetic activity and increase root surface area, their integration into soil-building strategies is clearly advantageous, but mycorrhizal fungi also help sequester the "liquid carbon" in the soil by building soil structure via the release of glomalin and other fungal glues. These compounds gives soil its ability to hold water and nutrients, thereby supporting so much of life and making it absolutely critical for soil health.[45]

Carbon sequestration is soil building, and mycorrhizal fungi are responsible for sequestering approximately a third of all the carbon found in the soils globally. Mycorrhizal fungi are a primary mechanism for building soil, restoring organic matter levels, and sequestering atmospheric carbon. Soils, plants, and seeds can all be inoculated with mycorrhizal fungi to accelerate and support healthy growth.

Crop Rotations

Crop rotation is an ancient practice found all over the globe. There are numerous iterations of crop rotation throughout all cultures. Legumes are the most common companion and crop rotation choice, but the concept of leaving a field fallow, to rest, is a form of rotation as well. In ancient times when fields were still surrounded by forest and vibrant, wild ecosystems, the fallow fields were revitalized by the surrounding wild soil food webs and ecosystem elements like deer, rabbits, etc. In the middle ages, a common rotation was a grain then a legume then fallow for a growing season.

---

[44] Jones, Christine and Frisch, Tracy. *"SOS: Save Our Soils."* 2015.
[45] Ibid.

The same concept is at play in the Victory garden systems of WWII Britain where the chicken yard and the family vegetable garden switched locations seasonally. The chickens would tear down the leftover garden patch, manure in it, and lightly till the soil for next season meanwhile turning their efforts into regular eggs, chicks, and meat.

Today we have entire books dedicated to the practice and study of crop rotation. More on Crop Rotations in the Food Forests and Gardens chapter.

Green Manure and Cover Crops

Green manure and most cover crops create a relationship in the soil with nitrogen-fixing bacteria but not all—some, for example alfalfa, are superb at accumulating nutrients in general. The main idea is to provide biomass and to cover the soil during the offseason, and then, usually, to till that biomass into the soil in early spring though it can be flailed in place as well. Nitrogen fixation, which focuses nitrogen in the leaves and seeds, accelerates the decomposition of the carbon in the biomass and creates humus more quickly, providing more sustenance for plants and the soil food web. Rye grasses, buckwheat, clover, and other non-leguminous cover crops also fix nitrogen. Green manures can be a combination of legumes and non-legumes depending on your goals and available resources.

> **"One gram of soil has the nutrients for growing an entire acre [of food]"**
> –Dr. Elaine Ingham, Sustainable Design Masterclass, 2016

Compost

Composting creates large carbon chains in a reaction that binds many exotic heavy metals and other undesirable pollutants, making them inert, though it does release some greenhouse gases in the process. The final product is a humus-rich soil medium that is made of decomposed organic matter. There are two types of compost: hot and cold. Thermophilic composts are hot composts that rely upon soil life and turning the pile repeatedly for 18-26 days. Cold, or mouldering, composts are slow processes that take a season or more to decompose fully. These require soil life to decompose the organic matter, and there is little to no disturbance.

Compost is often called "black gold" by gardeners. It is so fine that it sticks between the ridges of our fingerprints and can be hard to wash off. It's this binding quality that soils utilize to improve their structure when compost is added or when compost tea is applied.

Commercial operations use windrows, or large compost berms in rows, and turn them with large machines. This can happen on-site on a farm as well, using farm waste and a tractor.

Thermophilic or "Hot" Compost

Thermophilic composting is a way of decomposing organic matter to create a safe, pathogen-free, weed seed-free, parasite egg-free, and humus-rich soil amendment that is saturated with beneficial aerobic soil microbes and a broad spectrum of nutrients. Humus itself doesn't contain nitrogen or phosphorous; it is comprised of carbon compounds that are continuously breaking down into smaller, finer compounds. Hot composts are a quick process: if you aren't observing, measuring the temperatures within it, and turning the compost when it needs to be turned, you will lose fertility, nutrients, and time—it can also catch fire! It will also exhaust greenhouse gases into the environment instead of sequestering the carbon, nitrogen, and more.

For a compost to be effective, the ingredients must be as precise as possible. The initial pile by volume is made of a third brown organic matter (high carbon, dead, dried plant material that has already gone to seed), a third green organic matter (these are plant materials that haven't gone to seed, so they still contain enzymes, nitrogen, protein, and sugars—they can be dried or fresh), and a third manure or another nitrogen-heavy component. This even split is the ideal composition for a pile that is at minimum one cubic meter in size. The larger the pile is, the easier it will be for the microbes to go to work but the harder it will be to turn. Large operations use heavy machinery to turn their piles. (In the summer heat, smaller piles are possible).

Compost piles need more carbon when you use hotter manure sources. A pile that is a third chicken manure will have a higher nitrogen content than one that is a third cow manure simply

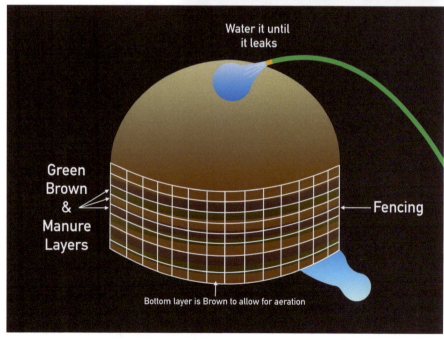

because of the nature of the waste itself. Wood chips likewise contain more carbon than straw; they are higher in density. Hotter piles burn quicker and provide less compost as their end product, so if you notice yourself turning it more often, mix in some saw dust or well-shredded, dried plant material for more carbon to balance out the excess nitrogen.

When we arrange the materials it is best to start with a carbon layer on bottom to provide aeration pathways while it heats up initially. In layers, add brown, green, and manure layers, and then water it all until it leaks. Wrap it in a wire fence to encourage aeration but contain the pile efficiently. Cover it with a tarp. Wait 3 days and then turn it every other day from thereafter, as long as it maintains a 131°F–140°F (55°C-60°C) temperature range. If it is too cool, either don't turn it as often, so it builds heat, or add more manure and possibly greens to speed up the reactions and resultant heat.

For perennial systems, a woody compost is ideal because it will require fungi to breakdown the wood lignin, and the resultant compost will be more fungal-dominant. If we add extra fresh cut grasses or weeds, we can easily create a bacterial-dominant compost which is ideal for more bacterial-dominant garden soils.

Don't forget—it's hot! It should stay between 131°F-140°F (55°C-60°C) for 15 days, turning at least 5 times. The heat kills the pathogens, parasitic eggs, and the weed seeds. The pile also needs time to cool off unless you are going to burn the plants or weeds in place by

Dr. Elaine Ingham with a handful of material that is 6 days into composting.

adding hot compost to the garden beds—some choose to do this. Dr. Elaine Ingham prefers to let hers cool and then use it on the garden or to make compost tea or compost extract.

Vermicompost

Using earthworms and compostable kitchen waste, we can create a bacterial-dominant compost of mostly worm castings. Together the worm juice (a soil leachate) and the soil itself can be used to bacterially inoculate soils or to start a bacterial-dominant compost tea.

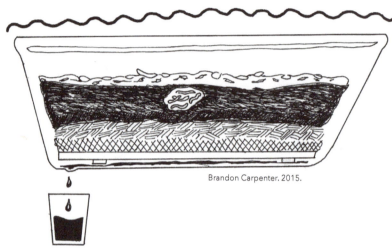
Brandon Carpenter. 2015.

Using an old tub or a waterproof container that allows liquid and air to pass through freely, we can add kitchen scraps to a base layer of manure populated with compost-loving worms and over time (three months) the container's contents will be transformed by the worms into a finely processed earthworm compost.

Jean Pain's Compost-Power

Jean Pain was a Frenchman who used woody compost heaps to heat his home and water and to power his truck. His simple ingenuity captured the methane from the compost as it gassed off for his vehicle and cooking range, used the heat from the pile to heat water for using in the kitchen, bathroom, and radiators, and then used the rich compost at the end of the decomposition in the garden. Instead of burning the fuel wood, perpetually choking the countryside around him, Jean Pain slowly released the potential inside of the wood to create numerous, stacked benefits. The process is anaerobic—you never turn it, so it takes longer and does not have the many benefits of aerobic compost (in fact some might say Jean Pain's anaerobic woody heaps do not qualify as official compost because it is not an aerobic process).

Pain, Jean and Ida. The Methods of Jean Pain Archive.org, 2016.

Methane from the compost is stored in the inner tubes.

Compost Teas and Foliar Sprays

Compost teas are solutions of concentrated aerobic soil life that have been proliferated in water from a sample of well-developed compost. Separating the soil sample with a micro-perforated bag lends the practice it's name, *compost tea*, because it looks like a tea bag in boiling water, and it also ends up being similar in color to tea. Compost teas are aerated with air pumps, usually the kind for aquariums, that have the smallest surface area because they need to be cleaned between each batch and sterilized. The aeration keeps things aerobic, but when biofilm from the last batch remains, anaerobes can still ruin a batch. Compost tea can be fungal- or bacterial-dominant (depending on the compost sample), and it can be used directly on the soil or as a foliar spray. Using a variety of composts creates the most diversity in your compost tea: use vermicompost, hot compost, EM, and more. Compost teas are usually diluted to a water-to-tea ratio of between 2:1 to 4:1. A little compost can go a long way in the form of compost tea.

Brewing can take 1-5 days depending on the volume of batch, aeration rate, and temperature, the last being the most important factor after aeration. In the heat of the summer, brewing compost tea can happen in under 48 hrs. Use the brew within 6-8 hrs of

A repurposed feed barrel can be used to brew compost tea as well.

brewing completion. It takes time observing and smelling regularly to know when a tea is ready, but anyone can identify the bad-smelling anaerobes that indicate a brew that is no longer beneficial for plants and soil. The margin for error is quite large, and soil life is vigorous to begin with, so dramatic effects can often occur with even amateurish attempts at compost tea.

A 5-gallon (19L) bucket can be used to aerate compost to make compost tea.

There are many other beneficial foliar sprays we can create at home that are natural and non-toxic. Some act to nourish, some to repopulate, and others act to protect, as in the example of a physical barrier like diluted clay on fruit trees; however, compost teas are the most versatile and beneficial overall.

Compost Extract = Humic Acid

Sold in stores at exceptionally high prices in stagnant bottles, humic acids can easily be generated at home with native soil biology. After the compost has cooled, water can be gently run over some compost and through a sieve, the brown water, the compost extract, that trickles out of the sieve is rich in humic acids which are fungal foods.

Using Effective Microorganisms (EM)

Effective Microbes (EM) are a consortia of selected microbes discovered by Professor Teruo Higa in Japan, and they are a trademarked brand owned by EMRO. These microbes are lactic acid bacteria (LAB), yeast, purple nonsulfur bacteria (PNSB), and other naturally occurring microbes that enhance the behavior of indigenous microorganisms in both soil and water. It is not a strictly aerobic group of microbes but rather a mix of facilitative microbes that encourage growth and health in aerobic food soil and water webs (they can also produce foods for each other within the consortia symbiotically). Yeast breaks down the organic matter using fermentation to keep nutrients bioavailable while lactic acid bacteria (an active ingredient in pickling) keeps pathogenic microbes at bay and the pH at a low, acidic, level. Purple non-sulphur bacteria (PNSB) is a remarkable bacteria: it can feed four ways. It can feed off

Make your own EM at Home
a Cuauhtemoc Villa recipe with LAB recipe from Chris Trump

Ingredients:

Yeast - from home-brewed kombucha or kefir

PNSB - from worm castings

LAB - from rice wash-water that has sat covered for 3 - 5 days (until it smells slightly sweet), and then combined 1 parts fermented rice wash water to 10 parts milk, and then covered for a 48 hrs until there's a clear separation of the milk curds from the whey which is a lactobacillus-rich liquid. Remove the cheese curds and then strain off the whey serum. Store the LAB in the fridge and make cheese out of the curds.

Molasses - organic blackstrap un-sulphured molasses

Water - ideally water from a thriving ecosystem: avoid chlorinated water from the tap.

Combine equal parts of the kombucha, LAB, and worm castings and then add molasses equivalent to the mixture amount. From there, dilute the mixture 1:20, mixture to water. Culture the mixture in an airtight container - it should be an anaerobic condition. pH should drop within 7-10 days to below pH 4 (ideally pH 3.5).

To Extend the EM:

1 part EM

1 part Molasses

20 parts Water

1 tsp Super Cera Powder (a bioceramic powder) or Sea Salt

---

of heat, toxic materials, light, and carbohydrates (sugars). PNSB is a thermophilic bacteria that can thrive in cattle manure lagoons, the edges of ponds, the digestive tracts of worms, and toxic waste dumps, but it can also consume CO_2 and release it as oxygen, amino acids, and folic acids. Like AMF, it is ancient and defies convention, leading many to theorize it to be one of the earliest microbes one earth. EM's amazing team of microbes do their work and then bow out of the food web after they finish their work: they stimulate the IMOs and then get consumed by them. EM is used to purify water, improve soil, enrich aquaculture systems, invigorate animals, and help plants grow.

"Lactic acid bacteria, yeast, and phototrophic bacteria contained in EM have the ability to ferment organic substances and prevent putrefaction. Therefore, for example, when making compost with EM, putrefying bacteria will be suppressed and, due to the fermentation action of EM, it is possible to manufacture compost with less turning than usual. Also, compost fermented by EM is rich in amino acids and polysaccharides compared with compost produced by the usual methods. EM prevents the production of ammonia during protein decomposition, metabolizing proteins in such a way that amino acids are produced instead. These amino acids can be directly absorbed by plants. Also, under normal circumstances, cellulose will be decomposed and broken down to form carbon dioxide. However, due to the fermentation action of EM, low-molecular polysaccharides will be produced and these will be absorbed by microorganisms and plants. Generally, proteins are synthesized from nitrogen. However, if the plants can directly absorb amino acids from their roots, they can repurpose the energy that would have gone into producing amino acids and proteins, thereby producing fruit with more sugar."

- How EM Works, EMROJapan.com (2016).

With Bokashi

Bokashi means fermented organic matter, but more specifically it is organic matter inoculated with EM. Bones, dairy, and anything organic can be digested by the bokashi EM fermentation. It is commonly used on kitchen counters in Japan to digest organic matter of all kinds. Two weeks after it has fermented in the bucket, it can go into a hole in the ground and covered lightly for two more weeks, and after that, it is nearly completely decomposed.

How to Make Biochar Bokashi

There are many methods, but this is the method taught by Cuauhtemoc Villa in *The Advanced Permaculture Student Online* and synonymous with the TeraGanix recipe on their site. Combine 1:1 biochar to organic matter (this can be compost). You can add ingredients like wheat or rice bran, insect frass, brewer's spent grains (beer mash), compost, seabird guano, cow manure, worm castings, agricultural waste, nut husks and hulls, and more to add in unique biology.

- *50 lbs (22.7 kg) bag of wheat or rice bran*
- *Equal parts biochar to bran*
- *4 gal (15 L) of water*
- *2 - 6 oz (59 - 177 ml) EM*
- *2 - 6 oz (59 - 177 ml) Molasses*

Combine the ingredients and mix them until just moist all the way through - 25% moisture, so you can squish out just a bit of liquid when you squeeze the soil. In an anaerobic condition like in a sealed barrel, ferment the mixture 21 days. You can then apply this to the soil or use it in a bucket system to digest your kitchen waste - including meat and dairy.

With Korean Natural Farming

This is an entire farming fertility, waste, and pest management strategy that relies upon indigenous microorganisms and biomass, usually agricultural waste. Chris Trump is a Korean Natural Farmer in Hawaii that uses KNF methods on his family's farm for fertility, soil health, and to fight ubiquitous problems other macadamia nut farmers face in his area - he has an entire series of freely available KNF videos showing step by step how to create all the KNF preps at home using resources almost anyone can access or replicate. The basic concept is IMOs are captured in thriving wild ecosystems to be scaled up using a variety of methods

How to Make KNF IMO Preps

IMO-1 - using a woven basket or a box with holes in it (that are covered with screen), we can inoculate undercooked rice with IMO (primarily mycorrhizal fungi but an entire spectrum of microbes will be present) by filling the box 2/3rds full, and placing it in a healthy soil area where mycelium is present and visible. Cover the box by stapling on breathable paper or use the basket lid. Check back in 4-5 days to look for white fluffy mold. That's IMO-1.

IMO-2 - combine and mix the harvested rice and mold with equal parts brown sugar in a large glass jar. Seal the jar with two layers of paper towels and let it ferment for a week. Now you have IMO-2!

IMO-3 - similar to bokashi, two ounces (60 ml) of IMO-2 are mixed into 60 lbs (27 kg) of biomass like flour, bran, or even macadamia nut kernels and shells. Compost the mixture for 7 days at below 110°F (43°C), and you get IMO-3. IMO-3 can be used to make a compost tea for foliar application and soil soaking.

IMO-4 - combine two parts IMO-3 with one part field soil and one part native ecosystem soil with 5 gallons (19 L) of water. Compost for a week, and you will have IMO-4.

IMO-5 - combine IMO-4 with a nitrogen-rich biomass like manure or the leftover biomass from producing black soldier fly larvae with 4-5 gallons of water depending on the moisture levels of biomass. Not too much moisture - it should just hold and then breakup easily in the hand. It will get up to 130-140°F (55-60°C) and then cool off quickly, and be ready to use in the garden in approximately a week. This prevents the nitrogen from gassing off, trapping it in the fungi, and prepares the fertility to be more bioavailable to the plants and the soil food web. This is IMO-5!

These are just the basic preps - inputs can be added along the way that aren't described here. There's a lot more specifics to KNF and even more abbreviations! Chris Trump's Youtube channel covers his practice and all the preps in-depth. Tune in to his channel or attend his courses for more information.

akin to mushroom cultivation but without the sterilization or pasteurization usually required, but there's even more to it: there's ferments and other products used, and it's not just about soil. KNF pig operations have odorless pig pens, and KNF agriculture is always no-till.

Mulch Plants as Nutrient Accumulators

Mulch plants are plants that accumulate lots of biomass. Artichokes, banana plants, comfrey, dandelions, alfalfa, nettle, lamb's quarters, and annual legumes are great examples of mulch plants, but every plant type also accumulates its own spectrum of nutrients, and to add even more complexity, each individual plant accumulates a unique spectrum of nutrients that is dependent on its microclimate and rhizosphere. Plants that accumulate the most nutrients or the highest concentration of a specific nutrient usually have extensive root systems or unique soil associations. These allow them to gather a specific spectrum of nutrients from the soils to return them to the top layer of soil through the decomposition of their roots, leaves, stems, and even seeds. The top 2" (5cm) of soil is the most fertile; this is why most plant roots are found in the top 6" (15cm) of soil. Because mulch plants accumulate so much biomass, they also tend to accumulate specific nutrients in larger quantities than other plants. Often by just using mulch plants alone, we can remediate soil nutrient deficiencies through regular chop and drop of those plants.

Daikon radish was preferred by natural farming pioneer Masanobu Fukuoka, whose own brand of permaculture relies on using only the minimum amount of work and effort to grow his farm's food—with no inputs other than plants already on site and using human labor for harvesting and planting. Daikon radishes can grow as deep as 5-6ft (1.5-1.8m), gently aerating and tilling the soil without machines, and they accumulate large amounts of phosphorous and scavenge and release a significant amount of nitrogen when they decompose. When they are left to rot in the field, they create vertical compost towers that penetrate deep in the soil, allowing infiltration of water and providing food for other plant roots and soil food web members. Buckwheat gathers even more phosphorous, making it an incredible green manure because it also fixes nitrogen. What you use where depends on your site and goals.

By understanding what our soils lack and what nutrients plants accumulate, we can apply strategies using just seed mixes of mulch plants to build and enrich our soils quickly. There are databases and studies on individual plants, but primarily they focus on N, P, and K levels.

Sheet Mulching

Sheet mulching is an easy and effective way to turn a lawn into a garden. It's also a great way to deal with waste products like newspapers and cardboard though any paper or even wood chips will do. It is a mulch to cover the soil, a way to smother weeds, and part of the ingredients for in-situ composting. Remember: these amounts depend on your site and your ingredient types (an inch of chicken manure is very different than an inch of rabbit manure). Aerate the soil, chop and drop the plants in place, and prepare the area for the sheet mulch. Soak all ingredients to begin with and use what you can - a simplified version can accomplish the same goals in some contexts.

Add onto the soil surface:
- 1-2" (2-5cm) Composted manure or fresh manure that lacks seeds and has a balanced C:N ratio.
- 1/4-1" (1-2cm) newspaper or cardboard—test this on your own site
- 1-2" (2-5cm) composted or balanced manure, preferably with no seeds—from something like a rabbit or goat
- 6-18" (15-46cm) of organic mulch like straw, seedless yard waste, wood chips, or any source of carbon lacking seeds.
- 1-3" (2.5-7.5cm) compost—if you have the compost, use it. The more compost on top the faster the process below occurs. This is where seeds may be planted, ideally legumes.
- Scatter seedless mulch, like straw, lightly atop to protect the growing seeds.

Jar Soil Test

Using only a jar, a soil sample, some water, and some observation, we can see the composition of our soils. Fill a jar halfway with soil and the rest of the way with water, leaving room for shaking. The sand, silt, clay, and organic matter settle out individually when you shake soil up in water and wait. The heavier particles settle out first and the lightest settle out last. We can measure the individual thicknesses of each layer, note them individually, and then add them together. If we divide the each individual section thicknesses against the full soil's height in the jar and multiply by 100, we can calculate each section's percentage.

VII

Organic matter is the most critical component, the top layer, and it can often float. We should aim for at least 10% if not 20% as the minimum organic matter levels for our soils.

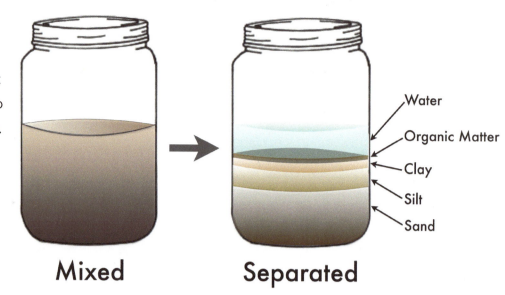

Mixed   Separated

Reading your Soil Results

Once you know your percentages you can mark them on a soil chart (see example). Find the intersection between all three percentages—each percentage has only two lines extending away from it. For dam building, knowing your soil types and placement on the site is critical to knowing what is possible using only on-site materials.

Determining Soil Texture by Hand[46]

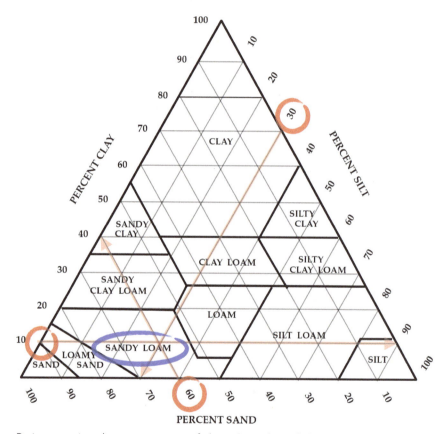

By intersecting the percentages of clay, silt, and sand, the soil type can be categorized which makes planning for earthworks, gardens, dams, and more considerations easier.

1. Wet the soil until it is like workable putty.
2. Rub between forefinger and thumb to ascertain its feel
3. Squish the putty between your forefinger and thumb to create a smooth length of soil, or ribbon.
4. Measure the length of the ribbon

---

[46] *Determining Soil Texture by Feel Method*, 2016.

What did it feel like when you rubbed it?
- *Gritty soils are sandy.*
- *Sticky soils have a lot of clay.*
- *Smooth soils have plenty of silt.*

How long was the soil ribbon?
- *Good Ribbon - 2"(5cm) or longer is correlated with 40% clay content or higher.*
- *Fair Ribbon - 1-2"(2.5-5cm) lengths of ribbon are correlated with 27-40% clay content.*
- *No Ribbon - less than 1"(2.5cm) is correlated with less than 27% clay content.*

Test for Soil Life

By Hand

Get a shovel full of dirt and sort through it; look for arthropods, earthworms, and nematodes. The last will be visible only if you can see down to a mm. If you sample on a warm, damp night you may get a better sampling of the current population.

Pitfall Trap[47]

Pitfall traps catch large soil organisms, allowing us to count them to get an idea of the populations present. Dig out a hollow in the soil for a small dish or cup to be level with the soil surface (yogurt cup to a soup bowl size), and pour non-hazardous antifreeze up to 1 cm deep in the bottom of the dish or cup. Cover the dish if you expect rain and allow your bugs to populate the dish—the antifreeze preserves all the bugs and doesn't allow a higher level bug to eat all the lower level bugs present.

Burlese Funnel[48]

This test focuses on sampling the smaller soil organism population. Pour ethyl alcohol or antifreeze into the bottom of a cup, so it just covers the entire bottom 1-2mm deep. Using a funnel or the cut top of a two-liter bottle inverted, place a small screen in the bottom of the funnel, and then fill it with soil. Suspend an incandescent lightbulb over the soil 10cm (4"). Within four days time in a sheltered area or indoors, the heat from the lightbulb will drive out

---

[47] Lewis, Wayne and Lowenfels, Jeff. *Teaming with Microbes*. 2006. p. 106.
[48] Ibid. p. 106.

all the smaller soil life down into the cup. Use a magnifying glass or a microscope to view and count the organisms.

For further study on soils:
<u>Teaming with Microbes</u> by Jeff Lowenfels and Wayne Lewis
Dr. Elaine Ingham's <u>Soil Food Web Courses</u>, books,
and Certification program.

Test for Plant Health

Refractometers or BRIX meters

Refractometers or BRIX meters are devices that refract light through plant leaf or stem sap, honey, or fruit juice to detect the concentrations of starches or sugars in that liquid. Industrially they have many applications. They don't require batteries, just the sun's light. Its reading of sugar/starch levels indicates how efficiently a plant is producing exudates, interacting with the soil food web, and photosynthesizing. To avoid inconsistent data, do not take measurements in the early morning or on cloudy days—try to test at the same time of day when sunlight and temperature are similar as a form of control on the experiment.

BRIX readings are variable per plant per soil per location. Exudate, mineral, and overall nutrient offerings differ from place to place, so nutrient content is varied in both soils and plants, as is BRIX reading. It should be noted that refractometer readings correlate to overall plant health and nutritional density, but they lack the sophistication to give us the nutrient profiles and the concentrations of each individual nutrient. It is still a great tool for measuring overall plant health, fruit nutrition, and ripeness as it is used by viticulturists, beekeepers, and orchardists globally.

Designer's Soil Checklist

- Test the Soil: Analyze its components, ratios, and biology.
- Decide on a regenerative strategy to manage the soil: compost, compost tea, legumes, etc.
- List site goals, identify soils and decide which strategies will be needed to maintain and build those soils, and to achieve those specific goals.
- Build Soil Wherever Possible

VIII. Fungi
edited by Peter McCoy

Before there were animals, before there were trees, before there were plants, there were fungi, the first **eukaryotes**. It has been argued that the first plant cell formed when an early fungus incorporated a photosynthesizing **cyanobacteria** into its cells. Lichens, a mutualistic relationship between fungi and algae, were first to venture onto the shores of the early terrestrial earth. At one point in the development of life on land, fungi dominated, with some fungi being as tall as buildings! Their actions and decomposition created the first soils for animals and plants to flourish thereafter.

Fungi are also intrinsically part of all animal and plant physiology, often making their categorization difficult. Fungi are the primary decomposers, re-distributors of nutrients, and ecological catalysts that guide and support plant and animal development in all biomes. They are found in the air, water, soils, and likely in the tissues of all plants and animals. They are found on all continents and even at the bottom of the ocean. Lichens can even survive the vacuum of outer space and successfully reanimate when returned to a suitable environment. All this evidence has led some to hypothesize a fungal genesis for all life on earth.

Fungi also exhibit unique abilities and qualities that

complicate or even defy much of our established understanding of many branches of science. Fungi alone can break down wood lignin—a development that came only after a long period of tree wood lignin piling up! They can also dissolve rock itself (*geomycology* is the study of rock weathering by fungi). Phosphorous is unlocked by fungi from decomposing organic matter, as well as from rocks and soil particles. Different strains of fungi can be adapted to feed on crude oil or to even grow in radioactive soils! With that said, change has to be incremental for the organism to tolerate it—fungi are powerful but still sensitive.

Lightning is associated with fungal flushes in nature, especially in relation to desert truffles. Commercial mushroom-growing operations in Asia are now discovering that shocking the mycelium with high-voltage electricity can also stimulate larger mushroom flushes! Fungi force those who study them to holistically reframe their understanding of the world; they are the "managers" of all natural systems and the facilitators of life itself. These characteristics place fungi at the very center of all natural systems.

Fungi are working everywhere in helpful and sometimes surprising ways. We can increase fungal cycles, harvest the abundance of these cycles, and, in turn, remediate toxins in our environments and bodies with this abundance. Research into fungi is still in its infancy, with mycology as a field of study largely ignored by American and European academia until rather recently. However, a large grassroots movement combined with the advent of the internet has led to a revolution in cultivation practices for home-scale growers and researchers.

The Phyla

The "queendom of fungi," as coined by <u>Radical Mycology</u>'s Peter McCoy, is vast and so diverse that it is impossible to categorize it with our current methods. The reproduction processes and life cycles of fungi differ dramatically across the fungal phyla, and even between species within a given phylum. When we compare fungal genetics, things get even more interesting.

When speaking of growing mushrooms, we are usually referring to species in the **Basidiomycota** (excepting truffles and morels). Basidiomycotan mushrooms start out as singular spores that travel alone and inoculate a substrate (their growing medium) where they begin to form mycelium, a **monokaryotic** network of hyphae that are comprised of strands of fungal tissue that are one-cell thick. These hyphae grow and branch through their substrate seeking food, water, and a mate. Once they find a compatible mate, the two networks will fuse together and combine their genetic material, forming an extended **dikaryotic** state wherein two nuclei are contained in one organism. Extended dikaryotic states are unique to fungi. Other fungi in the group known as the Glomeromycota (a.k.a. Arbuscular Mycorrhizae, or AM fungi) can contain 800-35,000 different nuclei in their spores. These nuclei can be from divergently different fungi, making glomeromycotan fungi difficult to categorize due to our inability to know which nuclei influence the fungus' growth.

Once the substrate has been completely consumed by a basidiomycotan fungus (this is done by releasing digestive enzymes from the tips of the hyphae), the fungus growth will become restricted, causing it to create a hyphal knot that will swell into a *primordia* and eventually into the mature fruit body, or mushroom. Inside the mushroom, meiosis and mitosis occur, creating new nuclei which are taken up by the mushroom's spores. Using condensation, the mushroom spores collect moisture and use this weight of water to induce an explosion into the air at astonishing speeds of one meter per second. This explosive process is thought to be why mushrooms are cooler in temperature than that of the surrounding forest. Every breath and gust of wind has spores in it.

The substrates commonly used in commercial mushroom production include wood, grains, manure, or other agricultural or urban wastes. The different growing media represent different ratios of carbon to nitrogen. Fungi digest complex carbohydrates and organic polymers like cellulose and wood lignin. Fungi can digest leafy mulch, used cigarette filters, various types of manure, old pairs of jeans, cardboard, paper, and even coffee grounds too, but most prefer a specific diet. In all instances, the mushrooms are grown on the "wastes" that we don't know what else to do with. Mushroom cultivation provides an abundance of food, medicine, soil, and more—all while reducing our pollutant loads and helping to more efficiently close loops in human systems.

Mushrooms don't develop well without proper food: they need a balanced diet that includes plenty of carbon (such as wood) as well as nitrogen from things like soybean hulls,

Oyster mushrooms grown in a commercial system.

wheat bran, or manure (and sometimes more specific ingredients). Mushrooms are like the fruit of a tree in that they contain the means for reproduction. Their "seeds" in this case are spores that are released in a fantastic burst as the caps flare outward and are carried away on the wind. The mycelium is like the tree's roots, trunk, branches and leaves.

Glomeromycota are among the oldest types of fungi on earth. They are found in soils across the globe, and do not form mushrooms. Their life cycle is quite different from that of Basidiomycotan fungi. Glomeromycotan species dwell in the soil around and inside the roots of plants—that is, they form **endomycorrhizae**: mutualistically beneficial relationships similar to that formed between rhizobia bacteria and legume roots (discussed in the prior chapter). With these fungi, the structure formed is known as a mycorrhiza. Glomeromycotan species are often referred to as Arbuscular Mycorrhizae Fungi (AM fungi or AMF) and

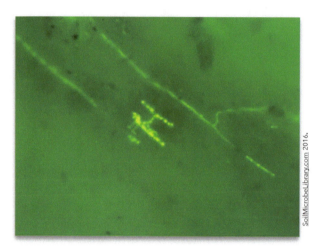

Arbuscular Mycorrhizae fungi Entering a Plant Root

form this relationship with at least 90% of all plants. This relationship is formed when a spore germinates near a root, sending out specific exudates that beckon the fungi to extend its hyphae into the plant root. The hyphae form tree- or "arbor"-like structures that penetrate the plant's cell walls from within the root, creating a direct link between the mycelium outside the roots and the cells of the plant inside the roots. As with the rhizobia bacteria, the plant provides sugars in exchange for nutrients and water. AM fungi magnify the efficacy of roots as well as their reach and surface area.

The life cycle of AM fungi is an ephemeral process that cannot be seen from an above-ground perspective. The arbuscules form in a matter of days and only last 4 to 15 days—though what is produced in that time period can last decades. Discovered in 1996, glomalin is a sticky protein exclusively produced by the mycelium of AM fungi. It significantly contributes to soil structure and sequesters nearly a third of all soil carbon. As much as a third of all soil carbon is sunk by mycorrhizal fungi as both glomalin and fungal tissue (mycelium). Most of this sinking is done by AM fungi. Glomalin is the sticky carbonaceous material that holds the soils together and gives it the fluffy, loam texture that is sought after by all farmers and gardeners everywhere.

> **"With such important influences on the soil environment, whole plant communities, and animal diversity of an ecosystem, the Glomeromycota may be the most ecologically significant of all fungal phyla"**
>
> –Peter McCoy, *Radical Mycology*, 2016.

The **Ascomycota**, one of the other seven fungal phyla, includes a wide array of fungal forms: yeasts, mildews, lichens, morel mushrooms, underground truffles, and other complex fruit bodies. Alcohol, cheese, bread, and antibiotics are all made with ascomycotan fungi. When reproducing, the compatible mycelium of compatible ascomycotan fungi do not initially fuse together as with basidiomycotan fungi. Rather, they meet and co-habitate in their substrate, only fusing right before the fruit body is formed. Ascomycotan fungi–and many other types of fungi–can also reproduce asexually (meaning they can self-clone without a separate genetic partner). Depending on various environmental conditions, these fungi will switch from asexual to sexual reproduction modes.

The other four phyla–the **Chytridomycota**, **Neocallimastigomycota**, **Blastocladiomycota**, and **Microsporidia**–represent single-celled microscopic fungi. There is even a group of uncategorized fungi called **Zygomycota** that has yet to be officially categorized.

The science of mycology is still quite young with fascinating new discoveries constantly arising. Part of the confusion in this field of study rests with the fact that in terms of genetics and appearance, fungi defy reduction and categorization. For these reasons and more, mycology is one of the best natural sciences for exploration–with so much to be discovered, anyone can add to our understanding of the importance of fungi in the world.

Mycorrhiza

There are seven types of mycorrhizal fungi which are fungi that associate with plant roots. The most common are the Arbuscular Mycorrhizae fungi (AMF). AM fungi augment the root structures physically, increasing surface area by up to 10,000 times to greatly increase the plant's nutrient absorption. Fungi can make nutrients available to the plant that would otherwise be completely unavailable.

> **"[AMF] are some of the only fungi that can perform the energy-demanding process of reducing nitrate into a form of nitrogen that can be metabolized"**
>
> –Peter McCoy, *Radical Mycology*, 2016.

Mycorrhizal fungi either penetrate the root cell walls or sheath the root cells in various ways to exchange nutrients easily with the plant. Endomycorrhizal fungi penetrate the cell walls of plant root cells while **ectomycorrhizal** fungi do not. Rather, they usually just surround the cells within a root. There are even endo-ectomycorrhizae fungi that form both structures! Many plants form various mycorrhizal associations that can change over their lifespan, and some plants, like orchids, are entirely reliant on mycorrhizal fungi for their survival for at least part of their lifespan.

These mutualistic relationships are the ecological foundation of our forests, grasslands, wetlands, and even aquatic habitats. Incredibly, at times this mutualistic relationship even can appear altruistic as the fungi may not receive any perceived benefit for all that they do for a particular habitat or plant. Plants would not cover the earth if it were not for fungi, but fungi do not control plants. Plants maintain autonomy and use their exudates (and perhaps bioelectrical signals) to communicate with fungi. The nutrient exchanges between these organisms range in content from sugars to carbon to nitrogen to phosphorous, but many are surprising in their behavior, at times exhibiting complete role and dependency reversals.

Lichens (Fungi+Algae)

Lichens are an incredible ecosystem of mutualism where algae (a plant) and/or cyanobacteria (a microbe) work in symbiosis with fungi to exchange nutrients and gases in a shared microbiome adapted to niche climates and environments. The photosynthesizing partner (the **photobiont**) lives in what is essentially a protective greenhouse that the fungus constructs with its mycelium. Lichens are found nearly everywhere and in all climates (though air pollution sharply reduces their populations). They can grow as biocrusts in deserts, locking soils together to protect them and raising organic matter levels. Lichens can fix nitrogen where it is scarce. In cold temperate climates where nitrogen-fixing plants and trees (like alder or redbud) are sparsely distributed, lichen decomposition can account for 50% of the nitrogen input in soils (though lichens can fix nitrogen both near and far from the equator). In deserts, lichens are often the main source of nitrogen.[49] Lichens are often overlooked as both a form of fungus and as a nitrogen-fixer, but they are critical to the vast majority of habitats.

Spore Prints

Spore prints are one of the only ways to be sure of a mushroom's identity. To collect a spore print, harvest the mushroom in question before it has released all of its spores (usually

---

[49] McCoy, Peter. *Radical Mycology Webinar 1: Seeing Fungi.* 2016.

this is before the cap is fully flared out). This is done using either white and black bi-colored paper, glass, or tin foil to make prints; spores can range in color from dark to white, so those surfaces allow you to better see the complete print. Make a hole for the stem of the mushroom to pass through the paper or tin foil—so that the cap's gills rest evenly on the surface—or discard the stem and put the cap directly on the surface. Place the mushroom and paper/tin a place free from breezes (spores are easily blown by the wind) and allow 12-24 hours for the mushroom to release its spores. Or, just cover the mushroom with a large bowl to protect it from the breeze.

Once you have a spore print, you can compare its color to descriptions or pictures in a mushroom hunting field guide. Often, the combination of the mushroom's spore color, habitat, and the visible features are all you need to know which species it is. This can be subjective at times and even confusing as some guides are better at describing mushrooms than others. It is always best to source local mushroom mentors for guidance on identifying wild mushrooms found in your area.

Agrocybe Cylindracea spore print on tin foil.

William Padilla-Brown, 2016.

Products

CO_2 is created as fungi digest a substrate: the fungus consumes oxygen and releases carbon dioxide, just like an animal. When growing fungi indoors, it is important to have enough indoor plants, ventilation, and space in your growing area. Heat is created as the fungus' digestive enzymes work, so be sure to provide enough airflow and space. Otherwise, the mycelium may overheat or go anaerobic and become moldy! Fungi need water to be present in the soil or substrate to perform their life cycle as they are constantly taking in water as they extend their hyphae. After a mushroom stops producing flushes, the final product is exhausted substrate that is ready to be composted—that is, it's spent in terms of nutritional value for the fungi and can now be digested by red wiggler worms and then used in the garden.

Compared to the estimated seven billion metric tons of CO_2 annually released through human activities, fungal decomposition dwarfs that figure with 85 billion metric tons of CO_2 released annually.[50] That accounts for nearly the entire 87 billion metric tons required annually by earth's plants. Animals release the final two billion metric tons through respiration. Though man-made CO_2 emissions, which are causing climate change, are by no means insignificant, they are manageable when seen from this perspective. This is especially true when we consider how much room there is in the soil for sequestration. Fungi sequester approximately a third of all carbon in global soils and are responsible for 40-70% of the carbon found in specific soils - mycorrhizal fungi, specifically AMF, are a critical tool in carbon sequestration efforts worldwide. [51]

Asian Reishi grown on the oxygen exhaled by Spirulina and Chlorella at MycoSymbiotics in PA, USA.

Fungi as Food

The most obvious food sources in the Fungal Queendom are mushrooms, but they are far from being the only fungal product we eat. Alcohol and bread, which are consumed all over the world, are both made with yeast. Yeasts are fungi. Sourdough bread and wine are both examples of ancient fungal foods that many people still consume today.

Tibicos or Water Kefir, a home-brewed probiotic drink that has over 36 symbiotic probiotic bacterias and yeasts (fungi) that improve gut health.

Tempeh is made with by growing a mold on soybeans (or other legumes), and many probiotic foods and drinks are combinations of both beneficial bacteria and fungi like water kefir (see picture). Spent growing medium—a byproduct of mushroom cultivation—can be used as livestock feed, and mushrooms can be grown on all paper and cardboard waste indefinitely and consumed if toxin-free materials are used.

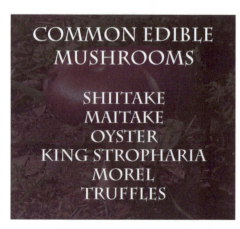

COMMON EDIBLE MUSHROOMS

SHIITAKE
MAITAKE
OYSTER
KING STROPHARIA
MOREL
TRUFFLES

---

[50] McCoy, Peter. *Radical Mycology*. 2016. p. 38.
[51] Grover, Sami. *Fungi sequester more carbon than leaf litter*. 2014. treehugger.com online.

Fungi are not just consumed by people. They are a primary food source in the soil food web. Ants and termites farm fungi inside their underground complexes of tunnels. Various flying insects (e.g. beetles, wasps, and bees) feed off mycelium, and wild animals like bears, foxes, deer, squirrels, and more also eat mushrooms—even psychedelic ones! Always keep in mind that many edible mushrooms have poisonous lookalikes in nature, so be careful and don't eat any unidentified mushrooms!

Fungi as Medicine

Though hundreds of studies have been done to look at the medicinal benefits of a few mushrooms species of interest, there are still thousands of studies that need to be done on the medicinal applications of fungi. Many mushrooms contain anti-cancer, anti-inflammatory, anti-cholesterol, anti-oxidative, anti-diabetic, anti-fibriotic, anti-fatiguing, anti-microbial, and anti-viral properties. Many mushrooms—most notably the woody polypores—have been found to stimulate the immune system, target specific cancers, and help heal and mitigate some of the negative effects of chemo-therapy, radiation, and other harmful cancer treatment therapies.

> **"Coriolus spp (Turkey Tail mushrooms) is the most studied natural agent against cancer with over 900 studies of which about 125 are human clinical trials. Most mushrooms have adaptogenic effect on immune function (e.g., boosting when low, reducing when excessive)"**
> –Jeanne Wallace, Phd, CNC

In Chinese medicinal practices, mushrooms have played an important role for thousands of years, but they are just beginning to be recognized in the Western world. The first US Food and Drug Administration clinical trials using Turkey Tail mushrooms to treat cancer only began in 2013. Mushroom prescriptions may yet be in everyone's future. Many companies offer medicinal fungal supplements in stores across the US. Peter McCoy's book, Radical Mycology, clearly details how to create much more potent and inexpensive medicines than those that are available in stores.

Colony Collapse Disorder (CCD), a condition that causes entire bee colonies to die-off suddenly, may be

Turkey Tail Mushrooms

addressed in part by feeding bees mushroom mycelium or extracts, a practice currently being investigated by Paul Stamets[52]. Paul noticed that his bees were feeding on exposed soil mycelium one day and then noted that in nature bees make hives in semi-decomposed trees, which readily have fungal activity close to the hive for easy access. Following this observation, Paul worked with Washington State University's Entomology department to test different types of medicinal fungal sugars in bee feeders, using captive bees. The fungal sugars were tested for their effects on viral counts in the bees and on the bees' average lifespans. Currently, it appears that the sugars from certain polypore mushrooms can help reduce viral counts and prolong a bee's life.

COMMON MEDICINAL MUSHROOMS

SHIITAKE
MAITAKE
TURKEY TAIL
CHAGA
REISHI
AGARIKON
LION'S MANE

> "Strongly anti-tumor and containing potent natural antibiotics[,]... Polyporus umbellatus [Umbrella polypore] contains powerful immunomodulating compounds and has also been implicated in the limiting of leukemia (1210) cell proliferation"
> –Paul Stamets, <u>MycoMedicinals: An Informational Treatise on Mushrooms</u>, 2002.

Methods

- *Dried Mushrooms* - These are dried and ground caps and stems used in food and drinks or packed into capsules. Mycelium can also be used this way by drying and grinding the substrate and can often create a more potent product.
- *Mushroom Tinctures* - These are made by submerging the mushrooms in grain alcohol, vodka, or vegetable glycerin for six weeks (daily shaking optional).
- *Mushroom Teas and Soups* - These are made by cooking diced or minced mushrooms in 70-80°C (158-176°F) water until a dark-colored tea or broth is produced (this can take several hours). This is why eating mushrooms in a soup is medicinal: it's essentially a hot water bath! Reishi or Chaga mushroom teas are popular medicinal choices. It should be noted that a hot water bath does not kill undesirable microbes as alcohol does.
- *Dried Mycelium* - Mycelium inoculated grains serve as a nutrient-rich and pre-digested food. Dried pure mycelium from a liquid culture is the purest and highest quality form of dried mycelium.

---

[52] Christensen, Ken. *Could a Mushroom Save the Honeybee?* 2015.

Poisonous and Psychedelic Mushrooms

Many traditional cultures revere mushrooms, including those of the psychedelic variety, often working with them for both spiritual and medicinal purposes. Archeological evidence suggests that this has been going on for quite some time. That said, there are many deadly mushrooms that look like edible mushrooms or may even taste wonderful, such as the Death Cap mushrooms (*Amanita phalloides*). It is very important to always learn to identify mushrooms with someone trustworthy and experienced.

Today we have studies being proposed and conducted that are exploring the medical applications of psychedelic mushrooms for extreme cases of addiction, depression or PTSD. While clinical research is ongoing and still in its infancy, humans have long consumed psychedelic mushrooms and plants for cultural, medicinal, and personal reasons. This practice continues to this day in many cultures both traditional and modern.

Fungi as Fiber, Fire Starter, and More

Mushrooms like Amadou (*Fomes fomentarius*) have been recognized for their utility for millennia. This mushroom was used by early man as a fire starter, as a means to carry fire from location to location, and as a fiber for paper and even as hats, like those traditionally made in Transylvania. Today, dried mycelium is being recognized as a building material of the future with the potential to replace much of the plastic we use to today.

Our ancestors depended on fungi for their survival and progress. Mushrooms like Inky caps (*Coprinus spp.*) are used for inks and dye. The mold-, water-, fire-, and rot-resistant qualities of dried and exhausted Reishi mycelium blocks are garnering attention as both a natural building material and as a means of utilizing paper, cardboard, and woody waste products. Some fungi, like sea algae, have bioluminescent qualities, and entrepreneurial mycologists globally are working out how to grow them as a living nightlight.

The ink was made from Inky Cap Mushrooms. It was painted with cat whiskers, a wild turkey feather, and an osage orange twig. More at NickNeddo.com

A hat made from Amadou mushroom fibers

> "Many fungi that attack and kill insects (e.g. the mold Beauveria bassiana) have been studied for decades as a natural pesticide. This notion of 'mycopesticides' has been argued to be a more natural alternative to chemicals when attempting to manage problematic insect population. However, this approach often overlooks the underlying ecological problems that lead to the imbalance in the system or the conditions in which these insects proliferated. A better approach to dealing with this issue would be to cultivate plant guilds that deter negative insects and to bring in reptiles and predator insects to naturally balance insect populations"
>
> –Peter McCoy, 2016.

Fungi as a **biofuel** is also gaining momentum as a concept. The future is seemingly limitless in terms of what working with fungi and integrating them into human-designed systems can offer to solve a range of pressing global problems.

Composting

Paper, cardboard, coffee grounds, and nearly any complex carbohydrate source can be digested by fungi (to produce mushrooms and the end-product can be made into mycelium-rich compost). We just have to get things moist and introduce enough inoculated substrate to seed the pile. Once it is all inoculated, you have a large pile of starter for an even greater amount of substrate (if that is your goal). Otherwise, it can be allowed to fruit until exhausted and then composted. Vermicomposting this material with red wiggler worms is an even more efficient and effective way to use this material, thereby generating worm castings for the garden or for brewing compost tea.

Dog and Cat Waste

Spent mushroom substrate can serve as kitty litter and, afterward, can be easily processed by vermicomposting. Dog waste can be combined with inoculated wood chips and then vermicomposted as well. The carbon in the wood chips absorbs excess nitrogen while fungi can digest the toxins and pathogens in the animal

waste, helping form a more balanced compost. If the pile is too hot or too toxic, the earthworms leave. For urban and suburban cat or dog owners, this can be a consistent, easy, and regenerative source of soil for balcony or rooftop gardens.

Mycoremediation

The negative impacts of many toxic wastes—from cigarette butts, to nuclear fallout, to chicken manure, and oil spills—can all be managed by fungi. This process is known as **mycoremediation**. Fungal hyphae exude digestive enzymes that have the ability to decompose complex compounds more efficiently than any other life form. As nature's greatest chemists and recyclers, fungi are critical to cleaning up the most complex pollutants in the environment. King Stropharia fungi form such a tight mycelial web that they can temporarily filter out microbial contaminants from flowing water systems. The sea floor is largely a fungal decomposition system.

There is also hope that fungi can digest the world's plastic pollution. It is logical that oil and its derivatives (such as plastic) which originated as organic matter, would be consumable by fungi. For many decades, however, plastics companies have been researching how to make plastics that do not degrade or decompose.

Preliminary research suggests that mycoremediation may be one of the fastest ways to deal with the issue of global pollution (including nuclear waste). In remediating an area affected by radioactive fallout, fungi expert Paul Stamets suggests laying down 1-2" (2-5cm) of wood chips, chipping the radioactive trees on-site and then "[planting] native deciduous

King Stropharia Mushrooms are excellent mycoremediation fungi to use in a wood chip bed for graywater remediation before it enters the garden beds or swale paths.

and conifer trees, along with hyper-accumulating mycorrhizal mushrooms, particularly *Gomphidius glutinosus*, *Craterellus tubaeformis*, and *Laccaria amethystina* (all native to pines). *G. glutinosus* has been reported to absorb—via the mycelium—and concentrate radioactive Cesium 137 more than 10,000-fold over ambient background levels. Many other mycorrhizal mushroom species also hyper-accumulate."[53]

These radioactive mushrooms can be harvested by people wearing full nuclear hazmat gear and composted somewhere isolated, so that the composting process, if given enough time, will turn the mushrooms into soil that is eventually non-radioactive. Paul Stamets suggests burning the resulting radioactive mushrooms and storing the radioactive ash. It may however be more pragmatic and more efficient to harvest these mushrooms and compost them in an isolated location, where they would have the time they need to naturally cycle through their radioactivity (for instance, Cesium 137 has a 30-year half-life).[54] Even more pragmatic would be to leave all nuclear power and weapons behind.

Fighting Fire with Fungi

Inoculating downed wood, wood chip piles, and stumps with fungi can help build soil and decrease the amount of fuel for wildfires. It also builds humus, traps moisture, and encourages the growth and diversification of soil life as the inoculated wood breaks down further. All the punky wood is fire-resistant, and all the fuel that left alone would take years to break down is processed at an accelerated rate by the fungal decomposers.

This fallen tree is completely inoculated with fungi, so now it is starting to fruit. This wood is punky and will not easily burn, making it a fire suppressant instead of a fuel source.

If the excess wood choking the unmanaged national forests of the American West were instead turned into inoculated wood chip piles covering water-harvesting earthworks) the resulting decomposition would aid in sequestering carbon, preventing wildfires, and encourage a regeneration of wildlife flora and fauna. This would all be achieved by

---
[53] Stamets, Paul. *How Mushrooms Can Clean Up Radioactive Contamination - An 8 Step Plan.* 2011.
[54] Ibid.

stimulating the foundational layer of the forest ecology; the soil life which is predominantly fungal.

Fungi and Plants

Almost all plants form mycorrhizal associations in the soil. These may be endomycorrhizal, ectomycorrhizal, or both. It should be noted that a small group of plants prefer neither, thriving instead in an actinobacterial or alkaline, bacterial-dominant soil environment though even these contain **endophytic** fungi which may be involved in root behavior.

- **Endomycorrhizae** - Preferred by annuals (e.g. beans, corn, peas, and carrots) and perennials (e.g. avocado, apple trees, citrus trees, olive, hemp, cacao, rye grass, and willow). This is by far the largest group of mycorrhizae associates at nearly 90% of identified fungi.
- **Ectomycorrhizae** - Preferred by trees (e.g. douglas-fir, alder, pine, poplar, manzanita, beech, and chestnut). This is a small group at around 5% of known fungi species.
- **Both Endo- and Ecto** -Preferred by trees (e.g. eucalyptus, poplar, willow, aspen, cottonwood, and alder). This is an even smaller group.
- **Neither** - This group comprises both the Brassica (e.g. broccoli, cabbage, kale) and Ericaceae (e.g. blueberry, azaleas, huckleberry) families as well as other plants like beets, and rushes[55].

Grow Your Own Mushrooms

Oyster Mushrooms on Paper and Cardboard

Using grains or sawdust inoculated with oyster mushroom mycelium (grain spawn or sawdust spawn, respectively), we can easily grow mushrooms on a roll of toilet paper or even cardboard and coffee grounds. Grain or sawdust spawn can be ordered online.

First, sterilize the growing medium by either pouring boiling water over the toilet paper and letting it cool or by soaking the cardboard in non-chlorinated water and then letting it drip out. Fresh coffee grounds are a great substrate as they are essentially

---

[55] Fungi.com. *Mycorrhizae-Compatible Plants.* 2016.

pre-treated during the brewing process. Please note that Oyster mushrooms are one of the few types that can readily outcompete bacterial and fungal competitors on these non-sterile substrates! Observing this happen can be both fascinating and educational.

With clean hands, stuff the spawn into the center of the toilet paper roll, and then seal the roll up in a plastic bag that has many small holes poked into it. Alternately, cardboard, coffee grounds, and grain spawn can be layered in a clean yogurt container or bucket until it is full. The container is then placed in a cool, dark area with its lid loosely applied. For either project, the mycelium should grow over the material in two to five weeks.

What causes fruiting?
- Increase in Humidity
- Increase in Oxygen
- Decrease in Temperature
- Increase in Light

To initiate fruiting, place the bag or bucket in a cool area. Mushrooms will start to form from the opening in the container and they need to be lightly misted with water to keep from drying out. A bucket of mycelium can produce at least two to five flushes of mushrooms (each one to two weeks apart) before it is exhausted. The inoculated cardboard can be used two to three times more to inoculate more cardboard, but after three to four flushes of fruit, they tend to lose their vigor.

Shiitake Mushrooms on Logs

Using wooden dowels inoculated with Shiitake mushroom mycelium, we can grow mushrooms on a hardwood log. Inoculated mushroom logs can fruit for 5-10 years! Logs should be cut from a living tree while the tree is dormant or at the end of the growing season. They should be 4-8" (10-20 cm) in diameter. Using a drill bit the size of the dowels, holes that are 1" (2 cm) deep are made a palm's width away from each other. The dowels are gently but firmly tapped in with a small hammer. Avoid damaging the bark. Hot

beeswax is then applied over the pegs, sealing the hole to protect the fungus, keep out competitors, and retain moisture. In 12 to 18 months, try soaking the logs overnight in cold water to induce fruiting. You can also hit the logs with wooden mallets to stimulate growth! Shiitakes grow in a wide variety of climates outdoors. Sawdust spawn can also be used as well as other methods to grow medicinal and plentiful Shiitake mushrooms.

King Stropharia in the Garden

Like Oyster mushrooms, King Stropharia mushrooms do not necessarily need sterilized substrate: they can outcompete competing bacteria and other fungi. King Stropharia, also called Wine Cap Stropharia, is a delicious mushroom that can grow on cardboard or wood chips in and amongst garden plants. It is incredible easy to establish and continuously feed and harvest from. A mushroom bed can quickly be made anywhere large or small, in any shape or design, by laying down alternating layers of inoculated substrate and wood chips, finishing with wood chips on top. Just keep it moist and it will continuously generate large, delectable mushrooms.

Liquid Inoculation Jars with Airport Lids

Using just mason jars with modified canning lids, we can cultivate liquid cultures of mycelium at home or almost anywhere. The liquid broth is primarily comprised of sugar water, but other ingredients such as yeast, peptone, or gypsum are also often added (for exact recipes and guidelines consult _Radical Mycology_). Sugar types can range from honey to malt to dextrose to even corn syrup. 500 ml of water is used in a standard quart canning jar. With an airport lid and a sterilized jar with sterile liquid culture inside, we essentially have a portable clean room (or sterile field) inside the jar. This means that liquid inoculation jars can be worked with in almost any circumstance with minimal concern for contamination–though you still have to sterilize your syringe and port site.

Liquid cultures are an inexpensive imitation of industrial practices that use large vats instead of mason jars. Mycelium grows best in liquid because it can grow in all directions rapidly.

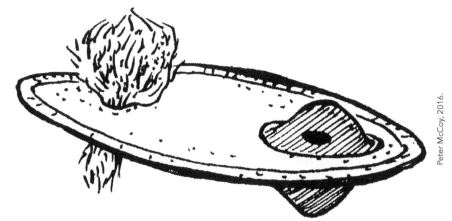

This method is also more time-efficient with less preparation required from the cultivator, and it is more efficient at inoculating substrates over traditional lab-based methods. Inoculated agar or grains can be used to start liquid cultures—even samples of mushroom tissue can be used!

Liquid culture inoculation in jars is spreading in the mycological community like wildfire not just because of its efficiency, low-cost, and easiness to work with, but also because of its greater efficacy as a medicine. Liquid culture mycelium can be dried on a plate of glass and turned into pure powdered mycelium. This is drastically different from what's offered in stores currently, which is mostly freeze dried and powdered inoculated brown rice. These products are mostly just rice and have very low mycelium content. Liquid cultured mycelium offers a pure medicine that is many times more potent that can be made at home for a fraction of the cost of inferior, store-bought products.

How to Make Airport Lids

A breakthrough in home cultivation, airport lids are modified mason jar lids that create a clean, sterile environment for the mycelium to grow. Two holes are made in the top of a mason jar lid (5/16" (0.8cm) in diameter). One is generously covered with a blob of high temperature RTV silicone on both sides of the lid. The other is stuffed with synthetic fiberfil (synthetic cotton) or covered with something breathable and disposable like micropore tape. This allows sterilized syringes to inject mycelium through the silicone port, and it also allows filtered air to flow through the fiberfil port.

Designer's Fungi Checklist

- Explore, list, and consider the possible mycorrhizal relationships between plants, soil, animals, people, and waste streams on your site
- List site goals and decide upon a strategy or set of strategies to enhance and support selections from these possible relationships that align with those goals
- Inoculate soils as well as perennial and annual plantings with mycorrhizal fungi to support growth and fruiting
- Avoid tillage to preserve mycorrhizal fungi
- Grow mushrooms with wood, waste, and more resources on-site
- Grow mushrooms in gardens and food forests
- Partner with fungi wherever possible

For more in-depth reading on Fungi and Mushroom Cultivation consult Radical Mycology by Peter McCoy or enroll in Mycologos, the first mycology school for online and in-person students.

MYCOLOGOS

IX. Earthworks & Earth Resources

In permaculture, the term "earthworks" refers to working with earth—literally! Working with earth is as ancient as humanity—we've used earth or soil in innumerable ways: from using clays to bond with toxins or poisons in our digestive tract, to using it as a building material, to a medium in which to grow all our food, and as a spiritual focus. It also should be noted that earthworks can be as damaging as they are positive. The strip mining, mountaintop removals, open-pit mines, and damage associated with most industrial earthworking operations makes most environmentally conscious folk concerned about disturbing the soils—and rightly so. Earthworks in permaculture are a powerful set of actions that can be taken to foster an amazingly positive and regenerative change; however, these works need thorough planning and consideration before implementation.

Earthworks prevent erosion, manage excess water in flood events, provide areas for catchment, provide areas to grow food, shelter buildings from light, provide habitat and living spaces, dampen and block sound and wind, provide walls and insulation for building, and provide the structure for a sustainable site.

Always remember...

Plan carefully

Taking into account the largest flood or precipitation event, the wind, the clay fractions of the soil, legal requirements, the social climate, and more site-specific details is vital as we plan our earthworks. Do careful water accounting and reflect thoroughly upon your ideas. You don't want to redo your earthworks —you want to get it right the first time; the less disturbance the better. Draw it out, discuss it, and get feedback before you cut, or experiment with scalable test plots or sections. Re-doing earthworks is a last resort and can be prevented with enough observation, experimenting, and planning.

Don't mix topsoil with subsoils

Always keep the layers separate, and know that the top 2-6" (5-15cm) are the most fertile. When digging a swale, remove and reserve topsoils upslope from the the cut while you shift subsoils onto the berm. Afterwards, topsoils are added to the top of the berm. Sod can be flipped and undesirable plant roots exposed to the air and sun especially on the edges of the berm. It should be noted though, if you live in an area that has not seen native plants in decades or even hundreds of years, if you mix your subsoil with your topsoils, you will reintroduce seeds from the past, sometimes the very distant past. This can both add diversity to the system and more work to re-establish the soil layers, but it is something to consider or test on your own site as it won't work for every site. In effect when we mix subsoils and topsoils, we are undoing the settling and development of soil that time and natural cycles have already set in place.

Don't Guess Contour

Contour is very difficult to see in a landscape without a tool to measure the level. Use stakes, colorful flags, and ropes to keep it clear where to cut and where not to cut! Being exact with earthworks will only enhance their beneficial qualities. It will also make building, tracking, and maintaining them easier if they are well-designed initially. Especially if you are creating a guideline for your keyline design, you want your contour to be precise. In general, it's better to be precise!

Plant Immediately!

Ideally soaked and inoculated seed and transplants should be applied immediately to the finished earthwork with its returned layer of topsoil, and then it all should be covered with a light mulch. If it is on a slope, or if needed due to wind or bird-pressure, small sticks and branches can be laid over the mulch to hold it all in place. Cowpeas work well in nearly all climates and work both as a nitrogen-fixer and as a biomass accumulator. Sunchokes, comfrey, artichokes, clover, all legumes, radishes, mustard, shrubs, perennial grasses, and many more choices are appropriate for quick recovery and soil coverage. If a plant grows quickly, is beneficial to all, and can be chopped and dropped, it is a good choice. Stacking functions is ideal, but not all areas need to be dedicated to producing food for people—especially if you are in a dry climate diverting most of the water to the garden and orchard areas. Some areas can be focused on native, drought-tolerant species that cover the soil, keep moisture, and create shade.

Compaction over Time

Moving soils for earthworks increases the air in the soil and visible mass, but over time, the soils lose that air and compact down. Anticipate that soils will settle as earthworks age; build them bigger than you intend them to be.

Fire

Fires can kill the soil life in the topsoil and even turn clays into impermeable pottery in the soil! Mudslides are common in Mediterranean climates, a landscape dominated by **pyrophytic** plants, if rains are too heavy in areas still recovering from wildfire. Rebuilding after a fire is delicate work and requires extensive restoration and intensive **silviculture** because fires are occurring more frequently than even pyrophytic life cycles can handle. Using earthworks and strategic plantings, we can bring back the hydrology in the soil and protect against perpetual megafires.

Always Consult Experienced Locals

Whether you are hiring them or not, consulting with someone who has made lasting, high quality earthworks in the local area is a wise course of action. Landscapers that make ponds and dams would understand the concepts necessary for any rehabilitative earthwork even if they aren't well versed in permaculture—excavation professionals can follow directions extremely well. Vetting local professionals is as easy as visiting the sites they worked on that are mature and talking to the customers about their experience with that company, service, and the final product.

Building Sites

Foothill pediments, midslopes, or lower non-eroded slopes allow for homes and settlements to be built between upper forests, lower forests, and plains. Above, the forest belt condenses water, slows or stops erosion, and prevents and delays snowmelt. Below the midslopes, productive animal systems, gardens, and forests can be created, fed by the waste products of the home system: graywater, compost, manure, paper waste, etc. Water storage in upper slopes can provide gravity-fed irrigation or even provide electricity, and in lower plains, water is easy to store in water-retention dams. Building a home in the mid-slope positions it above the frost line and shelters the home from winds in all directions. This can also be seen as: the house is positioned just below the keyline in the landscape.

Slope

Slope is expressed in several different ways: the rise over the run or "grade" (rise:run), the % of verticality, or a scale of degrees from 0° which is flat to 90° which forms a right angle. 1:4, 25%, and 14.04° are all the same slope expressed in three different ways. The slope tolerance of a soil depends on the soil type. In general, gravels can withstand very steep slopes at 1:1.5 (66.6% or 33.6°), free-draining clays 1:2 (50% or 45°), sands 1:3 (33.3% or 30°), and wet clays and silts 1:4 (25% or 14°). All these numbers are subject to stretching and shrinking depending on site specifics and soil composition. Soils on slopes also tend to sag in the middle, so measuring in a straight line tends to skew the actual layout of the land.

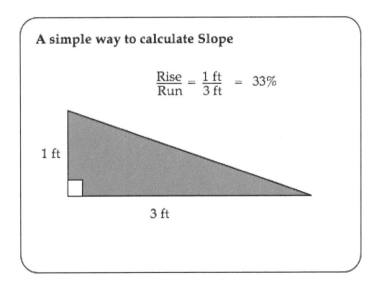

A simple way to calculate Slope

$$\frac{\text{Rise}}{\text{Run}} = \frac{1 \text{ ft}}{3 \text{ ft}} = 33\%$$

1 ft

3 ft

A General Guide to Slopes

If a slope is 50% or higher, it should be maintained as permanent forest, untouched and pristine. If it is 0-10% slope, it is easy to cultivate food, harvest water, and control erosion. 10-20% slope requires more work to stop erosion and cultivate soils regeneratively. Between 30-50% we can carefully forage, cut timber, and hunt. As with almost everything though, it depends on your site and goals.

Levels and Leveling

The ability to use levels and create a level area is critical to many components of design including swales, ponds, gutters, house foundations, overflow sills on dams, and more. If we cannot pacify water, slowly spread water, or evenly distribute the weight of a house, designs can be ineffective or even dangerous. There are many ways to make an area level, but most are simple and can be accomplished using on-site materials.

Water Level

A water level is as simple as it sounds: use water to see if an area is level. Fill up your leveled area with an inch or two of water and measure it in multiple areas to see if it measures evenly. This can be done with a ruler, a finger, or a notched stick just as easily.

Filling up the area with more water, observe closer: Does the water run off to one side? Does it slow, soak, and spread? How fast does it soak in? Is the soakage evenly distributed? All these signs give you important information.

The A-frame

A must for every home gardener or designer. Used for making swales most often on the home-scale, these ingenious and simple devices use gravity to measure how level an area is and help create perfectly flat areas. A-frames are made with three sticks or poles forming an A—they do not need to be of equal length. Hang a weight from the pinnacle of the A with a rope or string down to just below the horizontal crosspiece of the A.

To calibrate, simply mark the spots on the ground where you initially place your A-frame and allow the weighted rope to still—mark that spot on the crosspiece. Flip the A-frame around so that the feet go into exactly the reverse spots they originally were. Once the rope stills, mark the new intersection between hanging rope and crosspiece. The exact middle between those two marks on the crosspiece indicates perfectly flat ground or a contour line (which are always perfectly level). Anyone can make an A-frame and use it.

Calibrating an A-frame

1. Place the A-frame where you can easily see the foot marks, so you can flip it to exactly the opposite spots.
2. Mark where the still weight hangs
3. Flip the A-frame around and put the feet exactly where the opposite foot was. Mark the new line. You should have two lines now.
4. Mark the exact middle point between the two lines you've made—that's the level for that A-frame.
5. Your A-Frame is calibrated and ready to find contour!

Once the A-frame is level, stake the spots where the feet are. Then with one foot of the A-frame on the last staked spot; find level again. Drive a stake in where the other foot is. Now you have three stakes in a row on contour. You can tie a string to each stake and make the contour line more visible or start digging in sections. You can continue staking out the contour line as well, one stake at a time, to make sure that no complications arise. Often we cannot know what a area will look like on contour until we stake it out first or use Google Earth or a drone GPS mapping system to make a topographic map.

Using an A-frame works well at the home and garden site level, but an A-frame cannot remain accurate over long distances, large projects, and for precision farming or construction operations.

Hose Bunyip Level

Using a capped, clear section of hose with water in it (leaving enough air in it to measure on both sides), we can create a flexible and cheap level based on the same principles as the store-bought kind by attaching the hose ends to sticks that are marked out like a ruler, or the sticks themselves can be yardsticks. It can easily work faster than an A-frame in finding contour.

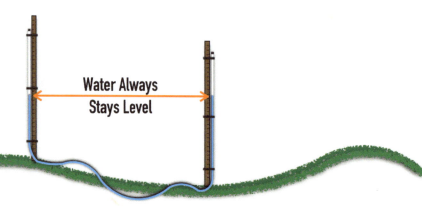

A Bunyip Water Level

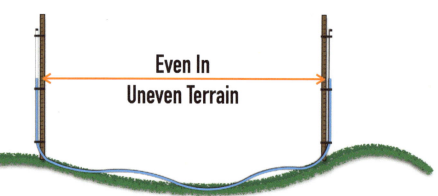

Plane Table, Theodolites, and Dumpy Levels

Plane tables are effective and inexpensive precursors to theodolites and dumpy levels which, though more precise, aren't necessary for most earthworks. Plane tables come in all sorts of permutations and can use a U-shaped water level mounted on a tripod like a mini-bunyip, a bowl of water with a piece of floating wood in it mounted on a tripod, a table with a sight, or a flat table on a tripod with a bubble level. The level creates a sight to view across valleys and see common contours. Theodolites are telescopes designed to measure vertical and horizontal angles while dumpy levels are the classic automatic levels with scopes used by surveyors the world over. These can be used as a sight for looking through or as a way to measure the ground the tripod is on.

Ethical Earthwork Applications

- To reduce energy, resource, and water usage
- To provide habitat for wildlife and people
- To rehabilitate damaged or degraded landscapes
- To spread and store water in the landscape

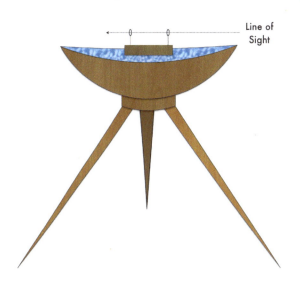

An Example of a Plane Table

Types of Earthworks

Berms or Banks

These are un-compacted mounds of earth that are stable. They can be used as windbreak, for flood control, as sound or visual barriers, as garden or perennial beds, for raised pathways, for building or animal shelter walls, for loading or unloading ramps, or in a swale system. Often the most inexpensive of earthworks, they are also the most common.

When air is lifted by a bank of earth, the sheltered space immediately behind it is ideal for a microclimate, animal shelter, or housing site. Banks are extraordinarily useful and versatile in their application. These can even be used to make water tanks or to enhance wind tunneling to improve wind power generation. Soil is endlessly moldable and adaptable.

During the Great Depression, President Roosevelt famously put Americans to work repairing the damage caused by their expansion westward, calling the program the Civilian Conservation Corps (CCC) or "Roosevelt's Tree Army" (1933). Today we can still see evidence of their interventions in the Tucson area. Earth berms were put in place there in the 30's and then abandoned. They serve as a powerful example of what earthworks and manpower alone can accomplish. The area is thickly vegetated, a stark contrast to the sparse desert surrounding it. Berms can be used on a large scale to mitigate drought, to adapt agricultural lands in a time of changing climate, and to stop erosion of topsoils. Images of the berms can be seen on the next page.

Benching

Cuts into the hillside that are nearly flat or on contour for roads or house sites provide benefits for the ecology as well. They slow water, runoff, and erosion. They also allow steep

While commonly referred to as the Tucson Swales, the CCC constructed flood mitigation berms to trap some water while they mostly divert it. It still created a vibrant series of microclimates by pacifying water, creating a windbreak, creating shade, holding in organic matter, and by stacking the berms one after the other. These berms can be still seen from satellite, on GoogleEarth (below), or visited in person (above).

slopes to be accessible (up to 30-40% grade can be benched if accessible initially by machinery). Valuable fruit trees as well as hardy, self-seeding annuals and perennials can be planted in the loose lower side of the bench as soon as it is cut to prevent erosion.

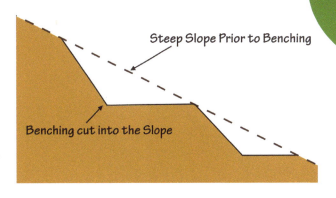

Culverts, bridges, gabions, or piping of some sort are used when benches encounter passing water—otherwise the water would erode the bench in that intersection quickly.

Terraces

These are large level areas separated by embankments. They build from the lowest terrace upward, reserving the topsoil from the first terrace for the last terrace. Once the first terrace is dug, the topsoil is pulled off the second site above it and used to finish the first terrace. This pattern repeats until the end when the first terrace's top soil is carried to the top and finally used. The level nature of the terrace pacifies rain and flood waters, makes flood irrigation easy and even, and prevents nutrient loss.

Terraces are not advisable on very steep slopes or in soils that cannot withstand the slope and water pressure. Trees are used on the embankments between terraces to not only provide leaf litter and nutrients but to hold the terrace embankments together and keep the terraces level and intact. Living fences or hedges are ideal for protecting the embankment slopes from grazing animals. Ponds and aquaculture in general can be utilized on the terraces for great effect, as Sepp Holzer's Krameterhof demonstrates elegantly. Staggering terrace spillways also prevents erosion over time. Terraces are the longest-lasting and oldest agricultural artifact still being used in the world today—while monuments of stone weather away, Asia's terraces endure—albeit with maintenance and good management.

Earth-Sheltered Greenhouses and Homes

These are structures that use earth to mitigate outside environmental conditions. They can be as simple as earth banked on the northern wall of a greenhouse in the northern hemisphere, or a southern wall in the southern hemisphere—this prevents heat loss and makes a thermal mass storage in the soil that continues to warm the structure through the cooler night. It also works to keep structures cool in the heat, even though they are under glass in the sun.

The *walipini* is a submerged garden (8-10ft/3m) facing the sun path with a ground-level, glass or plastic roof oriented perpendicular to winter solstice's sun at noon (the time of

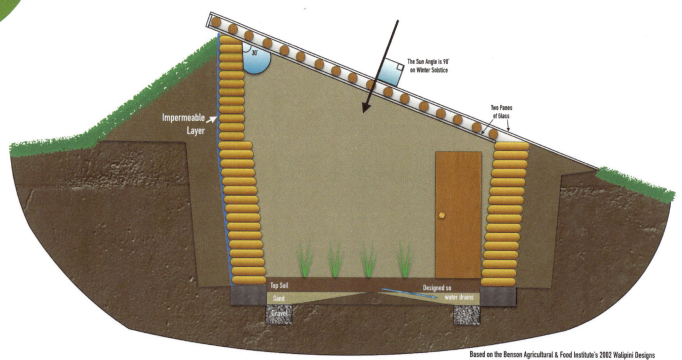

Based on the Benson Agricultural & Food Institute's 2002 Walipini Designs

least solar energy)[56]. Topsoils are reserved during excavation and returned over a graveled and graded floor to prevent water logging and to give plants healthy soils (since the subsoils at 8 ft (2.4m) deep are far from ideal for plant growth). This makes for a warm winter garden even in sub-zero temperatures.

This design is based on a general thermal constant of the earth. All over the world starting at 4 ft (1.2m) deep in the dry soil it is 52-54°F constantly. When we cap that heat and add the sun's energy to it, we easily create enough warmth to grow food through the winter.

Ha Ha Fence

This is a submerged garden with steep sides that serves as an inverted fence to prevent large herbivores like elephants or cattle from grazing in areas where fences are unfeasible or timber is scarce. Cows can climb up stairs but cannot climb down them; the same principle is being applied here.

Hugelkulturs

Mentioned earlier, *hugelkulturs* build soil by imitating the natural process of turning wood into fungal-dominant soil. They retain water and trap released carbon and nitrogen. They are great for large-scale and small-scale annual and perennial plantings. *Hugelkulturs*

---

[56] Benson Agriculture and Food Institute. *Walipini Construction (The Underground Greenhouse)*. 2002.

are a great alternative to burning and an easy way to build more soil with excess woody materials.

In sandy soils, the wood should be sunken into the earth. In soils with more clay, semi-sunken or even on top of the surface is viable. How deeply or not deeply embedded is also dependent on the context—if you are on a slope or get a lot of rainfall, you will want to bury the wood at least enough to give it a stable resting point in case of flooding. Wood floats, so *hugelkulturs* on undisturbed hardpan in flooding events have been known to float and then roll downhill, which can be dangerous.

Above all *hugelkulturs* are a waste management strategy for putting an abundance of woody material to productive use and should not be a strategy for building gardens from scratch. It is much easier to sheet mulch with already processed woody material wastes that we have in abundance, such as paper and cardboard, without cutting down more trees.

The Crater Garden

Another Sepp Holzer innovation, it is an area dug into the earth to shelter it from wind, to trap heat, and to collect and condense moisture.[57] A lake may form at the bottom, or it may just stay more moist than the surrounding area. It is a unique microclimate technique. Often it is dug out into terraces on contour that are deepest at the center, or it can be dug like a snail shell in a downward spiral as well. It doesn't have to be circular, just lacking straight lines and corners. The crater's depth brings the cultivation closer to the groundwater as well. Subsoils that are removed can form berms around the edges while reserved topsoils can be returned to the interior of the crater's surface. These are especially powerful in arid climates.

Swales

Mentioned earlier, swales prevent erosion, pacify water, infiltrate moisture into the ground, and provide shade and wildlife habitat. The best places for swales are always the highest and lowest areas first.

Swales vs SubSoil Ripping

Swales are highly effective when used properly and equally visual - such that people see them working, but they are not always the best option for a site and, in some cases, are overused by designers and can lead to subpar results. Ripping soils on contour in contrast allows water to infiltrate almost upon contact instead of forming runoff flowing between

---
[57] Holzer, Sepp. *Desert or Paradise.* 2012. p. 135.

SubSoil Ripping

A Swale

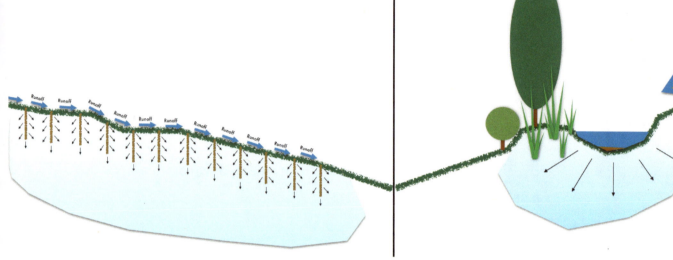

swales, but subsoil ripping requires machinery and an area that the machinery can navigate within. Ripping works best in a grazing, pastoral situation, but it's versatile too: it can be used between orchard rows and on slopes. Swales work superbly with alley cropping, silvopasture, orchard food forests, and reforestation projects with soils that can absorb the water at a rate that supports the site and ecological goals and not erode. Swales are excellent on hilly, rocky, or mixed forestry areas where machines cannot go since size can be adjusted from macro to micro, long or short.

Swales can be made with only a shovel, among already developed trees and rocks. They can restart springs, though in humid pastoral settings, areas over bedrock, or places with fast draining soils, too many springs can become an issue. Ripping is usually done on production farms and always with machines while swales are cut in all settings and can be done by hand on homesteads.

Net and Pan

This is a method for planting trees on steep slopes, especially in drier areas. Trees are spaced out evenly and planted in shallow pits (pans) then connected by even more shallow diversion drains. This allows all trees to be watered efficiently—all runoff is caught in

Brandon Carpenter 2015.

the net, and all trees are included. Organic matter is carried and deposited along with the water, making it a compost tea of sorts which is shared by other trees when water levels exceed the individual pans. When all pans and diversion drains are full they release water in a sheet down the hill evenly.

Mudbrick, Cob, or Rammed Earth Blocks

Mudbricks, cob (cobb), adobe, rammed earth, and many earth-building techniques are ancient and time tested. Rammed or pressed earth blocks are an easy way to generate your own building materials on site. Machines to press blocks are available to rent many places in the world. Cob can be made into any shape. Cob is "a traditional form of earth-building in which a moist, plastic mixture of earth, straw, water, and sometimes sand and gravel is piled on a wall while wet and worked into place."[58] Cob ovens and rocket mass heater benches are perhaps the most common application promoted. Adobe is just like cob but can also have manure in it. Both can sometimes include lime in their recipes. Please note: in some areas, the terms cob and adobe are interchangeable.

Earthbagging

This is a technological adaptation of mudbricks. Bags (biodegradable and otherwise) are filled with soil, predominantly clay, that is mixed to dry hard as cement. These bags are laid on top of each other while still damp, flattening them. Barbed wire or some other stabilizer is used between the bags to hold them in place while they dry and harden. It is similar to building an igloo: starting at the base and working your way up to the top. Domes are possible though many prefer a flat roof for easy rain catchment.

An earth-bagged water tank design under construction in Peru

Reading the Landscape

- **Levels** - These levels allow us to see contour on site in real time. While our eyes may fool us, these devices won't as long as they are working properly and the user understands how to use them.

[58] Evans, Ianto, Smith, Michael, G., and Smiley, Linda. *The Hand-Sculpted House: A Practical and Philosophical Guide to Building a Cob Cottage*. 2002. p. 330.

- **Topographic Maps** - These maps can be acquired from county record halls, the internet, Google Earth, by mapping with GPS coordinates, and with drones. They are incredibly useful with their aerial views of the sites and clear contour lines. Remember that not all sources are perfectly accurate which has led to GPS mapping drone operators being in high demand these days. Some designers like Sepp Holzer will even fly over a site to see landscape from the same perspective.

Earth Tools

Animals

Possibly one of the first places humans learned about earthworks was from other animals—in this way it is animals, rather than tools, that are our teachers. Pig wallows seal up, eventually forming permanent ponds. Herbivore manures improve the soil and often have seeds already in them, ready to germinate. Earthworms generate worm castings which improve soil structure. Omnivores like pigs and chickens can beneficially disrupt an ecosystem and provide an opportunity for designers to put in something more productive and useful for both the local ecology and people. Animals and animal partnerships have helped humans develop and progress throughout time.

A small hand-made rice knife from Permaculture Tools

Hand Tools

Though slower to make progress with and dependent on human power, hand tools are more reliable, efficient, adaptable, and precise than large machinery. From the rice knife to the hoe to the shovel to the scythe, hand tools are more nimble and sophisticated than machinery will ever be. When we work with hand tools on a site moving earth, we recognize indicators in the landscape more easily. An excellent example is found in the broadfork which aerates the soil without mixing the different soil horizons.

Fossil Fuel-Powered Machines

Though used primarily for destructive purposes, machines have a huge role in restoring our degraded and damaged landscapes and transitioning our urban areas into more sustainable cycles. With all the damage done with heavy machinery, it is only fitting that we will use them now to undo the damage they've done and regenerate and restore our landscapes.

While many will use only a Bobcat with a few attachments, a ripper on a tractor, or maybe an excavator with a bucket that can tilt, there are many types of industrial earth-working machines with many functions. These can be rented in many areas of the world. Machines can have buckets, blades, drills, hydraulic jets, levelers, rippers, rollers, and more, but primarily they use blades and buckets.

- **Blade Machines** - Tractors, bulldozers, graders, and more machines can mount a large blade in front for moving earth, leveling, or making ditches in a pushing or pulling motion. The blade can also be mounted in the middle for leveling or grading; leveling can also be accomplished with the blade mounted on the back. Almost all earthworks can be accomplished with time, patience, and skill with a simple blade machine. Often renting a small Bobcat with a few attachment blades can accomplish whatever earthworks need to be done on most home sites with little expense and time.
- **Bucket Machines** - These machines can lift and tilt but not swivel. They are excellent at gathering loose material, picking it up, and dumping it. They are essential for excavating certain projects like digging out a *walipini* or excavating silt from a pond.
- **The Four Way Bucket** - The four way bucket (lift, dig, push, pull) is the ultimate excavator as it combines the abilities of bucket and blade machines. These are sophisticated machines that can do what no other excavators can, from selective removal of objects and soils to elegant curves in earthworks.

Designer's Earthworks Checklist

- Observe and analyze the landscape thoroughly before working with any earth: topographic maps, historical accounts of the landscape, observing seasonal changes on-site, etc.
- Analyze and account for soil types and amounts on-site
- List overall site goals and identify where earthworks can help achieve those goals
- Make a plan with a measurable outcome
- Use earthworks to heal the land

X. Permaculture Processes and Frameworks

> "There is no one way and no one answer to any problem, and never can be. There are millions of answers and potential solutions, and these have to be worked out case by case, situation by situation, by people who are driven by a desire for something better"
>
> —Alan Savory, <u>Holistic Management</u>, 1999.

Concepts and Themes in Design

All Things have their Time and Season

We can't plant summer crops in winter, we can't run before we learn to walk, and we won't have any reassurance of safety or a positive outcome unless we know what we are doing before we act. We plant when it is time to plant. Plants and seeds grow when it is their time and place to grow. In all things, even social, timing is everything.

Self-Managed Systems

Perhaps the ultimate design concept is the self-managed system. Many dream of a farm that runs itself without human input or management: the farmer need only harvest and sell the products. This idea of foraging and living off the wilderness' abundance is actually possible to a certain degree—you would still need to gather the chestnuts and grind them to make bread; nothing is without some input on our part.

We can always design aspects of a system to self-manage. This can be accomplished through rotational grazing, larger aquaculture systems, food forests, perennial gardens, decentralization of authority, **unschooling**, or even restored wilderness. That being said, all thriving production systems require a human input of work, observation, and intervention at appropriate times.

The Unknown Variables

In natural systems we can influence the cycles in place, introduce new cycles, replace, damage, or remove cycles, but we can never know all the variables involved in these choices. Everything is in motion; everything is part of many cycles, and everything is at some stage in a process, usually in tandem with innumerable other processes in the ecosystem. We cannot control everything. Observation and careful reflection over time as we make changes, followed by careful adaptations, are always needed.

Longevity of Design

Systems we construct should last as long as possible with the least amount of maintenance possible. These systems should be fueled by the sun and produce enough abundance to support themselves, the creators of that system, and the wildlife of that area. Whatever energy used to create the system has to be regenerated or captured by the system itself and continue to do so as long as possible. Food forests, for example, can last thousands of years if designed properly.

The Big Pumpkin Fallacy

Striving for the biggest yield leads to more watering, more fertilizer, and more energy inputs for the targeted crop – while decreasing the overall yields of a system. It is preferable to create a resilient system with minimal inputs that has diverse yields.

Some tomato plants put on only two to three small fruits per plant, but those fruits are more vividly flavored than any store-bought tomato. The case is the same with dry-finished tomatoes in Sonoma county, California—it makes the flavors more intense when the plants only feed off the water table and farmers withhold irrigation. Bigger or more is not always better, especially when it comes to flavor and quality.

Cycles: A Niche in Time

A niche is a role in an ecosystem, and cycles are niches interacting over time in an ecosystem. Cycles have neither beginning nor end; all members in it play critical roles. Cycles trap energy, water, minerals, and nutrients in a system and cycle them through as many elements as possible, potentially for a very long time. We can increase this effect by adding in more elements, supporting or enhancing current cycles, or removing something that is blocking or hindering niches or cycles in the ecosystem. From the deer grazing in our meadows to the moles tunneling beneath them, they are all busy in the business of their niche and cycle.

Principle of Cyclic Opportunity

If we increase the number of cycles and speed of cycling in a system, we will always increase the yields in that system. If the system is unbalanced, the increased cycles could lead to problems like more chickens in a limited space could lead to unclean water and more aggression among the birds which could lower yields or decrease their quality. Stability within a niche and cycle can only increase by connecting it with more cycles, increasing its role in the greater system.

Resources

- ***Those that increase with modest use*** - Lemon balm and most herbs produce better if they are regularly harvested; otherwise they go to seed. They can give a near continuous yield if persistently trimmed.
- ***Those unaffected by use*** - Sheep's wool returns after it is harvested without affecting the sheep negatively if done properly and at the right time of year.
- ***Those which disappear or degrade if not used*** - Languages disappear if they are not used. Such is the case with many living things - they need a naturally occurring ecosystem to be active in to maintain their population. Perennial grasses will die if they are not harvested to allow their lower growth points to receive sunlight. Grazed grasses improve their root structures if managed properly.
- ***Those reduced by use*** - Finite resources (like fossil fuels and rare earth minerals) that are not actively being recreated by the environment like fossil fuels can be used up.
- ***Those which pollute other resources if used*** - Acquiring many fossil fuels like shale gas deposits through high-pressure hydraulic fracking poisons or permanently ruins aquifers,

rivers, lakes, streams, and wells. These are unethical resources or processes that must be

> Observe, Ask Questions, Hypothesize, Test, Observe, Reflect, Ask New Questions, and Develop Theories

boycotted or banned.

The Scientific Method

This cycle of actions is at the core of all the sciences, and it is no different in permaculture. However, academic science has a long track record of isolating subjects under study and then subjecting them to testing and experimentation in isolation. This leads to a warped perception of the way the natural world actually behaves. In permaculture, the approach is different because for permaculturists, it is a given that all things are part of a cycle, a greater whole, and that testing performed separate and in isolation from the original context (the cycle) will produce skewed results.

The scientific method also has no ethical component, which can lead down some ethically slippery slopes, especially in modern medicine and biotechnology. The scientific method alone cannot be the sole rudder for exploring our world. Permaculture provides the ethical lens needed for the scientific method to be fully effective.

Methods of design

Functional Design

Similar to stacking functions, this principle strives for each element to have multiple regenerative functions and for these functions to be supported by multiple elements in the system. This creates back-up cycles and systems for both inputs and outputs. The reality of working with living systems is that each component likely has innumerable daily interactions indirectly and directly with its environment, but they are too subtle and well-stacked to observe them all. It is enough that we recognize how to set things in motion for natural systems to begin stacking cycles.

Analysis

Make lists for every element of your system, analyzing their behaviors, their variety, their products, and their needs. This could be for a pig, a chicken house, a fence, an entire herd (see diagram), or even a planting technique. Anything can be analyzed for how it works, what kind it is, what it does, and what it requires.

Observation and Reflection

Extended observation of both natural systems and all design projects, prior to working on them, is critical. Ecosystems develop around open resources in nature and capturing them in stable cycles. We can observe and extend these natural systems to increase yield and

A Salatin Cow Herd

Polyface, Inc. 2016.

Variety
Beef Cattle
Mixed Herd
Chosen for Body Type, not Breed

Products
Plant Grazing
Plant Trampling
Plant Fertilization via Manure & Urine
Methane
Meat & Bone
Ingredients for BioFertilizer
Calves

Needs
Water that they can't get into
Protection from Predators
Mixed Pasture
Daily Rotation (or More Often)

Behaviors
Tramples, Manures, Urinates, and Grazes on Grass & Soil
Breed
Gains Weight on Diverse Pasture
Sequesters Carbon
Builds Soil
Improves Pasture

fertility. We can note the high marks on the stream banks indicating flood events. We can see tree flagging and feel loamy soil beneath our feet.

Reflecting back on our observations, research, and experiences allows for further connections to be made, and it also allows for adaptation to occur in response to environmental conditions. Observation develops during reflection into more informed actions.

Options and Decisions

This planning technique involves a simple listing of all possible options and then deciding by process of elimination which ones to utilize: what types of main crops to use, what types of perennials to plant, etc. All these decisions are improved by choosing from a list of all possible options. Often further overlays or questions are useful in refining the lists—always keeping in mind how things will be, not just how they are currently.

Data Overlay

Overlays are layers to a map, design, or a concept that help demonstrate an idea without completely altering the original map, design, or concept. It quickly provides perspectives on the same data set. Topographic maps, food forest layout overlays, energy, air, or water flow overlays, and more specific overlays all help visualize the options and their behaviors, interactions, and further possibilities, making decision-making and design implementation easier.

Random Assembly

Though simple, this strategy can yield interesting combinations. List all elements in your possible system (before you actually set them up) and possible connective strategies like above, below, into, beneath, behind, downslope, etc. By connecting two elements with random assemblies like "above" or "inside of," unique effects and interactions can be

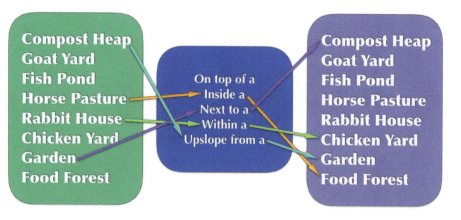

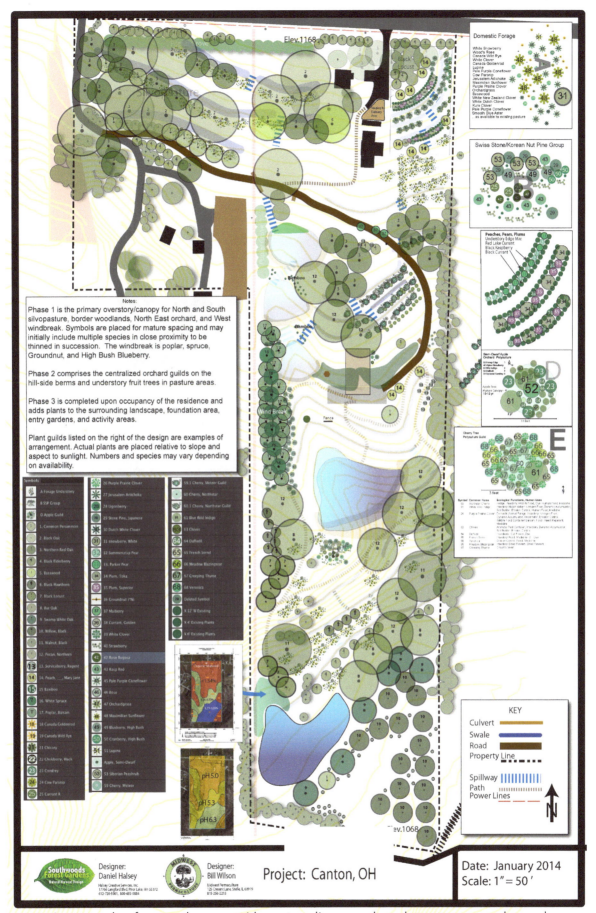

An example of an overlay map with contour lines, roads and structures, ponds, gardens, trees, spillways, property lines, and more. Often this can be an exercise in organization of information since there is so much information to convey—notice the smaller pH map!

discovered. While it can create useless or silly constructions, the possibility of discovering something we've never thought of before is much higher with this exercise. It's the most random and creative of techniques listed.

Flow Diagrams

These diagrams show the movement of energy, people, water, or any specific element in a system and can be as simple as a floor plan for a common workplace like a kitchen, studio, your desk, or living room. Flow diagrams can help improve efficiency in any work that involves repetitive movement in a finite space. Organizing resources and designing workspaces to be efficient and manageable is critical for long-term sustainability.

A Kitchen Work Flow

Soaking Kitchen Sink for Dirty Dishes	Soapy Kitchen Sink for Washing	Drying Area	Storage for Clean Dishes (also in cabinets above)
			Serving Area
			Food Preparation Area
Food Storage (the pantry)	the Refrigerator	Cooking Area	Cooking ware Storage Area (also in cabinets above)

Zone Planning and Analysis

Zone planning helps us arrange the elements of our system, so we don't waste time and energy. If it's an everyday task, having it close to home is ideal to minimize energy and time spent walking or carrying tools that distance. The exact layout of the zones will differ with each design based on the elements included, the geography, the people involved, the economics involved, and the designer's own style of designing.

- **Zone 0** - The home or house which can include an attached greenhouse, shadehouse, trellis, passive solar, etc. Seeds can be saved indoors, fungi can be prepared and grown, and numerous other regenerative plans and actions occur in the home.
- **Zone 1** - The immediate area around the home is ideal for elements that need continuous observation like tree nurseries, greenhouses, vermicompost, kitchen and herb gardens, edible and medicinal mushroom patches and logs, and quiet animals like rabbits. Other common elements: rainwater catchment, trellis, graywater systems, and greenhouses. This is an area you pass or visit daily and can spend time daily working in. This is also the area that usually has the longest growing season, being warmed by the house and closely attended.

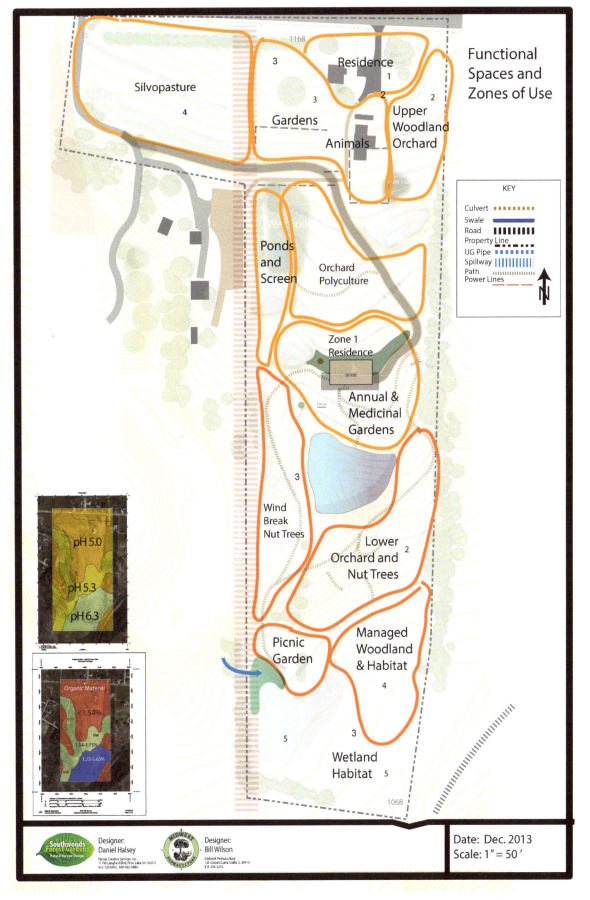

Using the same site featured earlier we can see the zones as an overlay as well. Organic matter and pH level maps are larger here.

- **Zone 2** - This area is less intensive and less visited than Z1. It includes small domestic stock, orchards, food forests, small pastures, broadacre crops, and animals shelters that connect to Z1 ideally. This area is visited every other day or just briefly once a day though it can occupy longer periods of time routinely as well.
- **Zone 3** - This is often a place of seasonal, annual, or monthly work, and it usually requires animals or machines to manage. It includes broadacre crops, larger animals, larger pastures, natural or low maintenance trees, large water storages, barns, feed-storage, windbreak, and hedgerow.
- **Zone 4** - This the wildest of the managed areas in the system; it borders the wilderness. It includes: timber, the largest pastures, firewood, native and non-native hardy trees, and large water storages. This is an area that is visited seasonally for specific tasks and lightly interacted with.
- **Zone 5** - The unmanaged or least-managed zone, Z5 is wilderness for hunting, timber, foraging, wild fungi, and observation. Though in general we are cautioned to leave it alone as much as possible, we must intervene to instill resilience in our native ecosystems: climate instability worldwide and atmospheric carbon levels are destabilizing these ecosystems, and they are in decline.

Zonal Development's Golden Rule

Develop the nearest areas first until they stabilize, and then expand the perimeter–start with Zone 0 and 1. Any soil will yield a good garden if prepared properly; save time and start close to home. This is very similar to making the least change for maximum effect.

It should be noted that the zone lists are just examples, and individual circumstances may shift these elements around in your system depending on your goals and ideas. This is to give you a basic idea of what works well.

Energy Flow

Sun, wind, precipitation, flood, and wildfire–these pressures shape our landscape and influence our homes and all designs. We can easily see where gardens, orchards, wilderness, home-building sites, shade, and windbreak can go when we consider the energy flow. We can map the sun path and the prevailing winds and their directions, and we can note the areas that are prone to frost, flood, and fire. We can use this information as a map overlay or to inform our design decisions.

Orientation

Orientation is the direction or angle of an element in relation to the sun path. If a house is perpendicular to the sun path and not parallel, energy (heat) will accumulate

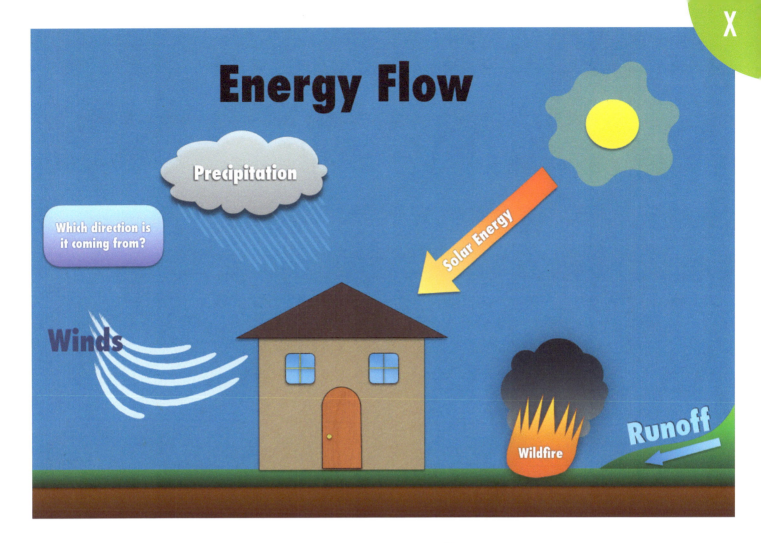

The Orientation of a House

Front Facing the Sunpath

Side facing the Sunpath

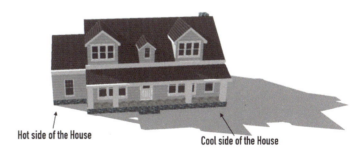

Hot side of the House   Cool side of the House

House model by Dilbert, 3Dwarehouse.sketchup.com, 2008.

unevenly, heating only the sunny side. If a garden is oriented away from the sun and into a wind tunnel, some crops will perform poorly. Homes should be designed to accept as much solar energy as beneficially and evenly as possible while sheltering the inhabitants from unwanted elements.

Slope

Nearly every site has some slope even if it seems flat. Understanding slope allows a designer to know where water will flow, where gravity is pulling the system, and where elements can best be placed. Slope can be seen with a topographic map easily. Water should always be stored at the highest possible point to gain as much gravitational potential energy as possible. Slopes at 20% or greater must be put into erosion-control tree planting. Flat plains need windbreak: copses and hedgerows. In cooler areas, orient everything towards the sun path while in more arid and hot areas orient elements to create, magnify, and share shade and channel cooling breezes. Refer to the next page for a simple way to calculate slope.

Aspect

Aspect is the slope's orientation. A slope facing the sun may get too much heat or, if turned away from the sun, not enough heat. This is critical in site planning for a home. Is it on the dark side of the hill or in direct sun? Will the trees shade the house in summer? How many hours of light reach the home in winter? Understanding aspect helps us answer these questions.

Shadows are shorter when the sun is directly above or when the land's aspect faces the sun path. Shadows lengthen when the aspect is away from the sun, or when the sun's angle is severe like during sunset and sunrise.

How to Calculate Slope

By imitating the picture or using posts with string and a level, we can measure the rise and the run to calculate the slope.

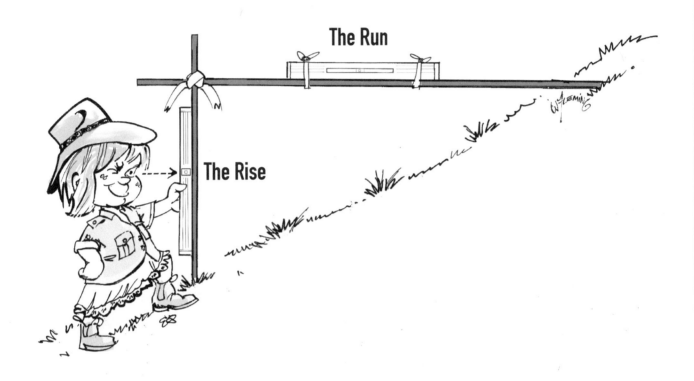

$$\left(\frac{\text{Rise}}{\text{Run}}\right) \times 100 = \text{Slope \%}$$

Sun Angle Equation

Vertical Height ÷ Shadow Length = Tangent
2m ÷ 3m = 0.67 Tangent

Tan^{-1} (Tangent) = Sun Angle
Tan^{-1}(0.67) = 34°

Tangent	Angle	Tangent	Angle
0.088	5	1.19	50
0.176	10	1.43	55
0.268	15	1.73	60
0.364	20	2.14	65
0.466	25	2.75	70
0.577	30	3.73	75
0.7	35	5.67	80
0.84	40	11.43	85
1	45		

Based on information sourced at DesignCoalition.org, 2016.

Sun Angle

We can see the sun angle easily in the early morning or afternoon when the angle is most acute. The shadows can be measured against the objects casting them. Knowing the angle as it changes over the year allows us to properly design roofs, shade height, plantings, windows, orientation of buildings, and more. If we do not want to observe and wait all year long to calculate the sun angles and sun path before finishing our plans, we can use one of the many online sun angle calculators.

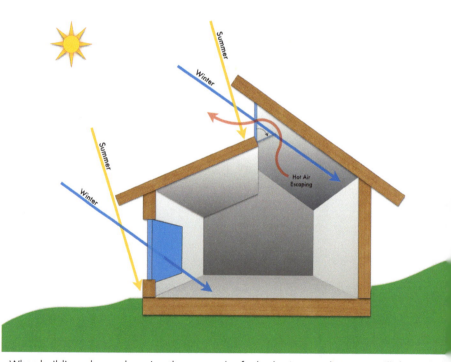

When building a house, knowing the sun angles for both winter and summer will determine the angles and length of the roof as well as window placement. Upper windows that let in winter light but not summer sun can be opened to let hot air escape as well.

Sustainable by Design hosts a free and precise sun angle calculator here: http://www.susdesign.com/sunangle/. Measuring on the solstices and equinoxes will give you an idea of the timing and the extremes of your sun angles throughout your year.

The Soundscape

While often overlooked, the sounds that we hear can affect us immensely. The sound of a highway or construction can steal the relative tranquility that would otherwise be present in well-developed ecosystems. Using sound-deflecting fences, earth berms, water features, and thick, absorbent vegetation like large and wide evergreen trees, much can be done to reduce or silence disruptive sound.

Timing

Timing is critical: the time of the year, the time of the season, the time of the day, the time in the life cycle of the organism, etc. Understanding when to do something, when not to do something, and how long to do something defines understanding proper usage or interaction with that thing; it can be cattle, chickens, a garden, or a person.

Incremental Design

Through small changes over time using observation and the extension of natural systems, we can continuously improve efficiency. This process often takes generations of development. It happens after a mainframe design has been put into place, and it manifests as continual adaptations for increasing benefit over time, sometimes generationally. This can also be seen in plant breeding; our modern, large-kerneled corn was bred from teosinte, a wild and small-seeded native grass. It took thousands of years of generational and incremental change. This strategy also mimics how ecosystems and organisms are perpetually adapting to the continuously changing environment.

Strategies that Create Yields

This is not a complete list as there can never be a complete list of strategies—these are simply to help suggest, guide, and improve your designs. Most of the concepts listed here are described elsewhere in this book in greater detail. Here they are generalized and gathered for an overview to aid designers in developing a more holistic lens.

Physical-Environmental

- Creating niches for new species by providing habitat or resources

- Soil-building and regeneration
- Water catchment, graywater, and stacking water usages
- Synergistic integration of structures and landscape

Biological

- Selection of low-maintenance cultivars and native species
- Investigation of other species for usable yields
- Supplying key nutrients through system arrangement and on-site input
- Plant and animal guilds and partnerships

Spatial, Managerial, and Configurational

- Stacking functions or nesting
- Tessellation or mosaic formation
- Edge and harmonic design
- Arrangement of systems to easily route energy for best use
- Zone, sector, slope, orientation, and site strategies
- Keyline Scale of Permanence or the Regrarians Platform
- Holistic Management

Timing

- Sequential nesting or patterning like interplanting, intercropping, etc.
- Accelerating cyclic frequency: chop-and-drop, compost teas, etc.
- Tessellation of cycles and successions, as in browsing sequences: cows followed by chickens followed by turkeys followed by rest.

Technical

- Use of appropriate or rehabilitative technology
- Design of energy-efficient structures, systems, and tools

Conservation

- Routing resources to the next best use
- Recycling everything
- Reusing or repurposing everywhere possible

- Safe storage of food products
- No-Tillage or low-tillage cropping
- Creation of very durable systems, structures, and tools
- Storage of runoff water for extended use

Cultural

- Removing cultural barriers to resource-use and regenerative living
- Making currently unusual resources, such as Humanure, acceptable
- Expanding choices in a culture or community
- Making resources available to stimulate and perpetuate cultural connections to nature such as heirloom seed swaps, regional and tribal seed libraries, etc.

Legal/Administrative

- Removing socio-economic, legal, bureaucratic, and governmental impediments to resource-use, regenerative living, and social development
- Creating effective structures to aid resource management and education—can be voluntarily achieved or through legislation

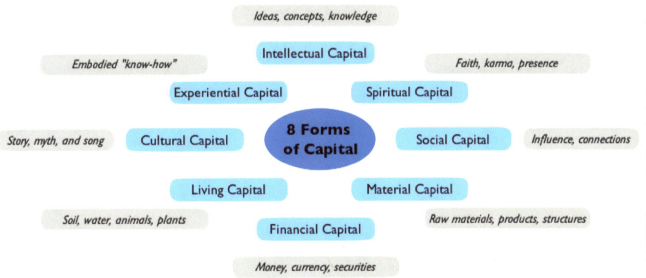

Social

- Cooperative endeavors, pooling of resources, sharing, trusts, land trusts, non-profit organizations, churches, local groups, online groups
- All eight forms of capital cycling within the community
- Positive social action to remove and replace impeding systems

- Cost-Analyzing and adjusting systems holistically for all energy inputs and outputs

Design

- Making harmonious connections between components and sub-systems
- Making choices as to where we place things or how we live
- Observing, managing, and directing systems
- Researching and applying new information
- Designing on paper first
- Setting priorities based upon resources available
- Developing Zone 1 first completely then expanding outward
- Expanding with observed successes
- Looking at the big picture to design but referring to the smaller picture perspectives to refine the overall design. Think in whole systems but be prepared to consider all perspectives of the systems including people within those systems at all times—be nimble!

Extending Yields

- **Early-, Mid-, and Late-season** - This should be applied to all our systems. We should have early, mid-, and late season apples, but we should also have all the seasons covered similarly with a diversity of yields.
- **Planting the same varieties in early or late ripening situations** - Like the sunny or shady part of a hill.
- **Staggering Plantings** - by planting seeds one section at a time every week or every few weeks, we can guarantee that we won't run out of lettuce or let any go to waste.
- **Long-yielding varieties** - These are usually perennials but annuals such as cherry tomatoes give continuous yields for many years if located in a suitable climate or greenhouse situation.
- **Long-lasting varieties** - Both the longevity of the plant and its relative products should be considered. For instance, olive trees can live for thousands of years. Dried grains, vegetables, and fruits remain edible for years.
- **Increasing diversity of yield** - Using polycultures in the garden, we can spread our yields out and increase them by having many different elements and products.
- **Improving preservation methods or expanding storage spaces** - Through root cellars, clutches, canning, curing, drying, smoking, fermenting, pickling, etc.

- **Trade** - By trading regionally, we can create bonds between communities while providing each with a service that benefits both and diversifies our yields by transforming them through trade.
- **Microclimates** - These allow us to concentrate or deflect on-site energies to create a different climate from the surrounding dominant climate. Using windbreaks such as a fence or a line of trees, thermal mass like a buried stone or a pond, and/or orientation to the sun path, we can shelter and warm an area to extend the growing season and, therefore, the yields in that area.

Frameworks

A Simple Task Framework

- **Technique -** How to do something
- **Strategy -** A technique or set of techniques applied to a cycle or system
- **Materials -** The resources on a site: trees, soil, rocks, people, etc.
- **Assemblies** - The arrangement of materials in design
- **Patterns** - The arrangements of assemblies

Holism

Holism, a term coined by JC Smuts but an ancient concept, is the idea that everything is a part of a whole and represents a whole in itself. The wholes individually are called holons. All systems in nature are holons that fit into larger holons and contain smaller holons within them, from the atom to the ecosystem to the solar system. Viewing the world and our interactions through a holistic lens allows for better management and understanding.

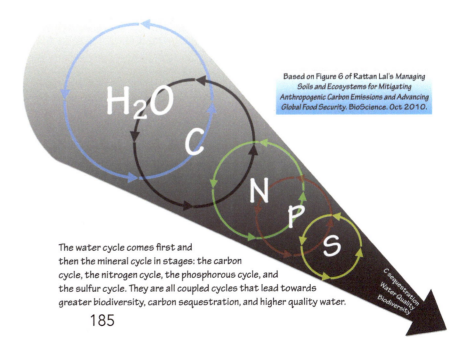

Based on Figure 6 of Rattan Lal's *Managing Soils and Ecosystems for Mitigating Anthropogenic Carbon Emissions and Advancing Global Food Security*. BioScience. Oct 2010.

The water cycle comes first and then the mineral cycle in stages: the carbon cycle, the nitrogen cycle, the phosphorous cycle, and the sulfur cycle. They are all coupled cycles that lead towards greater biodiversity, carbon sequestration, and higher quality water.

Holistic Management and Decision Making

Alan Savory discovered that the removal of large herbivores from desertifying land in Zimbabwe made the desertification accelerate.[59] He realized that even if we cannot recognize it immediately, ecological systems are holistically interdependent and complicated. Savory started considering all the aspects of a situation: the water cycle, the mineral cycle, energy flow, and socioeconomic factors.

Holism as a concept implies that components cannot be fully separated from the functions and interrelations of the whole. Created by Alan Savory, holistic management and the holistic decision framework both help designers take the arrangements and designs that permaculture generates and manage them holistically.

Forming a Holistic Goal

Forming goals alone is powerful—we articulate what we desire to see, have, or be in the future. The conceptual framework we design for accomplishing those big goals generates a series of smaller goals to reach each milestone. Without a goal, we are directionless. Without a holistic goal, we are bound to encounter conflict and make errors on our way to accomplishing our goals. By reflecting on the framework and the principles we can help shape, refine, and generate our holistic goals.

The Framework of Holistic Management

- **Define the Whole** - Define what it is in its entirety that you are working on.
- **Set Ethical Short-term and Long-term Goals** - Set goals for what you want and need based on how things will be in the future, not as they are now.
- **Observe and Document** - Careful observation for signs of degeneration or regeneration gives managers early indicators for course correction. Documentation helps locate patterns, extend retention, and deepen comprehension.
- **Use your Toolkit** - These are in no way limited to Savory's list for cattle management; your toolkit will likely have specific tools addressing your situation, but his list includes: financial capital, human ingenuity, herbivores, wildlife, soil biology, fire, rest, and technology. The concept can be adapted to any situation.
- **Test your Decisions** - For economic, financial, and social success over the long-term and short-term, choices must be tested and reflected upon from different angles and through different lenses.
- **Feedback Loop** - Without regular monitoring, reflection, and adaptation, productivity and the regenerative progress will inevitably decrease.

---

[59] Savory, Alan. *Holistic Management: A New Framework for Decision Making*. 1999. p. 16.

The Four Principles of Holistic Management

- **Nature is Holistic** - All are mutualists in a balanced ecosystem; disrupt one aspect of an ecosystem, and you disrupt them all.
- **Adaptation is Constant** - All successful plants, animals, and ecosystems adapt constantly to their environments, and so should any design that we implement.
- **Mimicking Natural Patterns can Restore Ecosystems** - Using herbivores behaving as they did under heavy predator pressure and at the herd densities per square foot they used to maintain, we can mimic the original patterns that created the grasslands and rich soils. This concept can be applied to everything and is most the basic tenet of permaculture: we can use natural cycles to restore the natural world.
- **Time and Timing** - Knowing how long to do something and when to do something—both these skills are vital to good management of any system whether they be relationships with people, animals, plants, or the land.

Keyline Scale of Permanence[60]

Created by PA Yeomans, this scale has influenced permaculture designers and regenerative agriculture practitioners all over the globe. The scale of permanence is a prioritizing framework, so it organizes the order in which we design and install a site. It was initially drawn up to enable Australian farmers to design their farms. Each step in the scale of "agricultural permanence" makes the next step easier to develop.[61]

1. *Climate*
2. *Land shape*
3. *Water supply*
4. *Farm roads*
5. *Trees*
6. *Permanent buildings*
7. *Subdivision fences*
8. *Soil*

The Regrarians Platform

The Regrarians Platform extends PA Yeomans' scale of permanence by adding elements from Holistic Management. Introduced in Darren J. Doherty's <u>Regrarians Handbook</u>, The Regrarians Platform is a more holistic take on the Yeomans scale of permanence.

1. *Climate*
2. *Geography*
3. *Water*
4. *Access*
5. *Forestry*
6. *Buildings*
7. *Fencing*
8. *Soil*
9. *Economy*
10. *Energy*

---

[60] Yeomans, PA. <u>The Challenge of Landscape.</u> 1958.
[61] Ibid.

Keyline Patterning in Design

edited by Darren J. Doherty

"Keyline planning is based on the natural topography of the land and its rainfall. It uses the form and shape of the land to determine a farm's total layout. The topography of the land, when viewed in the light of Keyline concepts, clearly delineates the logical position of on-farm dams, irrigation areas, roads, fences and farm buildings. It also determines the location of tree belts to provide shade and give wind protection. Keyline concepts also include processes for rapid soil enrichment. The shape of a landscape is produced by the weathering of geological formations over millennia. The processes are always the same. And so the topography of agricultural land has a basic fundamental consistency. It is the inevitable nature of land shape that river valleys collect water from smaller creek valleys. They in turn are fed their water from still smaller valleys, until finally the water derives from the very first, or primary valleys of the catchment area. In any country, anywhere, when rain shapes the land over long periods of time, it inevitably creates and determines the topography of that land. Ultimately, at the extreme upstream of any river system there always [exist] thousands of primary valleys. The only variation to consistent topographical shapes occurs where geological features, such as hard rock outcrops modify normal surface weathering"

—Allan Yeomans, <u>Priority One</u>, 2005.

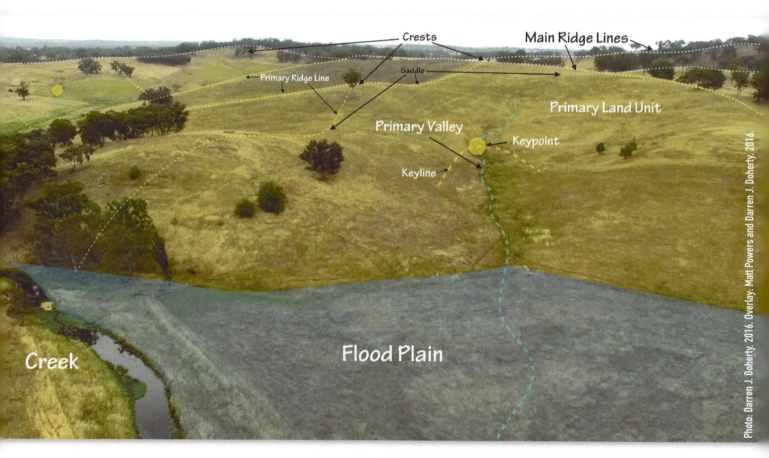

The Keypoint

This is the point after which the slope changes from an erosion zone to a deposition zone. Clay and silt particles settle here in greater concentrations as the water slows in its path on top of and through the soil. Keypoints occur some distance after the land coming down from the ridge has changed from convex to concave—how gentle or steep the slope is determines where the most settling occurs. The slowing of the water that leads to deposition only begins at the concave point, so at some point below that, deposition will begin and concentrate in the landscape. This observation and concept, coined by Australian author and farmer, PA Yeomans, are enduring. It is the idea that we can catch, soak, and divert water at these points in the land for better hydrology of the land. These are ideal sites for dams.

Sepp Holzer is much more free-form with his interpretation of these sites: any depressions or deep zones are areas where clays and silts have been collecting, and, therefore, they are all ideal pond or water retention sites.

Keyline

Keyline is a contour line that extends off the keypoint. A swale or diversion drain can be extended from off the keypoint at contour, increasing the water catchment of the keypoint. Using swales increases the catchment at both the keypoint and below the keyline; sealed diversion drains focus the water at the keypoint. Swale paths or roads can be ripped to increase infiltration. Where possible, terraces can extend off keypoints. It is important to note that the keyline does not extend to the ridge. It is contained in the primary valley between primary ridges extending out from a main ridge.

> **"The contour of the keypoint within the primary valley is called the keyline"**
> –Darren J. Doherty, Permaculture Voices: PV3, 2016.

Access

A key concept in both the Keyline scale of permanence and the Regrarians platform is Access (roads or pathways) in design. This is because when designing a site everything is first based off the climate, then the geography, and then water. Once you know where the water is, you are going to need to base your system around accessing and using that water. All trees, irrigation, housing, etc. are based off where the roads are placed. Improperly placed roads flood, erode, or wash away, but a well-placed road on the center of a ridge is ideal. Grading it to allow water to leave the road evenly will prevent erosion. This makes for a long-lasting, durable road. Roads along the ridges also allow tree systems to circulate the air and avoid frost pockets. Having access along the ridge means any contour off the ridge allows for easy access and road building. Roads serve as the skeleton for all keyline pattern designs with ridge line roads being the spines.

Often we see roads going down the center of valleys and homes lining it, but this is a recipe for disaster because the valleys are where all the water and everything it carries collects.

Different from "On Contour"

While some use swales on contour exclusively on their site with great results, many landscapes generate a lot of swale stubs, inconvenient narrowings or widenings, and dead end paths. Keyline patterning only uses the top or bottom contour of an area to determine the guide line for the rest of the rows, berms, roads, or fences. In valley systems, start at the top and then move down parallel to the original guideline (see diagram). In steep cultivation areas, start from the bottom contour and then make rows parallel to that contour guideline uphill at an equidistant spacing. This allows for standardization of access—the road or path is the same width the entire way and whatever equipment is being used can make turns easily at the ends of the rows or roads regardless of whether it is a wheelbarrow or a large tractor. Equidistant rows and the classic grid pattern is a time-tested system that can use contour as an initial guide but ultimately cannot follow the contours for every line or row. In a commercial setting, standardized rows give farmers the data necessary to more easily manage their farms and predict their yields, sales, water retention, carbon sequestration, and more.

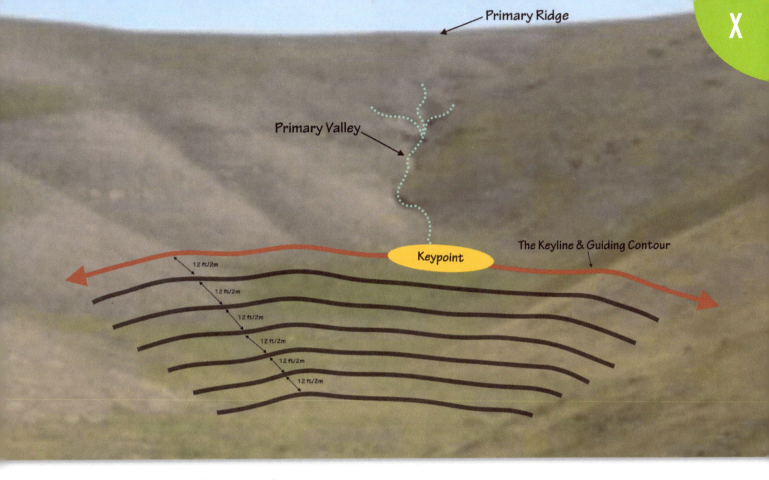

Pastoral and Beyond

While initially designed to help ranchers and farmers with hydrology and farm planning, the Keyline design method has been applied to forestry. Pastoral ripping is shallow 1-2 ft deep (50cm) while forest ripping is more often 1m. The water catchment from a forest is immense, so the ripping has to go deep for a fast penetration of water and easy access for roots to travel downward easily.

Keyline design, keyline patterning, the regrarians platform, and keyline subsoil plow "ripping" are all incredibly powerful concepts to source in design. The Regrarians Platform's addition of socioeconomic aspects allows for keyline thinking to find inroads into many different holistic contexts. For businesses, entrepreneurs, and consultants, the methodical approach, the detailed and advanced mapping, and analytical tools discussed in Darren J. Doherty's Regrarian's Handbook are an inspiration and a solid guide towards better practice.

Rewilding in Design

While it is often overlooked by designers, zone 5, the wilderness, is in serious distress. The natural mechanisms by which our annual and perennial cycles operated have been lost. The mastodons and their kind are gone. Large wild herds of grazers held in balance by natural predators are gone as well; their manures are also missing from the soils. Wildfires are out of control, and the deserts keep spreading. Critical components of our ecosystems are missing or out of balance, and that is why they are no longer operating like natural systems. Many landscapes have been unnatural, mismanaged, and damaged by human activities for hundreds of years; some of these areas are US national parks.

This has led to a movement called **Rewilding**, a call to bring back the wild again. As designers, we can help this process by regenerating natural habitats both in our cultivation areas and outside of them. In some cases we will have to replace the extinct megafauna with our own behaviors (silviculture), and where possible, we have to reintroduce natural predators. Reintroduction of large predators to their original habitat can lead to trophic cascades—where multiple levels of the ecological food chain are affected beneficially.[62] Wolves were reintroduced in the 1990s to Yellowstone National Park (USA) in an attempt to keep the elk at bay long enough for the willows and aspens to recover, in turn helping to bring back beavers and support riparian restoration.[63] Currently, while there is clear evidence of positive effects on biodiversity,[64] there is a debate as to how effective wolves have been on the specific task of supporting beavers[65]—the rivers' and creeks' lowered water tables have not all recovered; human intervention is likely needed to help repair the situation the original wolf extermination initiated. In whatever small or large way we can, we must bring back the wild, partner with nature, and, carefully, over time, learn how to support these ecosystems.

---

[62] Ripple, William J., Estes, James A., Beschta, Robert L., Wilmers, Christopher C., Ritchie, Euan G., Hebblewhite, Mark, Berger, Joel, Elmhagen, Bodil, Letnic, Mike, Nelson, Michael P., Schmitz, Oswald J., Smith, Douglas W., Wallach, Arian D., Wirsing, Aaron J. *Status and Ecological Effects of the World's Largest Carnivores.* 2014.
[63] Middleton, Arthur. *Is the Wolf a Real American Hero?* 2014.
[64] Farquhar, Brodie. *Wolf Reintroduction Changes Ecosystem.* 2016.
[65] Hollenhorst, John. *Are Wolves a "miracle" in Yellowstone? Science Seeks Answers.* 2016.

XI. Food Forests & Gardens

Usually the place where we fall in love with permaculture initially, a garden and food forest contain many of permaculture's most important lessons implicitly within them. The more you garden, the more humbling an experience it can become as we see what is possible when we partner with nature.

Plant Guilds

A plant guild is a harmonious and beneficial polyculture, focused usually around one or two central plants. An example could be an apple tree guild with onions or garlic for pest control, artichokes for mulch and phosphorous, nasturtium and strawberries for ground cover, and lastly, a small perennial nitrogen fixer like a Siberian pea shrub to pair with it. Plant guilds are creative assemblies that are limitless in their beneficial possibilities. There are lists of plant guilds, but many guilds are waiting to be discovered.

Likely the most popular and well known guild is The 3 Sisters: Corn, Squash, and Beans.

Interaction Matrix between Two Elements

Interactions between animals, people, and plants can be categorized into three separate types: those having positive effect, negative effect, or no effect. Sometimes it can be a positive or negative effect for only one of the two species, implying that one is negatively affecting the other. Species that negatively interact are rare. An abundant planting guild should focus on beneficial interactions.

Below is a table mapping the possible reactions between two species (plant or animal). Results can be positive (+), negative (-) or neutral (o).

		Species A		
		+	o	-
Species B	+	+ +	+ o	+ -
	o	o +	o o	o -
	-	- +	- o	- -

Plant guilds can be extensive even in a small space and a mix of both perennial and annual plants

Using this chart we can create patterns that help avoid or buffer negative effects between trees and plants (and we can apply this analysis to animal groupings too). For example, walnut trees have a root exudate (juglone or juglans) that is harmful to certain fruit trees. If we plant species that are not affected negatively by the exudate (mulberry trees, for instance) between the walnut and the vulnerable fruit trees, then the vulnerable trees will be protected from the exudate.

It should be noted that there are also many who feel passionately that you can disregard juglone effect with many fruit trees while others think that not even a buffer is enough separation. (Outside my window there is an apple tree that generated hundreds of pounds of Arkansas Black heirloom apples this year in Missouri, and it's planted next to a Black Walnut tree—both thriving). Soil and climate types may be a factor.

Succession of a Forest

Between the phases of being bare earth or sterile sand and that of old growth forest, there is a progression of successions. The process of succession moves soils from being bacterial-dominant to being fungal-dominant. Only when the system is more developed does it partner with a large fungal decomposer presence—though it should be noted that fungi are everywhere all the time. Succession can also be viewed in terms of soil disturbance or the length of time between disturbances. Annuals thrive on soil disturbance because they are pioneer species (we've just bred them to be super sweet, tender, and dependent on humans) while old growth forest relies upon a lack of soil disturbance.

Layers of a Forest

By including all the layers of the forest in our designs, we occupy any niches a weed or unwanted element might try to occupy. You can plant all at once, but when planting, keep succession in mind. The space will start out dominated by support species like annual legumes, green manures, mulch plants, and some fast-growing, nitrogen-fixing trees with small bare-root trees and perennial seedlings hidden among the explosion of growth. Use flags and colored sticks to keep track of your valuable tree and perennial plantings—they'll need intervention routinely as they develop to prevent them from getting shaded out or choked by vines. Once the valuable perennials and trees are established, the space will be largely occupied by valuable species with coppiced nitrogen-fixing trees and mulch plants found intermittently throughout the system.

XI

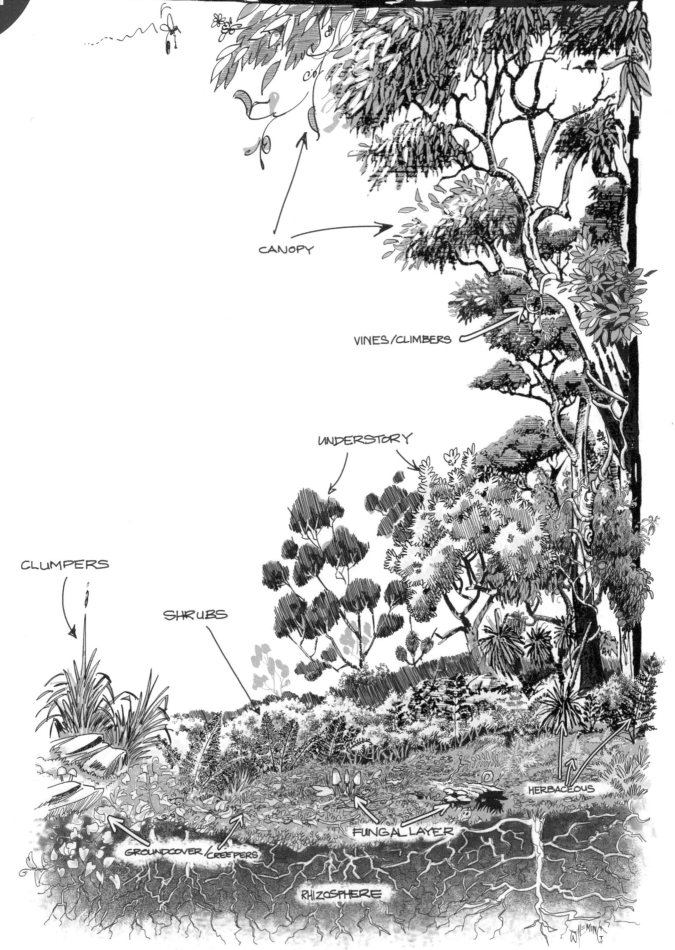

- Canopy, the tallest layer
- Understory
- Palms (some climates have both understory- and canopy-level palms, while some have none—such as in cold temperate)
- Shrubs (some climates have two layers)
- Herbaceous layers (in cold temperate climates there are two layers)
- Vines or climbing plants
- Ground cover plants or creepers
- Rhizosphere or root layer
- Fungal layer (both above and below ground)
- Clumpers or plants that spread by division

Planting Techniques

These are generally seen as gardening techniques or methods, but they flow just as easily into landscaping, forestry, land rehabilitation, ecology, and more. The more farmers and gardeners embrace perennials and native plants, the more diverse our planting techniques and methods will become; as this broadening of our understanding of plants progresses, we will also get a better understanding of our current and potential interactions with plants.

> **You can plant for a season, for succession, or for the next generation.**
> –Matt Powers

Plant Guides

There are many books and charts on when and how to plant everything. Few focus on the plant relationships or compile information on how plant guilds work. The Natural Capital Plant Database was created to serve this very purpose and features plants from across the globe researched by permaculturists for permaculturists. Free for anyone to use, the database shares research and data on plant guilds by region. The USDA has numerous databases that range in usefulness. Dr. Duke's Phytochemical and Ethnobotanical database (Phytochem.nal.usda.gov/phytochem/search) might be specifically the best USDA database of its kind. The USDA-GRIN database has rare germplasm available for teachers, researchers, and scientists. For basic gardening, individual plants can be looked up easily online, but for a thorough reference on seed saving, consult <u>The Seed Garden: The Art and Practice of Seed Saving</u> from the Seed Savers Exchange organization, edited by Lee Buttala and Shanyn Siegel.

Throw Sow

This is both a planting technique, sometimes called broadcasting, and seed characteristic—seeds that grow vigorously by throw sowing them. They do not need to be hand planted. They will grow faster than birds will eat them (or have another form of protection). Throw sowing for plant adaption or breeding generates extremely vigorous offspring. Save seed each year from the best plants in the worst soils, and their vigor will increase with each generation. You can also train plants to self-seed by shaking out their seeds all over the soil once the seeds have set. Over time, it creates a seed bank in the soil that outcompetes any native or non-native weeds.

A sample web page from PermaculturePlantData.com listing an herbaceous apple guild for zones 3-9.

Seed Balls

Masanobu Fukuoka used seed balls, a mixture of seeds, clay, and compost or manure to prevent birds from eating the seeds after they were sown and to provide them with the fertility they would need to grow once sprouted[66]. Seed balls also make planting easier. They can be used to plant out public space, abandoned lots, and other areas where birds or neglect are preventing recovery. They

Masanobu Fukuoka spreading seed balls on his farm.

---

[66] Fukuoka, Masanobu. <u>Desert or Paradise</u>. 2012. p. 161.

can be made and used by people of all ages and ranges of abilities because it does not require bending down to plant in the soil.

Direct Seeding

The most common form of planting in a garden or farm is direct seeding. This is where the seed is placed *directly* in the soil. Seed packets have measurements for seed depths and spacing, and these range in accuracy and necessity. Some seeds need to be directly sown solely to avoid being eaten by birds and can grow on the surface perfectly albeit vulnerably; others need a specific depth or their roots won't develop properly, and the plant will be weaker than it could have been.

Some plants are given the spacing we are directed to plant them at because of soil fertility, or, more accurately, soil infertility: they need to be able to work with a larger volume of soil in order to find adequate nutrients. Contrasting with this, there are examples of very close plantings in bio-intensive gardens with very high yields and large, beautiful plants. The soil structure and fertility are key components for determining the planting density and

success of your seeds. The seeds with the highest probability for success are either thrown-sown or directly seeded. Most direct-seeded plants have large seeds that can be planted deep enough below the surface of the soil to not be visible. Seeds like squash, corn, beans, okra, melon, peas, and other large seeds can all be covered with soil. Many of the finer seeds, seeds that birds love, or more tender seedlings are grown in a nursery and transplanted more often—though certain fine-seeded plants like carrots don't like transplanting, so they need to be direct surface sown and watered gently.

Some seeds like corn can be soaked overnight in water before planting to speed germination while other plants need to be soaked in hot water to germinate at all! Learning the needs of each plant and seed can seem daunting the first time you learn about it, but when you actually work with the seeds, it's hard to forget; they're just so amazing!

Pasture Cropping

As its name implies, this is sowing seed directly into pasture. Growing grains directly in the native, perennial pastures allows native habitats to be preserved, and the pasture can double as a crop field. When the grains are harvested, holistically managed herbivores like cattle can be used to harvest, process, fertilize, and trample the area—leading to more carbon sequestration, increased pasture biodiversity and soil depth, and higher quality products without any soil disturbance. This is a powerful system that Col Seis in Australia has been working with for over 20 years now![67] It requires timing and practice, but it is a proven and profitable method of regenerative cereal cropping.

Fruit Walls and Espalier Trees

Fruit trees and vines can be trained up walls and along fences creating fruit walls on buildings or surrounding properties. They are especially useful in areas with limited space but many walls—like in urban areas. The espalier technique trains fruit tree branches horizontal to control vegetative growth and encourage

Espalier fruit trees at Luther Burbank's Experimental Farm in Sebastopol, CA. 2016.

---
[67] Bradley, Kirsten. *Why Pasture Cropping is such a Big Deal*. 2010.

fruiting instead. This makes for smaller trees but more fruit and less if any pruning. It is an elegant way to control vigor.

Transplanting

While it is true that many natural farmers do not transplant, almost everyone else is transplanting seedlings, trees, and perennials into systems for a fast installation and environmental restoration. Using even a small plant nursery, one person can generate enormous numbers

of trees, perennials, and annuals that very quickly can occupy an empty space, even acres of land. Transplanting saves time and energy in different ways. By being able to focus our energy and time, we can more easily grow more, and then we can distribute them by hand and choose their exact placement, which can be very effective. Larger plants in place earlier in the growing season gives them a better chance of surviving the dormant seasons whether they be hot and dry or cold and dark.

When planting transplants, it is important to be aware of transplant shock. This occurs when plants feel that they've been moved or disturbed in relation to the soil, sun, moisture, wind, temperature, and/or their roots. Plants can take two to three weeks or more to recover depending on the type and extent of the trauma. Using compost tea and mycorrhizal inoculum, this largely can be mitigated. The transplant going into the ground accompanied by plenty of beneficial soil microbiology will have less shock if any. The plants can set up their root system associations according to whatever they desire since they have an abundance of options. Whatever exudates they put out, the appropriate soil life will be present to feed that plant.

Fungi in the Food Forest

At every step of development in a food forest, we can involve fungi. We can inoculate the legumes and other nitrogen fixers with rhizobia bacteria.

We can inoculate plantings with arbuscular mycorrhizae fungi (AMF). We can have wood chip mushroom fruiting beds or mycoremediation beds with oyster and king stropharia mushrooms. We can inoculate logs with mushrooms like lion's mane and shiitake and stack them in shady areas or embed them partially in shady *hugelkulturs*.

Micro Net and Pan

Net and pan is used in tree planting systems on steep hills to combat erosion, to help infiltrate moisture, and to establish trees. It is no different in a small-scale setting, especially if we are watering which is erosive by its very nature. Planting plants in depressions and connecting planting areas helps retain moisture and encourage deeper soakage. Our garden beds do not need to be perfectly flat—often plants need a microclimate or some protection to thrive.

Micro-Swales

Using the same idea of the micro net and pan, we can miniaturize the swale concept with a furrow on contour. When we plant in the furrow, we get more water retention and wind and pest protection, and it provides shade to seedlings, leading to a higher success rate. Age-old and time tested, furrows and swales can combine functions to make micro-swales.

Polycultures vs Monocultures

While some plants, such as corn, prefer to be close enough to touch or nearly so for the best pollination rates, even these can be grown in a polyculture. Crop rotation often allows for seasonal monocultures or at least a mix of plants restricted to the same botanical family. In nature, we have seasonal or temporary monocultures develop and pass, often in a successional event. A reparative plant, often called a weed, takes over an area and dominates it, only to die out, setting the stage for the next level of succession to take place. Depending on the specific climate and area, the reparative succession could take a few years or a few millennia. This is why human intervention is so critical; we need to help speed up the regeneration.

Keyhole, Double-Reach, and 30inch (76cm) Garden Beds

Alex McVey, 2016.

Keyhole garden beds are circular beds that surround the gardener in an C or U shape. By only having enough room to stand and pivot, this gardening technique saves space and effort. Double reach beds are garden beds or rows that allow you to comfortably reach to the middle from either side (to avoid straining your back). Even narrower are the **biointensive** 30" (72cm) beds that use Elliot Coleman's line of 30" (72cm) standardized tools. Smaller than that, we risk compaction from walking on the path, and wider than that, we have to enter the system, so we risk compaction there too. How you shape your garden depends on your space and preferences and can be, and often is, a combination of design techniques.

Annuals vs Perennials

Annuals grow in one season and die off leaving only their seeds and carbonaceous husk behind. Biennials are plants that take two years to develop and go to seed. Perennials are plants that are planted once, continue to grow or regrow each year for many years, and often spread without human intervention. Annuals prefer more alkaline soils pH while perennials prefer more acidic soils (though both need a range of pH in the soil). Annuals require more attention and input than perennials. As designers, we can choose how much and how often we will interact with the systems we design by choosing plants that support our goals.

For over a century, work has been done on breeding perennial grains. The main difficulty is that the roots and the seeds compete for nutrients which leads to small seeds. The closest we are to perennializing a popular grain currently is with rice, according to <u>The Carbon Farming Solution</u> (2016). Since it was once perennial, it makes sense that it could be bred back to that state. Work is also being done on sorghum, wheat, rye, kernza, and dozens of other grains in an attempt to make cereal grains that never need to be replanted.[68]

Crop Rotation

There are many ways that crops can be rotated, and there have been many different crop rotation methods throughout history. Initially it was based on resting the land, leaving it fallow, and that worked well while wilderness still surrounded the fields, but rest only works well in non-brittle climates with biology in place to remediate the land. Today it is an umbrella term for rotation with substitution plants or rest. Crop rotations from earlier in history involved resting the land or growing legumes, both of which would allow soil fertility to return in different ways. Rotations range from changing heavy to light feeders, to switching botanical families, planting guilds, and stages of succession. There are many rules of thumb that need to be considered. Often using slips of paper and rearranging them to see what works best for your system can be the most effective planning exercise.

For commercial annual production systems, crop rotation is invaluable at maintaining soil fertility and controlling pests and pathogens.

Medieval 3 year 3 Field & Crop Rotation

	Year 1	Year 2	Year 3
Field 1	Wheat or Rye	Rest	Peas, Beans, or Lentils
Field 2	Peas, Beans, or Lentils	Wheat or Rye	Rest
Field 3	Rest	Peas, Beans, or Lentils	Wheat or Rye

Crop Rotation Principles

- **Alternate Heavy and Light Feeders**
- **Alternate Root Crops with Leaf Crops**
- **Brassicaceae, Solanaceae, Liliaceae, and Cucurbitaceae** - Rotating these botanical families gives each area a three year break between growing the same family there again.
- **Green Manure (usually Legumes)** - Grown during Fall/Spring, this field crop is tilled in and allowed to incorporate into the soil before spring planting.

---

[68] Toensmeier, Eric. <u>The Carbon Farming Solution</u>. 2016. p. 108.

BroadFork Farm's 10-year Crop Rotation

	Year 1	Year 2	Year 3	...	Year 10
Plot 1	Solanaceae Compost	Greens & Roots	Garlic Compost	...	Greens & Roots
Plot 2	Greens & Roots	Solanaceae Compost	Greens & Roots	...	Early Cucurbitaceae & Brassicaceae Compost
Plot 3	Early Cucurbitaceae & Brassicaceae Compost	Greens & Roots	Solanaceae Compost	...	Greens & Roots
Plot 4	Greens & Roots	Early Cucurbitaceae & Brassicaceae Compost	Greens & Roots	...	Liliaceae Compost
Plot 5	Liliaceae Compost	Greens & Roots	Early Cucurbitaceae & Brassicaceae Compost	...	Greens & Roots
Plot 6	Greens & Roots	Liliaceae Compost	Greens & Roots	...	Late Cucurbitaceae & Brassicaceae Compost
Plot 7	Late Cucurbitaceae & Brassicaceae Compost	Greens & Roots	Liliaceae Compost	...	Greens & Roots
Plot 8	Greens & Roots	Late Cucurbitaceae & Brassicaceae Compost	Greens & Roots	...	Garlic Compost
Plot 9	Garlic Compost	Greens & Roots	Late Cucurbitaceae & Brassicaceae Compost	...	Greens & Roots
Plot 10	Greens & Roots	Garlic Compost	Greens & Roots	...	Solanaceae Compost

Based on the work of Jean-Martin Fortier, *The Market Gardener*, 2014.

Botanical Family Key

Solanaceae Tomatoes, Potatoes, Eggplants

Liliaceae Onions and Leeks, but not Garlic

Cucurbitaceae Cucumbers, Squash, and Melons

Brassicaceae Broccoli, Cauliflower, Turnips, Radishes, Cabbage

BioIntensive Gardening

Though you may hear different definitions depending on the person using the words, strictly speaking biointensive gardening is a method of increasing production per square inch by implementing advanced gardening methods, usually using hand tools or walking machines.

Some biointensive gardening methods start with double-dug beds which are not ideal. Double-dug beds are a method of making soils loose deeper than 12" (30.5cm) which involves digging up the topsoil, reserving it, pitchforking or broadforking the subsoil, and returning the topsoil back on top. For potatoes, carrots, and many root crops, it makes an immediate difference though it is energy intensive and negative overall for soil life in comparison to just broad forking. Using a broad fork, which is like a giant pitchfork, to loosen soils is a great way to avoid tilling and double-digging. Pitchforks can also be used.

Biointensive gardeners mix compost into their top soils regularly. This gives the soil high amounts of organic matter and soil biology. This allows for close planting which creates a miniature canopy that shades out weeds and keeps in moisture. In addition to a high density of plants per square inch, biointensive farmers focus on their density of calories per square inch, and in recent years, they have been thinking about the amount of carbon sequestered per square inch (or cm). Sourcing companion planting, crop rotation, new technology, mostly heirloom seeds, and biological controls of pests where possible, biointensive gardening has many valuable permacultural aspects.

Beneficial Insect and Pollinator Habitat

Providing habitat for honeybees isn't just about keeping your honey supply accumulating or helping the bees get through the winter. Bees require the pollen on flowers to survive. We, in turn, are dependent on bees for their pollination of much of our food. We each are part of a great **holon** with bees. Recognizing and supporting these pollinator holons supports the whole system.

Beneficial insects and pollinators might need specific plants to regularly visit or establish residence in a system. Plants like milkweed for monarch butterflies are superb attractors and critical for their survival. Many beneficial insects that do not pollinate are predators that eat the insects that feed on our plants. When insect populations are in balance, very interesting things can happen with short blooms of "pest" activity that target weak plants—followed by beneficial insects or birds feeding on those pests and leaving a stronger overall population of plants. This is similar to predator pressure improving the health of wild grazing herds. Allowing plants to go to seed regularly keeps a diverse offering of pollen available to attract and keep on-site as many pollinators as possible.

A Bug Hotel at the Krameterhof in Austria

A Bug or Insect Hotel can be made to accommodate beneficial bugs. Each one of these structures is unique to the location, resources available, and insects on-site. Bundles of hollow sticks, stacks of rocks, logs with holes in them, and more, all provide a home for a particular insect. Sometimes this can make for a buffet for passing birds, and managers need to put a screen over the face of the bug hotel which keeps out birds but let bugs in, as done on Miracle Farm in Quebec, Canada.

Predator insects, with such fantastic names as minute pirate bug, assassin bug, soldier beetle, mealybug destroyer, and damsel bug, can be attracted to the garden. There are many others with less fantastic names as well: earwig, green lacewing, wasps, some flies, ladybugs, praying mantises, and more. These bugs are primarily attracted to plants in the *Apiaceae* family (carrot, fennel, parsley, celery, cilantro, etc.) and the *Asteraceae* family (sunflowers, daisies, lettuce, artichokes, calendula, dandelion, dahlias, yarrow, zinnias, etc.) Planting a diversity of plants and letting them go to seed is important for pollinators and all the cycles they support and that support them.

Irrigation

The white film on this pear tree is from irrigating with well water.

Watering, using well water, always leads to the salting of soils—it often looks like a white film on leaves and fruit, or it can even be crusts of salt on the soil when it is severe. Watering with chlorinated municipal water bleaches the soils. Soils have developed in nature to prefer rainwater which is essentially distilled water (lacking in minerals). Plants have developed to anticipate rains (sensing the rise in humidity) and prepare themselves to accept the moisture. They receive moisture through condensation, precipitation, and groundwater. When we water, we fill up the spaces in the soil, making a temporary anaerobic condition. If we overwater, we can make our soils anaerobic for extended periods of time and stress out our plants. Water as little as possible. Instead try to create landscapes that gather and hold moisture. If you do have to irrigate, try to use dripline buried under mulch as much as possible to minimize watering and evaporation.

Seed Saving

While the many techniques and methodologies of seed saving could fill several books, seed saving is rather simple for the most part. To produce seeds with the highest germination rates, plants need to dry down completely and become "brown"—where all the energy from the plant has gone into the seed, and the remaining portion of the plant is mostly just standing carbon, ready to compost. All the sugars and nutrients were focused into the seed's production. The seeds, at this point, may still need to be dried down further before storing which can be done in the sun, inside, or very carefully in a dehydrator (and then only for large, dense seeds like beans and corn).

Seeds are kept in a cool, dark place for the next season, or in the refrigerator or freezer for years. Seeds that are centuries and even millennia old yet still viable are being found on archeological sites and being grown out at universities continuously, proving how amazing seeds are!

Some plants such as cucumbers and melons need to be isolated from other members of their botanical family—they cross! To keep plants true, hand pollinate, plan and time accordingly, and use physical barriers like screens or paper bags.

Potting Soil

We can make our own potting soil at home with just sharp river sand and sieved compost (though the finer we sieve the less fungi will survive the process). Sharp river sand is sand that lets water pass through it easily and can be found in the inside bends of rivers, streams, and creeks.

Most Annuals and Perennials
- 50% sharp river sand
- 50% sieved compost

Tropicals
- 40% sharp river sand
- 60% sieved compost (or more compost)

Fine seeds
- 90% sand
- 10% sieved compost

Rooting Plants

We can take cuttings of trees, perennials, and even annuals like Loche squash from Peru or tomatoes and encourage them to form roots, so we can grow an exact replica of that plant, sometimes referred to as "cloning". There are a number of ways to do this. Some plants such as willow, grape, and fig are naturally adept at rooting from cuttings, but others, like many hardwoods, are more stubborn and need coaxing and time. Using **willow water** and compost tea, we can sparingly water the cuttings keeping them moist in a greenhouse system with intermittent misters, planting them in a container of river sand that drains well. Keeping them warm, highly nourished, and moist but not wet will encourage them to put out strong roots.

A Rooted Comfrey Flower Stalk
To Root: cut a flowering stalk and place in a bucket with some water in the bottom that you keep refreshed for 2-3 Weeks.

Grafting

Grafting is the art of taking a cutting from one tree and growing it onto the cut branch of another tree. This can create an all-season tree where each branch has a different ripening time period, so you can have early, mid, and late season fruit all on one tree. It is ideal for small areas, orchardists, or anyone looking to turn pruning cuttings into new trees. Every pruning from a fruit tree can be grafted onto a root stock or an already established tree's branch to diversify yields. You can have one citrus tree with lemons (*citrus limon*), limes (*citrus aurantifolia*), and mandarin oranges (*citrus reticulata*) on it or one tree with a dozen different types of apple.

Rootstocks are chosen for their vigorousness, disease-resistance, and hardiness; they can also be linked to fruit-size and early fruiting. This is how it is possible to have dwarf and non-dwarf trees of the same variety: different rootstock. Pears (*pyrus*) graft onto quince (*cydonia oblonga*) rootstock to save time since quince trees don't take as long to mature as pears which can take 20 years! Plum (*prunus*) rootstocks can support grafts from all other *prunus* varieties: peaches,

nectarines, apricots, and other plums.

Trees don't all readily graft onto each other—most are done within the same family. It is both an art and a science. Some use grafting tools that work like scissors while other use traditional grafting knives that are extremely sharp. There are several different styles of cut though the main idea is to have the bark from the cutting, or scion, and the rootstock touch. The pieces can range in size from a 6-8" (15-20cm) section of branch or larger, to a small 1-2" (2.5-5cm) section of branch (benchgrafting), to just a cut out bud from another plant (budgrafting).

When finished with a new graft, you can tie on or leave a final top branch for birds to land on. Otherwise, your fragile grafting can be ruined by any passing bird. At Miracle Farms in Quebec, they use rubber bands to keep the grafts stable, and they seal off exposed wood with wood putty (as pictured).

Sepp Holzer has recently introduced a new concept called ReGrafting where commercial varieties are grafted with wild, native varieties for higher yields, better taste, and disease-resistance[69]. Recently, Sepp included this in a design in Southwest Spain with wild avocados which were outperforming the commercial orchards. It is like rewilding applied to grafting.

Stratification or Scarification of Seed

Seeds are protective containers of genetic material. Many seeds only sprout when they sense that conditions are exactly right. A soaked pea will sprout on concrete while a blueberry or nectarine seed in the same situation will stubbornly resist sprouting until it becomes non-viable.

- **Stratification** - Winter seed dormancy can be broken by spending a short period of time in colder temperatures in a refrigerator, soaking in cold water overnight, or even outdoors.
- **Vernalization** - Some seeds require a longer cold period in the refrigerator or outdoors to simulate a full two to three month winter.
- **Scarification** - Some seeds need to be stressed to germinate through either heat or physical action like sandpaper scratching, fire, or nicking a corner with a knife.

[69] Holzer, Sepp. *Desert or Paradise.* 2012. p. 60.

Pruning

While there is much debate over pruning in general, pruning can help control growth, determine shape, and stimulate fruit growth. When pruned, many fruit trees like apples, stonefruit, and mulberries respond with an increase in fruit production—they panic and start producing overtime. The timing of pruning is critical. Most fruit trees prefer winter pruning but not all, so do your research!

When pruning, a classic method is to remove intersecting branches to prevent them from rubbing each other and damaging their bark. This also allows for airflow. Small branches are removed from within 6-8" (15-20cm) of the trunk, leaving only large branches and a relatively open area. Some growers use wires to hold the branches down after pruning for a couple of months. After this, they may grow downward, below the horizontal, which stimulates fruiting. The orchardists make sure to not leave the ties on the tree for longer than that, or else, the tree may start to grow around the wire. Other growers are tying branches to

From Luther Burbank's experimental farm, espalier methods keep branches permanently laterally to promote a focus on fruiting.

the branch itself or to the trunk in early spring for a few weeks or a month to stimulate bud growth. There are many different methods of pruning and training fruit trees, and many new combinations and methods are being trialed regularly.

For those that do not prune, they commonly control vegetative growth by pulling the branches down below the horizon line and holding them there for a month or two in late winter/early spring. This stimulates fruit growth instead of overall branch, trunk, and leaf growth. It channels the energy that would have gone into vertical (vegetative) growth into the fruit. Unpruned fruit tree arms weighed down with fruit tend to develop a drooping habit as well, as do trees with vines on them or late winter snow. Unpruned branches can touch the dripline of a tree intercepting critters that would have eaten the roots or bark of trees with pruned branches. Many argue that parasites from the soil can be transmitted this way, but the growers using this method focus on parasite and disease resistant varieties.

Orchardists who do not prune often cite vigor as the reason why.[70] Vigor is caused by overwatering, over-fertilizing, and over-pruning. The sap that returns in spring is based on the quantity of branches, buds, and leaves that the tree had the season before. If the extra sap is a moderate amount, it can go into bud growth and form larger and more numerous fruits, but too much sap can turn into vigor: out of control vegetative growth which can lead to increased disease, weaker plants, and poor yields. This is why orchardists graft onto dwarf rootstock—it's non-vigorous! The branches that are grafted on are usually from vigorous varieties especially if they are heirloom. Trees want to turn water or nutrients into vigor and vegetative growth. In wet climates with heavy soils, no swales are needed for a line of trees on contour to act like a swale—a swale would create too much water and endless vigor!

Chop and Drop

Chopping and dropping weeds in place works wonderfully with weeds of all kinds; chop them before they go to seed to generate organic matter for the soil. Because weeds are reparative mechanisms, accumulating specific nutrients to remediate the soils, they are excellent mulch (if cut before they can propagate). They will continue to re-sprout from their root systems and provide more mulch for a period of time

---

[70] Greenman, Eliza. *A New Fruit Culture!* 2016.

before they get exhausted and die. They also don't require any room or spacing; they use every available space, profusely packing in. Pulling weeds up by the roots undoes the work they are doing and simply causes more aggressive weeds to show up.

Some weeds are noxious, poisonous, or have incredible thorns; these can be undesirable even as mulch. They can sometimes be called invasive though that term gets often misapplied. Because weeds are pH (and nitrate) dependent, using fungal-dominant compost tea applications will weaken weeds enough that chopping and dropping will remove them. Otherwise, using animals is a powerful reset for stubborn plants such as blackberries.

The Home Plant Nursery

A small plant nursery at home can save money, grow a greater diversity of plants, and add another source of revenue to the homestead through plant sales and eventually the products of a food forest. Rooting out cuttings from pruning is a great way to capitalize on a free resource, a natural return of surplus. Growing trees, perennials, and annuals from seed in a controlled environment, like a greenhouse, allows for homesteaders to generate more plants than they can use on their homestead within a few seasons. Small greenhouses equipped with misters connected to hoses on timers provides an easy way to start your own home plant nursery.

Plant Breeding

While it can fill several books with its intricacies, plant breeding can also be a very simple and profitable hobby. Many garden plants have obvious male and female parts (see pictures) that are easily hand-pollinated and then sealed off with tape or with a paper bag. When both a male and female flower are just about to bloom (usually the night before) remove the petals of a male flower and cut it free. Open up the female flower, use the male flower like a paint brush on the female flower, and then seal the flower shut. This can be done

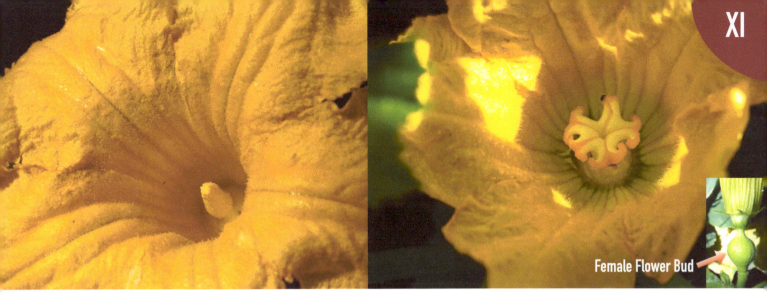

Male Squash Flower   **Female Squash Flower**

with corn, tomatoes, squash, and melons easily, and with the aid of a microscope, smaller flowers.

    While it is true that Mendel's inheritance chart works well with simplistic genetics like those found in peas, it does not work with other reproductive genetic arrangements. Potatoes, for instance, are much more complex in the way they reproduce genetically, and for that reason they can reproduce either asexually (vegetatively) from their tubers or sexually via flowers, bees, and small tomato-like fruits. Most commercial varieties of potatoes rarely flower, but heirloom potatoes can mostly be relied on to flower.

    One can have a lot of fun breeding and creating landraces by allowing for open crosses and then selecting the best each year. This is primarily how all plants have been bred for human use, but many are a bit more mysterious and complex. Studying the works of Carol Deppe will give you an in-depth perspective of the wild world of breeding plants for a purpose. Her most in-depth work is <u>Breed your own Vegetable Varieties</u> (1993).

1. Remove the Petals from a Male Flower   2. Paint the female pistil with the male stamens (transferring pollen)   3. Seal up the flower & watch the fruit develop

XII. Tropical Climates

Tropical climates are located between the Tropic of Cancer and the Tropic of Capricorn with the equator running down their middle. Subtropics are further away from these two boundaries, extending up to the 38th parallel and into mountainous regions of the tropics themselves. All areas of the tropics experience intense evaporation from vertical-sun exposure, and none of these areas described have a uniform range across the planet—they almost all extend throughout the equatorial zone. Disease, mosquitos, and perpetual growth make it a difficult climate for human habitation, but with proper design, it can be one of the fastest places to establish a fully-functioning, self-sufficient home site.

Wet Tropics

These areas are at and around the equator and are constantly bombarded with rain because the perpetual overhead sun is causing constant evaporation; the clouds, heavy with moisture, can't help but release it. The sun slowly wobbles in the sky without much angle change over the course of the year. Vegetation runs wild in an effort to cover the soil and tie up nutrients in the biomass. Most life and most of the biomass is in the canopy of the rainforests. The soils are almost all functionally infertile. When european colonists arrived in the tropics and applied temperate climate farming techniques, it was disastrous for the soils, ecologies, and the people. It should be noted that volcanic soils, some terraces, and flood plains in this region can retain their fertility and support more than rainforest-emulating food forest systems.

Because pathogens reproduce and spread quickly in the wet tropics, proper sewage treatment and isolation from groundwater is critical. Furthermore, water sources in general have to be regularly monitored for purity. Open water becomes a breeding ground for mosquitoes that carry pathogens. Natural insect controls, strict hygienic practices with water and waste, and raised homes with natural airflow and screens are all needed.

Wet-Dry Tropics

Further away from the equator, 5-10° and 15-20° latitudinal degrees away, where we are not immediately below the sun, we get a wet and dry season. The sun's path alters throughout the year to a greater degree, creating more differentiation in climate and ecology. An example of this kind of climate would be Hawaii.

Winters are the dry period in these tropics when they experience desiccating winds; heavier flooding and rains usually arrive in the wet summer season. The summer growth can be immense, creating both a challenge for farmers and an opportunity for designers. These areas constitute the tropical savanna and are home to many grazers: zebras, antelope, water buffalo, and more. There are also large predators to manage the grazers. Agriculture in this area is easier because the dry winter gives the soil respite and more fertility has been built over time, but over-grazing, erosion, fire, and over-tilling are all still real concerns in this climate. Windbreaks or hedgerows are needed against the winter winds. Constant mulching is needed to keep soil rich in organic matter.

Monsoon Tropics

A type of wet-dry tropics, monsoon tropics extend further than the other regions, 0–35°F (°C) degrees. Monsoons are extremely heavy summer rain periods preceded by building heat starting in spring and caused by a switch in the direction of the winds from inland to offshore. The soils are less nutrient-rich and in winter tend to be hard and brittle.

Tropical Soils

Unless it is in an area with volcanic soils or Amazonian terra preta, tropical soils are lacking in nutrients in comparison to the soils of the temperate climates. The constant rains leach nutrients out of the soil. Plants are the biological reserves of nutrients in the tropics, not the soils.

Some sandy soils are so well draining that plants have a difficult time establishing: legumes are vital in creating a living network for other plants to rely upon as they establish and create humus and more soil structure beyond sand; adding clay can also be helpful. For gardens in this region, plastic lined trenches, beds in sunken water tanks, or garden row gleys need to be employed to prevent water from leaving too quickly. Organic matter from trees, mulch plants, and even aquatic plants can be added to the soils. Composting in this region is difficult and growing annual legumes, like cowpeas or green manure crops, to chop and drop is more effective. Ants, termites, and also a few worms help develop and maintain the soils in this region.

The maximum sizes for growing spaces in the tropics vary slightly but in general the stereotypical temperate climate field is inappropriate—the average US farm is over 400 acres (165 hectares); most of these farms are managed in large open fields hundreds of acres in size. Smaller open spaces work better with tree systems around them to support them. Perennial polycultural food systems supported by legume or nitrogen-fixing trees are ideal for this region as tillage or any soil disturbance sets the ecosystem back sharply and recovery can take a very long time if damage is severe. Grazing cattle herds are not naturally found in most of the tropics outside the savanna, so graze carefully and lightly.

Start small and work your way up, use manure, annual legumes, and tree legumes. Grow perennials more than annuals, trees more than understory, and food forests more than fields, and return as much organic matter to the soil as possible.

Tropical Earthworks

Excessive water can be managed with terraces, swales, *chinampas*, and pits to drain water off of croplands, to prevent soils from being waterlogged (and then potentially anaerobic), and to turn the excess water into aquaculture. Terraces both wet and dry are common in the tropics for rice or taro (wet) and small seed grains (dry). Diverting and sinking water into slash piles for bananas, papayas, or other water- and nutrient-hungry plant guilds is also a passive and powerful technique.

Tropical Food Forests

In many tropical areas of the world, home gardens have traditionally been food forests. In some areas, local communities relied upon the natural tropical forests for foraged food, but now that these forests are largely gone, they have lost the traditions with which to restore them. Luckily, food forests are easily established in the tropics. This area prefers perennials since annuals work in disturbance, and tropics cannot handle consistent disturbance. The tropics are ideal for food forestry. Many exotic medicinals come from this region as well and can easily be grown.

Hans Eiskonen. Unsplash.com. 2015.

- **Tall canopy** - Jackfruit, breadfruit, coconut, avocado, cashew, and pecan
- **Palms** - Over- and under-story like coconut and chilean wine palm
- **Understory layers of trees, giant herbs, and large bushes** - Coffee bean, cacao, banana, and taro
- **Clumpers** - Cardamom, turmeric, bamboo, and sugar cane
- **Ground covers and vines** - Sweet potatoes and yams

Tropical Gardens

Tropical gardens are not tidy like the traditional temperate climate garden; they are larger and more rampant. Soil fertility is largely maintained through chopping and dropping mulch and green manure plants. High shade is vital to allow in early and late light but block the midday intense solar radiation. The gardens, though containing annuals which require soil disturbance, are more like microclimates in food forests than temperate gardens. They are small, roughly a quarter acre (1000m^2) size areas. Raised beds, *chinampas*, and berms can be used to decrease the moisture content of the soil.

Timing is essential in gardening in the tropics—growth and decomposition are moving so quickly that opportunities for intervention are brief and easily missed. The tropics can also have no dormant season—it can be constant growth year-round.

- 1-2 acre (4,000-8000m^2) fields for annuals, often using alley cropping (agroforestry)
- Mulch and pollard from alley cropping trees
- Chicken or duck tractoring between crop cycles

Tropical Chop and Drop on Zaytuna Farm.

- Terraces
- Perennials, bananas, pineapple, grains, cucurbits, and sugar cane

Banana or Papaya Circle

A classic permaculture design example where a mulch pit is lined with desirable plants that feed on the fertility of the decomposing mulch. Gardeners do something similar when they sink porous containers full of compost into their garden beds and surround them with tomato plants; it creates a passive feeding situation for high production. Slash piles of mulch and biodegradable waste can be used similarly but in any shape if there is enough moisture.

Pollards

By consistently pruning the top branches of vigorous trees that respond well to persistent pruning, we can get a consistent yields of straight sticks or canes for **biochar**, weaving, building, or stick-fire fuel. The pace of growth in the tropics makes it an ideal location for pollarding.

Chop and Drop

Chop and drop at the start of summer—winter is dry and spring is hot —you need the moisture of summer rains to properly break down the cut greens. Moisture contains the breakdown and keeps its products traveling down into the soil instead of oxidizing.

Biochar

Biochar is biologically infused charcoal. Charcoal is made at high temperatures in a low-oxygen environment (pyrolysis). Biochar is made when the charcoal is soaked in compost tea or combined with compost. When charcoal is created all the life and moisture is removed; what remains is like a porous carbonaceous vacuum. It provides habitat for soil life, but charcoal alone as a soil amendment will rob the humus of water and nutrients. It should be noted that biochar has a unique electrical capacity abilities such that it can store information like a silicone chip can - scientists are eagerly researching and testing biochar for new applications currently. We are also just learning that different specific heats for making biochar create different types of biochar - and the plant mediums used make a difference as well. The Redwood Forest Foundation, for example, makes superb biochar at a specific heat in a specialized kiln of 600°F/315.5°C, but that doesn't mean that superb biochar can't be made at other heats, each for a specific purpose.

Biochar was inspired by Terra Preta, a dark, rich, man-made soil from areas of the Amazon that was created using charcoal, fish bones, pottery, humanure, and bones. Biochar is gaining attention as a carbon sequestration method as it takes biomass that would be burned

and instead turns it into a durable carbon compound that improves soils. In some areas, terra preta is the best solution to depleted topsoils.

It should be noted that making biochar is a biomass-intensive process as it requires organic matter for both the compost and the charcoal. For those in the tropics or the temperate climates where biomass is plentiful and constantly generating, it can be a powerful tool for building soil. If the site only has its standing living carbon to be turned into biochar, it has to be imported - luckily, it's superbly lightweight and becomes a nearly permanent soil amendment. It can help in desertifying regions; they shouldn't use up the last of their trees to make it though. Many sites in contrast has an abundance of biomass that can be turned into biochar.

Biochar can be made quite simply by stacking your dried wood or organic matter so air can pass through the pile, small sticks at the bottom, bigger sticks in the middle, a bird's nest of sticks on top, and then lighting it from the top. Top lighting, or Conservation Burning, is a cleaner burning method than lighting from the bottom especially if you are a farmer burning agricultural waste already - top lighting produces much less smoke (estimated 98% less particulates); it is a clean burning method according to Sonoma Biochar and Blue Sky Biochar, and while it does create biochar, it is one of the crudest methods. Light from the downwind side, so the fire is both sheltered and fed by the wind. The conservation method burn converts unburned biomass into approximately 15% charred biomass. There are other methods where heat is applied around the wood in a kiln and the wood never catches flame but still turns into charcoal. Arresting the burning process is critical. Spray the pile down with water before the charcoal starts to burn into ash which is the 2nd stage of burning. Spraying it down with microbial infused waters also immediately imprints the char with life - Cuauhtemoc Villa of the Sonoma Biochar Initiative uses EM infused water to cool his biochar piles. The char should be very brittle and glassy looking - use a shovel to break up the char to create more surface area (though try to avoid pulverizing it since biochar dust is nowhere near as effective).

Tropical Animal Systems

Sanitation is a primary concern for containment systems with animals in the tropics. Wastes need to be washed away and cycled as soon as possible. Housing for animals must be designed to let waste flow downhill, to be easily washable, and to infiltrate and be absorbed into forage systems to return as animal feed.

These animals rarely free graze; they are contained and brought their feed, contained in animal tractors (open bottom pens for portable grazing), or monitored or tethered while grazing.

- *Rabbits*
- *Ducks*
- *Chickens*
- *Pigs*

Tropical Aquaculture

While aquaculture has the highest natural yield of all forms of cultivation, it is perhaps most prolific in the tropics because lacking a dormant time period and thriving on consistent high temperatures, waste cycles are fast and rich pond sludge quickly accumulates. *Chinampas* are ideal as are wet terraces for rice and duck systems. Ponds with herbivore fish can have forage plants along their edges for fish to feed upon; these plants can be food for fish, people, and animals. Rabbit, duck, or chicken manure can create temporary algae blooms in the ponds for fish to feed upon, but use with care as too large a bloom will suffocate the fish and other pond life.

Tropical Housing

In humid and hot climates, evaporative cooling is not an option—only dehumidification, air movement, and shade can be used. Buildings must be oriented to the direction of the prevailing winds, not the sun path. There is rising air during the day and falling air during the night, so orienting the home so that exchange passes through the building is critical. Avoid heat sources and thermal masses—even cooling thermal masses like caves create mold, fungi, and mildews. Thin walls and roofs prevent heat being stored in them during the day and quickly release whatever heat they've accumulated at night.

A solar pump can be used to draw heat out of buildings, and white or reflective paint can be used on the outside surfaces of buildings. Complete shading can also be accomplished with layers of palms or large legume trees which allow air flow beneath their canopy. Roofs are steep to avoid direct orientation to the sun and usually thatched to allow heat to easily leave. If in a flood zone and on stable soils, housing may be raised or sheltered by earth banks. Distance is maintained between the house and other structures and interceptive trees or plants to help air flow. Trellis is used to create more shade and extend the garden areas around the house without obstructing airflow.

Kitchens are outdoors unless there is a cold season, in which case there may be two kitchens: indoors and outdoors. Sealable, cold food storage is critical to keeping food safe and long lasting. It is also important that the kitchen area be designed to be cleaned easily.

Safe toilets

In the tropics, clean toilets are of utmost importance. They can easily become transmission sites for deadly pathogens if poorly designed. Septic tanks in the tropics tend to poison the groundwater. Dry compost toilet systems are ideal with a 1m drop which aerates the waste with the fall and impact, and then it tumbles at ideally a 45° angle and lands on a grate or mesh to be aerated as it decomposes. A black-pipe solar pump extending from the bottom of the toilet to two to three meters (six to 9 feet) above the top of toilet is used to siphon off the gases from the waste;[71] it works just like a rocket stove's tower. Liquid and solid are separated by the grating which allows for more aeration and for these materials to breakdown separately.

Safe drinking water

The continuous heat and relentless growth in the tropics leads to all decomposition and growth cycles working at their fastest possible speeds. That means in water and in soil, aerobic cycles work so quickly in the tropics that they can use up all the oxygen, and anaerobic cycles and organisms can contaminate the soil, compost, or water quickly. This makes composting difficult and timing crucial, but it also means that pathogens breed in our drinking water incredibly fast, so problems can easily spin out of control if not corrected promptly. Regular monitoring, screening, and safe storage are critical for keeping water safe in the tropical regions.

Insect screening and netting

Screening over rain gutters, water storages, and windows to keep out insects is critically important in the tropics where insects spread viruses and pathogens. Homes, bedrooms, beds, and fruit trees can be covered with netting. Habitat for bats, amphibians, and birds that feast on mosquitos and other insects can be built around the home for added protection.

---

[71] Lawton, Geoff. *The Geoff Lawton Online Permaculture Design Course*. 2014.

Graywater treatment reedbeds

Remember: testing water is critical both on the way in and on the way out. Multiple reed beds in succession made out of concrete basins may be necessary to filter out all the pathogens, toxins, heavy metals, and excess nutrients. You may even need to follow a reed bed with more diverse and complex biological filter plant and animal guilds as specific toxins can be persistent without the right plant in the system to cycle it. Graywater treatment reedbeds are even being encouraged and regulated by local governments in areas of Australia.[72]

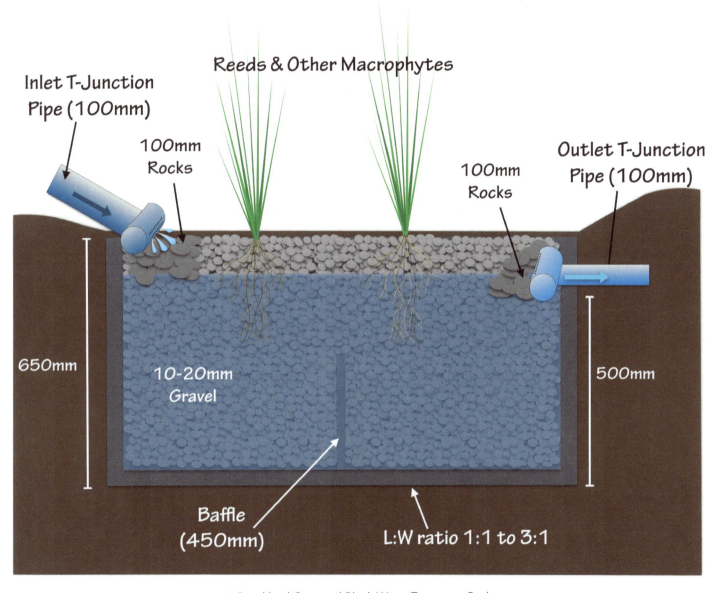

A Reed bed Gray and Black Water Treatment Bed.
Based on Lismore City Council's "Figure 2: Basic Reed bed Design (Lateral View)"
from *The Use of Reed Beds for the Treatment of Sewage and Wastewater from Domestic Households.*

---
[72] Alexandrina Council. *Environmental Health Fact Sheet: What is required with an application for a Reedbed (second-stage wastewater treatment) system?* 2016.

Bamboo

This incredibly versatile and useful plant grows especially well in the tropics. The tender shoots are edible. The mature poles are a sturdy and durable building material. It can be made into biochar or fuel. The power of bamboo to sequester carbon rests upon the fact that it is a perennial grass, and can be constantly harvested without damaging the vigorous root systems. It spreads creating areas of monoculture, so they do require management and planning. Wandering root systems can be controlled with subsoil plowing or hardscape like steel plating. Plan the placement of bamboo carefully. Ripping out the root system requires machinery most often but can be made into rich compost.

"Managed bamboo forests sequester more carbon than natural bamboo forests. This is because harvesting stimulates growth, and treatment of harvested culms allows for long-term sequestration in bamboo products"
–Eric Toensmeier, The Carbon Farming Solution, 2016.

XIII. Temperate Climates

Temperate climates extend away from the Tropics of Capricorn and Cancer toward the poles. This climate is typified by drastic seasonal change: *WINTER*! The tropics and hot drylands don't have winters with frosts or even snow (unless at high elevations). In this region, sun angle and season length are limiting factors to systems and need to be included in all design considerations. The northern hemisphere is predominantly a temperate region.

The Subtropics

This area is a blend of mediterranean and tropical. It is an edge climate, and it typically doesn't get frosts. A combination of plants from both regions can be grown in this area, but it is also home to much of the world's desert land. In the savanna regions, summer rains precipitate the start of the growing season.

Mediterranean

This area is characterized by mild, wet winters with very little to no snow as well as seasonal wildfire in dry, hot summers. The rampant growth in winter feeds the fire cycles of summer. This region refers to not only the area around the mediterranean sea, but also the

climate analogs all over the world. Both mediterranean and subtropical gardeners are the envy of gardeners in colder areas because of the length of their seasons and the diversity of plants they can grow.

Warm, Cool, and Cold

These regions have frosts, snow, and sometimes months of frigid temperatures. The growing season shrinks the further you travel away from the equator. The shortness of the season coincides with greater day length, so plants and animals have cycles and systems adapted to this and are very active during the spring, summer, and fall. This activity equates to an abundance of vigorous growth.

In Alaska where the growing season is only 150-200 days long, the sun wobbles in the sky for nearly a month but does not set. Alaska's growing season may be short but the constant sun and the rich soils grow gigantic vegetables that are the envy of other climates and are sometimes even met with skepticism.

Temperate Soils

Temperate soils are on average more fertile than the soils of all other climates, due largely to winter decomposition. Under a thick layer of snow, it is in general a constant 32°F/ 0°C degrees. This is usually much warmer than the ambient temperature, and beneath that snow, it is perfect for a slow, steady breakdown of all the biomass from last year's growing season.

Because of that richness, this region can support the largest field sizes and even allows for regular, light tillage since the winters heal the soil so effectively. This is essentially the garden that imperialism and colonialism exported to the rest of the world; it was damaging because it relied upon the rich soils and short growth cycles of the temperate climates.

Temperate Food Forests

These forests are characterized by their dormancy periods since that is when fertility is returned to the soil. There are deciduous trees that lose their leaves in winter and evergreens that never lose their foliage. In fire-prone mediterranean zones, fire-dependent, resinous shrubs and trees thrive. This is the climate of the double herbaceous layer and cane fruit like raspberries and blackberries. In the colder regions of the world, there are fewer nitrogen-fixing climax trees in comparison to other climates; instead, lichens provide the majority of nitrogen fixation in some colder climate areas (as well as deserts).

Pruning, pollarding, **coppicing**, and chopping-and-dropping are all powerful practices in this climate because all encourage acceleration and diversification of soil food web cycles. Active silviculture is vital, as is timing—since the season is so short, every day counts. Large *hugelkulturs* build soil quickly with winter's thorough decomposition process.

It is important to increase plant hardiness the further out from zone one you plant; we also need to mix in winter/fall hardy species as much as possible to spread yields out over the entire year. Strong sheltering windbreaks, hardy edge-species, and microclimates all work together to extend the growing season and increase yields. Animal systems help prune, control pests, build soil fertility, sequester carbon, till soil, seal ponds, make compost, and more.

Chestnuts at Luther Burbank's Experimental Farm in Sebastopol, CA. 2016.

- **Fruit and Nut Trees** - Apples, pears, plums, chestnuts, hazelnuts, and walnuts
- **Legumes** - Honey locust, black locust, western redbud, and Siberian pea shrub
- **Berry and bush layer** - Blueberries, raspberries, loganberries, and huckleberries
- **Mushrooms** - The constant generation of woody biomass is perfect for mushroom cultivation
- **Creepers and clumpers** - Strawberries, bamboo, aloe vera, and artichokes
- **Herbaceous layers** - Large, diverse, and medicinally relevant, there's a double layer of herbs in cool/cold temperate climates
- **Microclimates** - Used to extend seasons, yields, and diversity
- **Deciduous plants** - Deciduous vines and trees are plants that drop their leaves in winter and allow winter sun to reach further into a system—which warms microclimates while shielding and shading those same areas in the heat of summer
- **Windbreaks and Hedgerows** - To protect from wind and wildlife

Versaland, Grant Schultz's farm, features a keyline design and multi-species grazing

Silviculture

Temperate forests generate incredible biomass seasonally. Specific tree and plant varieties like willow or locust respond well to aggressive pruning methods. Coppicing (cutting trees or plants to the ground) and pollarding (cutting the top branches from trees) generate long sticks perfect for rocket stoves, biochar, basket-weaving, and more. Forests can easily become impassable or wildfire-prone if they are not managed with wildlife or human intervention. Many large forested areas in the cold temperate climate are not managed actively enough by either; these landscapes are degrading and need human intervention now that most of the wildlife has been removed or displaced by human activities. Proper management of forests can help aid the return of displaced keystone species; their reintroduction brings restoration and balance to the ecosystem.

For good timber, branches need to be removed before they are larger than 4" (10cm) in diameter otherwise it creates knotty wood that is less dense when harvested. If the energy and nutrition meant for that branch never goes out, it remains inside the the tree, making for more dense and smooth-grained wood (no knots). Removing branches as they form up 18 ft

(6m) in height using a ladder is critical in some fire prone areas.[73] Pruning to allow air drainage to prevent frost pockets or to allow light in during the colder and darker winter months can be vital in colder climates. When harvested, the logs can be sealed with wax on the ends to slow the drying out process, which prevents cracking.[74] Wood needs to be aged, even for burning, so having a place already prepared for drying down timber and fire wood is important.

All this work requires more than just a pair of loppers and pruning shears. One needs chainsaws of varying sizes, chaps, goggles, headgear, steel-toed boots, and even things like a portable mill that attaches to the chainsaw or a timber jack that allows one person to handle giant logs more safely. That said, in South America, they use giant chainsaws to cut timber boards by eye without guides and some folks use animals to move their timber for them.[75]

While removing all trees or too many trees is detrimental to a forest system, too many trees in place is also counter productive if sunlight cannot reach plant growth points. The forests of North America were widely spaced from regular controlled burns by the Native Americans. Today we can simulate the effects of fire and large-animal grazing under predator pressure by using good silviculture practices and holistic grazing of animals, from grazers to browsers and omnivores. The forests managed by the Native Americans were larger but there were fewer trees per square foot. The trees were also larger and of better quality.[76] Thinning out the poor quality trees, turning them into *hugelkulturs*, brush swales, wood chips, firewood, and anything ethical and creative, is a great way to improve the forest while netting another yield.

Burnt trees from fires are valuable biomass that can be used like biochar. Dead trees are habitat for numerous species. Good management of the forests takes all these cycles into account and finds balance. Each forest is unique as is the management it needs.

Temperate Gardens

Gardens in the temperate zone are busy in the spring, summer, and fall with winter gardens happening in the warmer regions only. Raised or double-reach garden beds are common as is a more refined, neat, and orderly appearance.

Temperate climate dwellers always marvel at the size of plants and insects in the tropics, but that is because the insects are filling a larger role in decomposition since the tropics lack a long, cold winter. In the temperate areas furthest away from the equator, insects

---

[73] Doherty, Darren J. *"Broad Acre Agroforestry Integration."* 2016.
[74] Ibid.
[75] Ibid.
[76] Catalyst. *Earth on Fire*. 2014.

are seasonal and nowhere near as diverse as in the tropics.

Ragged Jack or Red Russian Kale is an amazing temperate annual crop that can grow all year.

Despite the riot of annual growth in the temperate climate, it tends to die down or go dormant each winter. This climate has a double herbaceous layer, so herb gardens in this climate can display a great diversity of plants—ideally positioned close the kitchen. Berries, cane fruit, and hedgerows can all make effective garden fences for predators as well as distractions for deer and birds. Rockeries for snakes and lizards are also ideal in the garden to help keep down the populations of undesirable rodents and insects. Fences, trellis, and espalier trees help the many fruiting deciduous vines in this climate provide shade for gardens in the hotter regions, vines like kiwi fruit, grape, or passion flower.

One of the most enjoyable aspects of the temperate climate is that the work is seasonal. Humans experience a dormant season as well in winter. They take time off, they read, they sleep more, and they reflect. Unless you use poly-tunnels, cold frames, or greenhouses in winter, the season will not extend very far even with hardy crops in the colder temperate regions.

- 1-2 acre (4,000–8,000m^2) fields for annuals maximum
- Ideally on or close to contour, unless too wet
- Double-reach beds
- Hedgerows for compost, mulch, feed, bird habitat, and food
- Roots, Grains, Fruits, and Nuts

Temperate Animal Systems

Herbivores do extremely well in this climate and can help create change as fast or faster than they do in savanna drylands. The herbivores are a natural complement to the rapid annual growth. They also provide food for humans when gardens and food forests are dormant. Harvesting animals in winter is also cleaner since the cold slows down all reactions and acts like refrigeration—a time when insects are dormant. For many northern cultures,

A bunny tractor at Camp Singing Wind, a permaculture site and event center, in Toledo, WA. 2017.

traditional diets were light on fresh meat during the summer and heavier on fresh meat during the winter, with preserved meats being eaten and prepared strategically.

Animal systems in the temperate climates can be very creative, sometimes even involving humans and animals sharing buildings. In winter in Northern Europe, animals were traditionally housed below the home, so that their body heat would provide the house with passive warmth—some areas still use this method. If enough carbon is added in the form of straw or dry mulch as the animal manure accumulates, the carbon can balance the nitrogen, so there won't be bad smells rising up into the home. Deep-bedding also provides warmth as it composts in place.

Chickens can be found housed in chicken tractors, free-ranging like they do in the tropics, wearing hand knit sweaters against the cold, and living in straw-bale-house coops, warmed by their own bodies, bedding, and manure.

- **Silvopasture** - Running animals through orchards routinely is a great option for this region.
- **Cut forage and hay for silage and dry winter feed** - Fermented winter feed (silage) can be made with alfalfa and molasses for ruminants. Hay is also a great option, can be dried in the fields, and can be stored so cattle can have access in the field all winter long.
- **Holistic Management of Cattle** - Rotation of herbivores and omnivores in succession with rest and daily shifts through paddocks, recreating the natural patterns that built soil health.

Composting

Slow, mouldering composting happens naturally in this cool and cold temperate climate in winter. Compost piles can be easily built to decompose over winter months for springtime application—the plants that grow in that region are almost all designed to do this annually anyway. Gathering up the biomass only magnifies the effect and enhances the end product. Indoor animals provide bedding and manure for potential compost as well. Start pulling out the animal bedding about a month before spring planting, make a pile at least one cubic meter in size—fresh cuttings from early spring growth may be needed to make it more balanced. Enrich your soil just in time for your transplants or seeds to go in the soil.

Use *hugelkulturs* to compost woody biomass to build even more soil each spring with the added benefit of it all happening in place. Rather than compost being transported to the garden beds, the compost is made in the garden bed (which is similar to the banana circle mulch pit strategy). The *hugelkulturs* also hold water for long periods of time via the punky wood, the action of soil mycelium, the action of the soil food web, and the creation of more humus.

Temperate Aquaculture

Though very possible in the temperate climate, aquaculture has largely been ignored—instead this area relies heavily upon fishing. Much like those in the tropics accustomed to foraging from wild forests, those that relied upon fishing are witnessing ecological collapse. Because fish in this climate typically live in either large bodies of water or moving water, systems have to be either constantly moving, or large, to be regenerative. It is likely better to support established natural systems of aquaculture in this climate (for fishing) rather than to build small home aquaculture systems reliant on energy inputs—that is unless a system is situated in an urban environment, where the market is large enough to justify energy inputs. Keeping small bodies of water warm through the winter is costly while large systems hold thermal mass and take in the sun's heat constantly to maintain life, even under ice.

Temperate Housing

Buildings are oriented so that they face the sun path unless oriented slightly towards sunrise in the subtropics or slightly towards sunset in cold temperate climates. This keeps the subtropical homes cooler in summer and the cold temperate homes warmer in winter. By facing the sun, homes have a distinct front and back to the home. The front has natural light that keeps it warm while the backside is dark and cool. Bedrooms, bathrooms, laundry, and storage are usually on the cool side while daytime and social activities occur on the warm side in the sun. Homes have thick walls for maximum thermal mass to trap the heat and insulate against the cold. The center wall can also be used to to trap and distribute heat. In humid areas, air flow and sunlight are vital, just as salt-based dehumidifiers can be.

Earth-sheltered systems like the *walipini*, earth-banked walls, or earth-sheltered greenhouses are all ideal. They take the solar energy they get during the sunny part of the year and trap it in the earth for slow release during the winter. Sealed entryways, a two-door entrance, double-glazed windows, and a mudroom—these are all strategies to keep that warmth trapped inside the home.

The roof of Glenn Kangiser's Sierra Nevada underground cabin. California, 2015.

Glenn Kangiser has an underground cabin homestead nestled in the Sierra Nevada mountain range of California. It is a working example of an earth-sheltered home in a temperate climate. Combining Mike Oehler's ideas with his own contracting and construction experience, Glenn cut into the side of a ridge and built his home with large, uncut tree logs and covered it with earth to keep it warm in winter and cool in summer.

In cold temperate climates, graywater remediation systems like reedbeds, settling ponds, and biofilter ponds would all have to be under glass or inside a greenhouse to keep them warm all winter. When it is cold, all the biological filtration processes slow down and go dormant. Keeping the ponds and reedbeds warm keeps them cycling and cleaning the water.

Rocket Stoves and Rocket Mass Heaters

Rocket mass heaters are masonry heaters that use a J-tube to burn stick fires cleanly in a system that slowly radiates and conducts the fire's heat outward from a thermally charged mass. A rocket stove is a J-Tube without the mass; it has a focused point of heat that can be used for energy, cooking, or even blacksmithing. The intense heat and J-tube smokestack shape creates air-suction that pulls the fire sideways while gravity helps the sticks fall into the

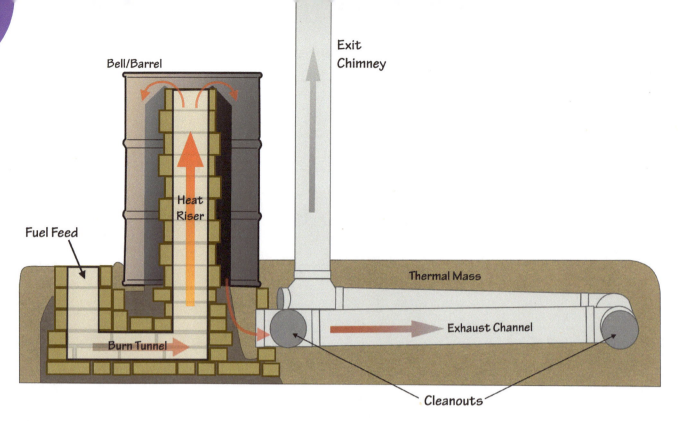

Based on diagrams from The Rocket Mass Heater Builder's Guide by Erica & Ernie Wisner.

J. This prevents smoke from flowing back into the house if properly done. It's so hot that the smoke inside the J-tube becomes fuel for a secondary burn—making for a clean burn. Once the fuel is completely consumed, the exhaust that exits is comprised of sterile ash and CO_2.

The most efficient form of heat and power in the temperate climates comes from stick fires. They burn at a higher heat than log fires because of the greater surface area. The intense annual growth of woody plants in the temperate climates, often seen as a nuisance, can satisfy our energy and heat needs, making the problem the solution.

This ends the need to leave any fires burning overnight because the heat is trapped in a large mass like a cob bench or brick wall which slowly releases it. Variations on these types of heaters were traditionally common in the colder regions of the world like Northern Europe (the Kachelofen) and China (the Kang bed stove). Today, they are the best option for decentralized, clean energy in temperate climates.

Burning wood cleanly is considered carbon neutral because tree regrowth sources that carbon dioxide to build its structure. It's a perpetual cycle: the sticks will either oxidize in the sun on top of the soil or oxidize quickly in a fire. Either way, it will eventually oxidize unless composted or buried in a *hugelkultur* (and both imperfectly sequester the carbon).

In very cold climates, the lack of sun makes solar panels and even greenhouse growing inefficient—wind energy may be used but wood is the most reliable and abundant heat source.

Rocket Stoves for Hot Water, Electricity, and More

The extreme heat created by rocket stoves can be harnessed in many ways: heating a home, cooking food, heating water, metal smithing, glassblowing, or even making steam—even waste disposal plants are using J-tube-style designs to cleanly burn waste. Building steam is definitely the most dangerous application of rocket stove technology since it can explode and be fatal to anyone nearby. Water heaters need to have steam releases like a tempering valve, and they also need to mix the hot water with cold water to guarantee it is within tolerable temperature ranges.

The biggest hurdle for many who desire a RMH with uniform readouts is that every rocket stove layout, layout space, time of year, type of wood, and style of loading that wood will differ, and any metrics taken from different systems or even the same system with different wood, different times of year, or a different person lighting it, and you will get variable levels

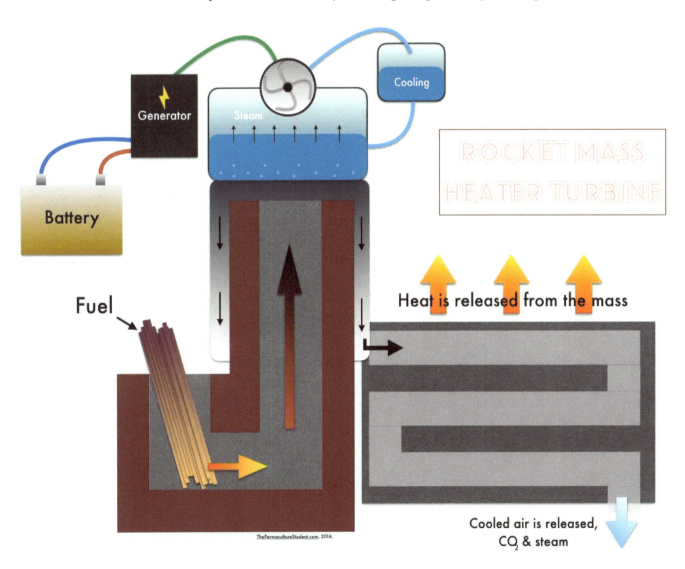

A concept map for a rocket mass heater steam generator. Notice the steam can vent off and return after cooling.

of heat which equates to varying levels of smoke in the exhaust. Rocket stoves and rocket mass heaters in general are each unique in their effectiveness at burning cleanly and heating a space. Using a RMH or a rocket stove requires a person to be present and monitoring its operation carefully. It also takes knowledge of fire and the physics of the RMH itself. Rocket stoves have limitless potential—there are already miniature wood-burning stoves that charge cellphones. Once we embrace this technology ubiquitously, we will see innovation bloom as it did with the automobile and the internet.

> *"Wood burns clean at 1200°F degrees [648.8°C], but on average most fires probably burn around 900°F to 1000°F [482–537.7°C]. A RMH burns around 2000°F [1093.3°C] and can go as high as 4000°F [2204.4°C] but, around 2100°F [1148.8°C], begins creating NO_X [Nitric Oxide]. We have a pretty narrow band to clean burning. We can do some real damage if we do not think this whole thing through. We really need as a society and global community to move in to the clean burning range because most things can be fabricated [to burn] in the clean combustion range rather than going over or under"*
>
> –Ernie Wisner, 2016.

To learn more about Rocket Mass Heaters consult Erica and Ernie Wisner's new book <u>The Rocket Mass Heater Builder's Guide</u>.

Greenhouses and Shadehouse Lungs

Heat rises into or out of a house, and cold air falls into a house. Greenhouses on the sunny side and shadehouses on the shady side helps accomplish this easily if designed properly. Humidity filters and plastic or impermeable seals between the house and the extensions are vital to stop the spread of mildew and mold. You can also have a shared thermal mass like a concrete foundation that the house and the greenhouse share to conduct the heat indoors—the same can be done inversely with the shade house.

Summer sun is blocked from heating the floor of the greenhouse by the deciduous vines and

Lizzie Guilbert. Unsplash. 2015.

surrounding trees, shade cloth, and/or using full ventilation and fans to shed heat. To increase the heat catchment, the concrete foundations can be extended downward with insulation on the inside edge to trap heat in the earth beneath the house. New types of greenhouses maintain their temperatures by pumping the summer's hot air into the ground through pipes with an inexpensive fan and then releasing it to heat the greenhouse during the winter. The same concept can be applied to a house as well.

A solar pump positioned in the wall between the greenhouse back-wall and the front wall of the house will allow easy siphoning of unwanted hot air from either area or from both, with the use of gates.

Low Grade Geothermal Heating and Cooling

By routing the hot air of summer down through the earth in winter, greenhouses can heat themselves through the summer. The system can be passively running constantly, or it can be a more sophisticated system, monitored by a computer program. This can eliminate or greatly reduce the costs of heating and cooling a greenhouse depending on the size and sophistication of a system.

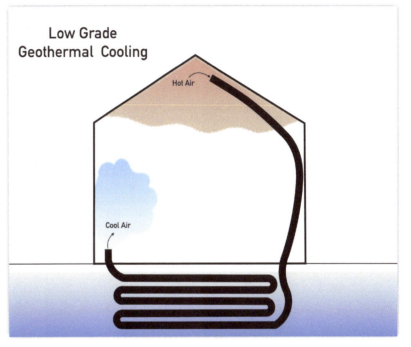

Hot air is sucked out of the top area of the greenhouse (or any house) and is routed below the surface at least a meter through a series of pipes. Running the air through pipes underground allows the earth to cool it. The air released is much cooler. The entire system is kept in motion by the single fan inside the feeding tube.

Condensation and moisture can be problems in these systems. De-humidifying the moist, warm greenhouse air before it enters the cooler below ground area will prevent or mitigate this but requires a stronger fan and, therefore, more energy input. Otherwise, allowing water to drain by design (having the piping slope to one corner) may help as it does with air conditioning units.

Root Cellars

These are earth storages that range from holes in the ground to entire rooms underground to store food through the winter and to keep food cool in summer. These, like the *walipini*, rely upon the thermal constant, the consistent temperature of the earth 4 feet below the surface. Sepp Holzer is a master builder of earthen shelters; he builds them for his animal shelters, storage, root cellars, and more. They are incredibly useful and easy to build since they are usually only the size of a room. Creating a safe, earth-covered roof for so small a structure is much easier to manage. The larger it gets, the more engineering is required to avoid weaknesses in the structure. The sides of root cellars should be also shored up in most cases to prevent soil cave-ins.

BSFL and Other Waste-to-Insect-to-Animal Feed Systems

Black soldier fly larvae are an incredible feed or supplemental feed for poultry or fish. They consist of more than 40% protein, reproduce quickly (laying 900 eggs in the 5-8 days they live), and feed on decomposing kitchen waste and manure.[77] This system turns waste into feed for animals efficiently. There are many insects that are edible and manageable that can be grown as feed; this list includes fly larvae, worms, beetles, and others. Mulberry trees over ponds have silk worms falling into them, feeding the fish passively. Recognizing and listing the insects we can attract and raise that our animals can feast upon is a first step to including them in our designs.

---

[77] Microponics. *Black Soldier Fly Larvae.* 2009.

    Arid climates are those where evaporation exceeds precipitation. Though typically we think of deserts and arid areas as being hot places, these can be in both cold and hot climates. These areas can suffer from prolonged droughts, extreme dryness, and high soil salinity. Most of the life in arid climates is beneath the surface, near oases of moisture, nocturnal, seasonal, rain-dependent, and even migratory. These are the areas Alan Savory has dubbed "brittle".

Neal Spackman's 10 Keys for Greening Any Desert[78]

- **Know your Original Climate** - Understanding what came before human or natural disturbance helps us plan. In historical and geological time, there have been many stable climates and ecosystems on every site. The most natural and beneficial changes we can make on-site are those that mimic or encourage these past, on-site ecosystems.
- **Know your Water Cycle** - Careful water budgeting comes from understanding precipitation, understanding the water path, planning for the larger 100- and 50-year precipitation events, knowing the recharge rate of the aquifer, knowing the longest time period possible between rains, and understanding what is driving precipitation in the area.

[78] Spackman, Neal. *10 Keys for Greening Any Desert*. 2016.

- ***Know your Mineral (or Nutrient) Cycle*** - Understanding how your nutrients cycle in your system is critical, especially in arid climates where a lack of moisture can prevent biological breakdown, essentially locking up nutrients.
- ***Find your Niche in your Watershed*** - Understanding where you are in your watershed is critical to managing water on your site.
- ***Precipitation and Evaporation*** - Knowing how much water you get in contrast to how much water evaporates on your site gives you a more accurate idea of what amount of your precipitation you can expect to infiltrate and store.
- ***Anti-Evaporation Measures*** - Plant mulch, rock mulch, shade, perennial grasses and trees, and windbreak. Converting evaporation from bare ground to evapotranspiration through plants can encourage a healthier upper water cycle.
- ***Species Selection -*** What type of ecology is your goal? Does that fit within your water budget? Using a climate analog, locate plants and animals that will work in that ecological model.
- ***Timing -*** Plan your planting and earthworks around the dry and wet seasons. Both plants and animals create disturbances that can be beneficial or detrimental depending on placement, timing, and management.
- ***Creating Microclimates -*** These smaller areas where more extreme climate features are mitigated can become stable microclimates that spread outward in fractal patterning to connect to other microclimate sites.
- ***Enabling Ecological Succession -*** Starting with physical infiltration of water and then following that with biological infiltration of water allows for the quickest and most stable ecological succession pathway.

Arid Soils

Decomposition is often suspended or seasonal in the drylands. The lack of moisture tends to preserve the nutrients in a dehydrated state—this leads to nutrient- and mineral-rich soils, but they often lack organic matter, moisture, soil life, and clays which are needed to make soil stable enough for plants to establish. Organic matter can compensate for any lack of clays especially if aided with aggressive pioneer species legumes that fix nitrogen, but use clay if possible because it accelerates the process and increases success rates.

Because rains usually occur annually or sometimes even less often, they are highly erosive events . The soils lack structure and root systems to hold them together, so they erode easily. Organic matter that has accumulated is dried out, so it is washed away easily.

Precipitation

During a few, brief weeks each year in summer or winter, the rains come, but drought in these regions can go on for years. Soils that dry for long periods of time can turn hydrophobic: they can repel moisture. For this reason, all deserts flood in large rain events. The severity of the floods is not just a factor of rainfall intensity, but also a factor of geography, plant density, and soil composition.

What needs to happen is the slowing, spreading, and soaking of the water into the landscape. People can encourage this through physical structures, such as earthworks, biological improvement (increases in soil carbon and soil life, as well as interception by trees and plant life), or chemical soil additives (though the last is not recommended). Precipitation can also be soaked in with large, flat areas that hold pacified flood waters for deep soakage.

For precipitation to occur in the desert, several things need to be overcome. The temperatures must be below the dew point. There must be cloud condensate nuclei (CCN). There must be moisture in the air. Desert dust heats up the air, reflects light at the height of clouds, and steals moisture—preventing temperatures from reaching the dew point. Trees are the key to reversing this trend; perennial grasses cannot go years without rain, but trees can. Trees provide shade and CCN, lower albedo, raise evaporation rates, raise infiltration rates, and increase water vapor density.[79]

> "We can afforest the whole west coast of Saudi Arabia, which is about 30 million acres [12 million hectares] in such a way that we're going to make it rain more frequently and at the same time create entirely new economies that don't exist now, agricultural economies that are actually going to increase water resources and sequester carbon and increase biodiversity and biomass."
> —Neal Spackman, Sustainable Design Masterclass 2016.

Drylands Earthworks

Incredibly long swales are often used like diversion drains in flooding events but still allow for soakage. They can concentrate precipitation into gardens, orchards, or homestead areas. Pits, gabion dams, and check dams can be used in conjunction with earthworks to slow waters down before they enter catchment. Any earthwork that allows the water to slow down removes silt from the water. Swales are soil-building and water-distribution tools in the drylands.

---

[79] Ibid.

XIV

Rock Check Dams at the Al Baydha Project in Saudi Arabia

Arranging stones for check dams, condensation traps, or even mosaics of stone-mulched pit-gardens, all can help to create niches for desert life. It doesn't take much moisture and windbreak for the desert to begin building soil.

Trees can go inside the swale bed, not on the berm only. These trees can help slow the movement of water though, in some instances, they can also be damaged by flooding. Use native pioneer species as much as possible to protect the valuable trees in your system. Microclimates are created with sheltering earthworks, native edges, sunken garden beds, and pits. Shade from the predicted canopy tree line defines maximum distance where the next swale or row will go—the more shade, the better; the higher the canopy creating the shade, the better, as well.

Rain catchment swales can go for kilometers in arid climates. The Al Baydha Project in Saudi Arabia maintains a water catchment to water usage ratio of 5:2 on their demo site, with plans for a 9:1 ratio on a 5000 hectare test site. They are putting more water into the ground than they are using to drip irrigate their trees—they are making a productive agricultural operation in an area considered unfit for agriculture while soaking in more water than they are taking out of their wells!

Qanats are irrigation systems that tap into the raised water tables of hills that are above agricultural areas. Using ingenious tunnels, they tunnel directly into the water table of the hill and form a channel to draw out the water. Wells uphill from the channel outlet can dip down and get water from the channel.

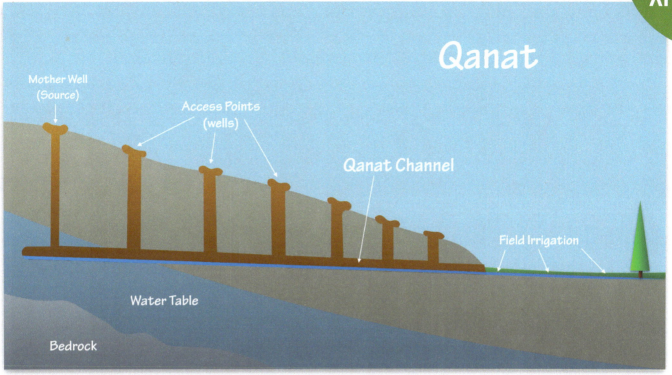

Dry Farming

Dry farming, once quite common, has become something inconceivable to many gardeners and farmers in arid regions. This is simply the practice of using whatever rainwater you receive for watering your system – instead of pumping from aquifer-fed wells – especially not from those that are thousands of feet deep, as is common today. Irrigating with solely the rain you receive in a year, or in some arid regions every year or few years, will determine plant choices and the length of the growing season. Very few places on the earth get regular rains throughout the growing season; the American midwest

Dry Farmed Huauzontle in Zone 8b in California, US. Note the planting density and the shade generated by the native wildflowers and grasses.

happens to be one of those regions. Farmers there can plant corn and never irrigate it; they actually fear it raining because their corn can get moldy! Dry farming in those areas is no great feat, but in dry regions, dry farming is very difficult. In desertifying areas, dry farming may be the only option for keeping farms profitable and ethical. At the rate we are draining the aquifers, we must be prepared to return to the idea of dry farming as a viable option for keeping farms and food production going.

Dryland Food Forests

With careful establishment, using shade and water resources wisely, and with careful monitoring during development, food forests in arid climates can be stable for thousands of years. World renowned permaculture educator, Geoff Lawton, has visited and documented a food forest in Morocco that is over three thousand years old. In Tucson, Arizona, the Civilian Conservation Corps' earth berms (mentioned earlier in the Earthworks chapter) harbor a vibrant ecosystem yet are surrounded by desert. Dr. Elaine Ingham has even created a dewfall palm ecosystem in Saudi Arabia that didn't need to be watered after six months of establishment. She also established a native Texas prairie landscape at the George W. Bush Jr. Presidential Library in Texas in just under a year using only compost and dewfall on soils that had been compacted for parking lots in many places. Stable oases in desert climates are

The award-winning Greening the Desert project site in Jordan developed by Geoff Lawton is a stark example of what is possible with food forestry in desert climates.

Geoff Lawton, 2016.

possible if the conditions are right and the site designed properly. Always time all plantings to coincide with the seasonal rains for best results.

Food forests in arid regions are characterized by high canopy shade, spiky legumes, and a noticeable lack of herbs or ground cover. It is primarily open closer to the ground. Organic matter is hard to come by, but mulch plants can provide consistent organic matter in place. Chop and drop happens after rain events. Dripline irrigation beneath 1/2-1m (1.5-3') of mulch prevents evaporation. Gravel filled, perforated pipes buried vertically next to trees for deep and sure watering can make the dripline system even more effective.

- ***High palm canopy*** - Date and wine palms
- ***Understory*** - Mulberry, tamarind, carob, moringa oleifera, moringa peregrina, zizyphus spinachristi, olive, fig, and apricot
- ***Legumes*** - Albizia lebbeck (palo verde), leucaena, acacia, prosopis (mesquite), parkinsonia aculeata
- ***Concave, water-wicking earthworks***
- ***Shade and Windbreak*** - *Causarina*
- ***Heavy mulch needed***
- ***Very few herbs***

Dryland Gardens

Dryland gardens are typically water-intensive and in need of organic matter—both of these are lacking in arid climates. A garden design must be strategic and creative for it to be successful in these regions. Sunken garden beds are used so water wicks downward into the planting area with paths raised; raised beds in this climate are heated by the air while sunken beds are cooled by the deeper soils surrounding them. Sunken beds can even be lined below the surface with plastic or naturally sealed to catch all irrigation and prevent it from leaving the system.

High shade is vital to block the midday sun and let in the milder morning and evening light. This is easily achieved with alley cropping—lanes of trees between perennial and annual garden beds. Ideally these trees are palms; they take up very little space on the ground while providing effective shade. The mulch from these alley crops help build soil and slow evaporation.

All gardens experience evaporation, but in the arid climates this evaporation is too precious to waste. With gardens directly around the home, cool moistened breezes flow into the home when garden moisture evaporates or plants transpire. Stone mulches can add to this effect by regularly condensing moisture onto their surfaces. Trellises in the gardens and around the home also provide shade and vertical growing space. These trellises can be created out of trees like mulberries with grape vines climbing them, or they can be permanent, constructed shade structures.

Summers are the time of dormancy for the garden in arid and hot climates, the opposite of cool and cold temperate climates. Unglazed clay pots or olla pots buried to their necks in the soil slowly weep water evenly. With a lid on top and insulated by garden soil, they let very little water evaporate out the top. Gardeners in almost any climate can use olla pots; they do not overwater since once they reach equilibrium with the moisture content within and without the pot, they stop leaking water. Olla pots can even be closed on top and connected by piping in a daisy chain to a water barrel uphill. The water pressure will keep all the olla pots full for an extended time period passively.

An Olla Pot

Wicking beds are used on the flat roofs commonly found in arid climates. Trellis and vines usually grow above these beds, further shading the house as well as the garden beds. Wicking beds are designed to conserve water by housing a reservoir of water below the soil in a gravel bed with an overflow spout just before the soil begins—this overflow prevents the soil from sitting in the water which would cause it to go anaerobic over time. Usually a pipe extends from down inside the gravel bed up to a few inches above the surface of the soil where water can easily be added to the reservoir. As the water evaporates in the reservoir, it keeps the soil above it moist. Some designs allow roots to enter the gravel areas while others do not—sometimes the roots find their way into the gravel area in spite of our plans!

Composting in drylands creates nutrient-rich humus. Though it requires water, it can be made in the ground to conserve moisture. Pigeons can

also be used in the garden to add nutrients and build soils. Pigeon lofts or pigeon pea trees can be placed nearby or above fruit trees. The bird manure and consistently falling leaf litter are excellent for the soil.

Salad mallow, chickpeas, corn, sorghum, amaranth, teff, and many well-known crops can be grown in arid climates. There are innumerably more crops that can be grown in arid climates, with much less effort than current commercial crops, though few people other than ethnobotanists or people living in traditional cultures know about them.

- Time planting to coincide with or precede seasonal rain
- A half acre (2000m^2) or less for field crops
- Sheltered inside a food forest's partial shade and windbreak
- Buried irrigation beneath plants

Dryland Animal Systems

Grazing animals like sheep can easily damage brittle, arid climates. Grasses are often too sparse to feed to grazers ethically. Browsers are preferable because they can be fed with trees and bushes. Carefully graze and browse with animals; brittle climates can be pushed into desertification easily. Many arid climate legume trees can tolerate regular browsing and have the added benefit of nitrogen fixation. Animals can be herded by people or fenced through contoured grazing alleys through the food forest in short cycles.

The Horticultural Solution

According to Joseph Simcox, the botanical explorer, instead of growing mediterranean or even temperate crops in the arid climates, "we should start growing, selecting and improving the myriad of food-producing desert plants that humans have already relied upon for sustenance the world over." Ethnobotanists travel the world finding, documenting, and sharing seeds from the rarest and most unique plants. There is no reason to genetically modify drought tolerance in crops when edible plants already exist that thrive on a fraction of the water that commercial crops need. For example, nearly all cacti are edible if not palatable; cactus orchards could be grown for their fruit or used as a vegetable in the near future.

In all climates it is wise to research native plants and animals for their potential for yielding food, fiber, fuel, tools, and medicine. These can form your outer edges, your microclimate support species, or your dry garden. With the work of ethnobotanists and organizations like Baker Creek Heirloom Seeds, the Seed Savers Exchange, and Seeds of

Preservation Independence, rare, historically significant, and extremely versatile plants are moving out of a place of myth and returning to backyards and farms all over the world. Studying their work can greatly expand our understanding of what is possible in our climate.

Desert Aquaculture

Though seemingly out of place in the dry lands, aquaculture is still possible. All dams and cisterns are designed to be deep and narrow. Tanks sunk in the ground with small, fully-shaded entries are perfect fish and frog habitat and can provide water for birds. They can even capture runoff from a hardscape and simply provide wildlife habitat.

Dryland Housing

People in arid climates tend to live in *wadis* or canyons where there is already natural shade. Roads in towns and cities that run east/west are narrow as are the spaces between buildings to avoid the sun, and roads running north/south are typically wider because they experience only a short time period of direct sun as the buildings shade them on either side. The taller the buildings, the longer the north/south roads are shaded. By clustering together and maintaining their shade, they create a cooling thermal mass.

Buildings tend to have flat roofs that double as a summertime sleeping area, since it's so hot indoors, and as a garden space with a trellis fully shading it. Evergreens and palms are used to shade the roof and home completely. The western sides of buildings need to be heavily shaded with no windows; there can be lighter shading on all other sides. Commonly buildings are underground or earth-sheltered to thermal bank the cool and insulate against the heat. Matte white paint is used to reflect heat without focusing it as glossy paint would.

Covered windows also shade and shutter out the heat while wind chimneys draw air in through buried pipes. Air should be drawn in from at least 20m away around 1m deep, so it is cooled by the time it enters the home. The mouth to the pipe can be shaded heavily, and evaporative cooling systems can be used to enhance the cooling effect. Damp charcoal trays or unglazed pottery filled with water can be used to slowly release their moisture into the moving air, cooling it further. This works like a passive air conditioning system. Drawing air in from deep wells can have the same effect. Curtains are also used to push airflow lower, cooling it, while insulating and shading rooms inside the home. Outdoor summer kitchens are vital as the heat a kitchen would generate inside the home would be unbearable. They are shaded by garden trellis, and they ideally share a wall with the inside winter kitchen to conserve energy moving everything from kitchen to kitchen seasonally.

- **Dry compost toilets** - They work very quickly
- **Graywater reed bed**s - This feeds into the garden
- **Solar energy primarily** - With some wind

Final Note on Climate

Some of us may find that we are in a combination of two climates such as a mediterranean winter with an arid climate summer, but the lens of climate still works in terms of what we can observe and what strategies we can apply. In fact being an edge climate can double or even triple the options you have.

Climate Chart

	Temperate	Tropical	Arid
Climate	Long, Cold Winters Mostly Humid Cold-Cool-Mediterranean-SubTropical	Summer Rains Drier Winters High Humidity when not a Desert Wet-Wet/Dry-Monsoon	Long, hot Summers Winter Rains Found in any Climate
Soils	Mostly Deep, Rich Soils Lots of Organic Matter Non-Brittle Most Productive	Unless Volcanic, Nutrient-Poor Soils Thin layer of Topsoil Mostly Brittle	High in Nutrients Low in Organic Matter Usually lacking Clay Little to no Soil Microbiology Brittle
Food Forests	Hardwood Overstory Extra Herbaceous Layer Seasonal Work Berries & Roots Seasonal Chop & Drop	Palm, Nut, & Large Fruit Overstory Double Palm Layer Constantly Growing Giant Fruits & Herbs Constant Chop & Drop	Palm Overstory Work when it Rains Extensive Earthworks and Soil Building Few Herbs & Many Seed crops Selective Chop & Drop
Gardens	Primarily Spring/Summer/Fall Growing Season Harvest before Winter Easy Composting over Winter Raised Beds	Year-Round unless too Rainy or Dry Harvest Before the Rains Composting is Difficult Raised Beds	Winter Growing Season Plant when it Rains Harvest before Summer Composting is Difficult Concave Beds Wicking Beds
Field Crops	Light Tilling Possible 1-2 acre (4000-8000m²) size fields without trees Hedgerows	No Tilling 1-2 acre (4000-8000m²) size fields with trees Agroforestry	No Tilling 1/2 an acre (2000m²) or less Sheltered inside a Food Forest or within a large Windbreak Buried Irrigation
Housing	Along the Sun path Earth-Sheltered Thick Walls Heated in Winter	Oriented to the Cross Winds Above Ground, even Raised Outdoor Kitchen (or 2 kitchens) Thin Walls Screening or Netting Special Attention to Safe Toilets & Drinking Water	Underground, Shaded Thick Walls Rooftop Gardens Dry Toilets
Aquaculture	Systems have to be Large Enough for the Biology to Survive the Winter Chinampas & Canals	Ponds can be Ubiquitous Wet Terraces Chinampas & Canals	Minimal, Sheltered, & Underground in Narrow, Deep Cisterns Mostly for Wildlife
Animal Systems	Ideal for Grazing The Cold & Composting provide Sanitation Silvopasture & Open Pasture Nearly all Animals	Little if any Grazing Sanitation Concerns Silvopasture Chicken, Pig, & Duck (other animals can be used as well)	Careful Minimal Grazing Weathering & Dryness provide Sanitation Silvopasture Camels, Sheep, & Goats (other animals can be used as well)

XV. Aquaculture

The only ethical response to the collapse of the ocean, lake, and river ecosystems across the globe is to stop eating wild-caught aquatic plants and animals and start raising our own in clean environments while we restore the wild habitats. It will be far healthier for our bodies and better for the environment. Almost all the animals and fish we take out of the current ocean and river systems are tainted with toxins. They need to finish their life cycle, decompose or be eaten. As for the mercury, atrazine, etc., they need to be returned to a place where they can do no harm, ideally trapped in the long carbon chains of ocean, waterway, and wetland soils. All the wild fish populations must be encouraged to exponentially increase either through seaweed farming or other creative means, and fisheries need to refrain from harvesting any fish for several seasons. The cycling will clean the oceans, the coastlines, and the wetlands along with all the organisms dependent upon them as well.

Orientation and placement

If ponds are placed against the prevailing wind, allowing it to blow over them, they get passive aeration naturally. This also has a cooling and evaporative effect. Placement so that the sun shines on the pond for long periods of time each day turns the pond both into a reflector and magnifier of that energy as well as a thermal mass that will even keep frost off of the trees planted around it. It should be noted that oxygen levels drop as the temperature rises, so shade is often important to keep waters within tolerable temperatures during summer and spring, especially in hotter climates.

Size and Depth

Whether small or large, pond aquatic systems that are less wild (and thus less self-sustaining) need a lot of tending and inputs, and they tend to have more problems. The wilder they can be, the better. Pond depths can range: some people recommend they be 2m (6.5ft) or shallower, with deeper areas of 4-5m (13-16ft) to retain fish if the pond has to be drained or for cold temperate climates to help fish survive through the winter.[80] Sepp Holzer, on the other hand, commonly creates ponds, or lakes, 10-15m (32-49ft) deep.[81] Deep areas allow for certain fish to spawn, protection in winter, and higher oxygen levels. Shallow areas allow for spawning of other fish, lower oxygen areas, and temperature and pH differentiation. Shallow and deep zones also support different species guilds. Overall differences in the depth allows for constant movement of the water as well, which keeps it fertile and oxygenated.

Yield

Aquaculture has the highest possible yields per unit of surface area because gravity isn't pulling on the organisms in the water in the same way that it exerts this force on organisms out of water, and organisms are surrounded by nourishing water and nutrients similar to mycelium in a liquid culture. The ponds act as a nutrient trap, capturing silts and all organic matter that breaks down in the water, forming a rich sludge that can be removed and used in gardens.

> "A properly built and managed pond can yield from 100 to 300 pounds [45-136 Kg] of fish annually for each acre [4000m²] of water surface"
>
> –<u>Ponds: Planning, Design, Construction</u>. NRCS, 1997.

Plant Layers

- **Floating plants**
- **Edge plants**
- **Shallow-water plants**

---

[80] Lawton, Geoff. *The Geoff Lawton Online Permaculture Design Course.* 2014.
[81] Holzer, Sepp. <u>Desert or Paradise.</u> 2012. p. 77.

- **Deep-water plants** - though no deeper than 8 ft/2.4m

Trophic Layers

- **Algae**
- **Zooplankton**
- **Plants** - Chinese water chestnut, kangkong, taro, cattails, water hyacinth, Indian water chestnut, lotus, arrowhead
- **Shellfish, crustaceans, mollusks, and echinoderms**
- **Fish** - Tilapia, Catfish, Bluegill, Bass, Carp, Trout, Perch
- **Small to large water mammals, reptiles, and amphibians**

Food Chain

Algae feeds on sunlight and nitrates in the water. Zooplankton, a diverse population of microscopic animals, feeds on the algae. Crustaceans, macroarthropods like crayfish, crabs, and shrimp, feed on zooplankton. Fish feed on plants and crustaceans. Animals like birds and amphibians feed on plants, fish, and crustaceans.

pH

Ideally large systems should have a range of 6-8 pH. Fish, mollusks, freshwater crayfish, and shrimp prefer hard alkaline water while food organisms that feed into the aquaculture food chain need more acidic conditions, so areas of different pHs are desirable. This can be achieved with deep areas, shallow areas, still areas, and actively aerated areas. Finding the balance for your system between all those variables can take some tinkering and will in the end be unique to every situation.

The Nitrogen Cycle

Distinct from the soil nitrogen cycle, the aquatic nitrogen cycle is equally important. Plants, fish, and microbiology are all interdependent as they transform nitrogen into different molecular arrangements. Oxygen concentrations determine the types of nitrogen available much as pH does in soil. In an aquaculture setting it can be easily managed and controlled by adding plenty of plants and monitoring fish densities.

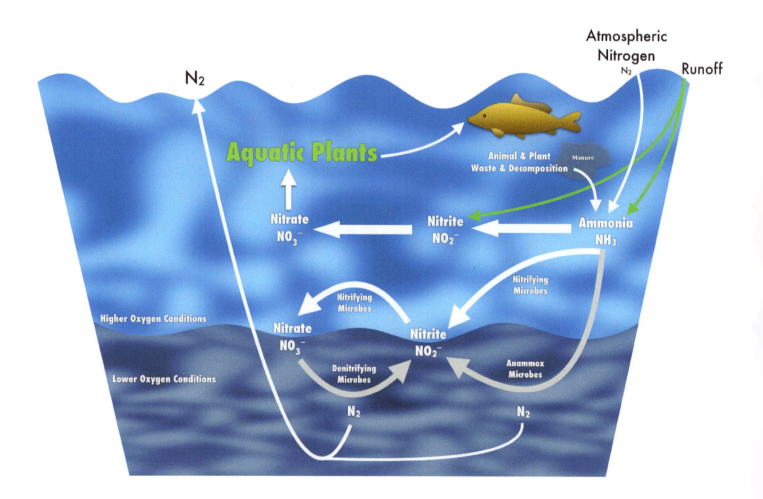

Stocking

Some fish can be stocked one per gallon (3.7 liter) and some one per cubic cm or inch. Knowing your fish will usually tell you how much you can fit into your system at maximum capacity. If fish breed, they tend to overcrowd quickly. Be ready with predator fish, perhaps in a caged area in reserve to help keep populations within oxygen levels. Predatory fish that are roughly the same size as their prey fish will not feed on them. A rule of thumb is that the predator fish need to be at least three times as big as their prey for them to feed on them.

The maximum stocking rate is a limit – not the ideal – nor does it take into account all systems, each of which is unique–keep that in mind as you observe and learn from your aquaculture systems.

The Pond's Edge

The more edge, the better–especially in pond design because it provides more areas for plants and sheltered habitats. A wavy meandering pond edge has nooks and crannies for frogs and other animals to hide from predators, to lay eggs in, to provide more organic

matter input, to provide more food for fish, and to allow for different pHs to develop for more complete cycling of pond waste and nutrients.

Fertilizer and Soil

Rich and abundant nutrient accumulation occurs in ponds and all aquaculture systems over time as organic matter and silts settle at the base of the pond or tank. When our systems become choked with plants or filled with sludge, we can harvest the plants and the sludge and aerobically compost them, adding the finished product to our garden or food forest soils. In turn, ponds can be fertilized themselves with inputs from land, such as animal and bird manures. This fertilizer causes a flush of growth which is harvested before it occupies too much of the system and chokes off the oxygen levels—this can be composted or used as direct mulch. Mussels can also be used to clean and add phosphate to the water, and subsequently to enrich the sludge and plant growth in the pond. Fertilizing a pond and using a pond to generate fertilizer is easy!

Aeration

Water aeration can occur in a number of ways: oxygenated water input, water pumps, falling water, or orientation to the prevailing wind.

Always have a way to aerate your water. 1ppm oxygen is too low for almost all fish while 5ppm oxygen is a good minimum for overall pond health. On warm summer nights,

heavily vegetated ponds may experience a dip in the oxygen level. Having an low-oxygen-detection system and automated aeration system may be desirable.

It is always best to aerate and filter water through gravel and plants as it enters a system. Gravels and other stone filters develop bacteria that digest nitrites in the water as it passes through.

More oxygen = more life
Warmer water = less oxygen.
More oxygen = more fish
More fish = less oxygen.

Cage area

Having fish in a caged area allows food and natural elements to grow and flow through their area without letting those fish free to be eaten or eat other fish. It is a great way to control population growth, protect fish, and protect plants. This can also be useful to grow smaller fish or organisms to feed to larger fish, yourself, or other animals in another system.

Restricted area

Having a restricted area for filtering plants, smaller fish, and delicate habitats to remain protected from larger fish protects the entire ecology's health. This can be done with screens or even a simple rock wall that doesn't allow those larger fish to pass through.

Pipe releases

Warm surface water in summer as well as cold water from the bottom of the pond in winter can both negatively affect pond life—both can be siphoned off a pond with pipes and valves. This can help keep temperatures tolerable for pond life during the extremes of the year. A pond can even be drained quickly with a valve just above

Geoff Lawton using a pipe release on a pond at Zaytuna Farm.

the base of the pond.

Fish Feed

Fish can be fed using the plants in the ponds, plants on the edges, crustaceans in the pond, and insectary plants like mulberries hanging over the pond to drop insect-covered fruit to the waiting fish. Sweet potato and black soldier fly larvae together make great fish, duck, and chicken food as is done on Zaytuna Farm[82]. Duckweed and other floating plants will feed fish as well. Fish can be fed other fish using screens, cages, or through multiple ponds with different stocking variations.

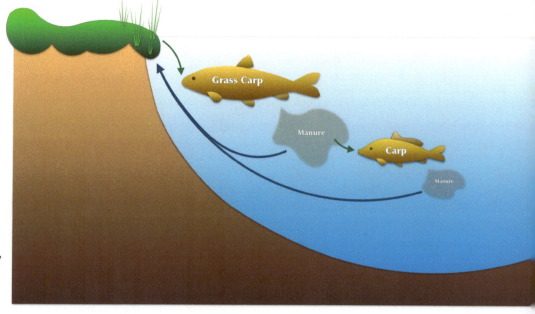

Grass Carp feed on vegetation along the edges of ponds while other carp feed on their manures, and plants feed on the manures of fish in general.

Different forms of waste attract different forms of fish food. Worms are attracted by kitchen scraps, ants by bones, and grasshoppers by the color yellow. Termites are drawn in by woody biomass, snails by moist areas of the garden, and even cockroaches by rough mulch. We can grow our own fish food and save money and energy.

Even a simple light over the pond at night can attract mosquitos for fish to feed on. Placing rocks beneath the light allows for smaller fish to exclusively harvest the mosquitos.[83] This is a win-win for farmers and fish! Rabbits and other animals can be housed over fish ponds as well to provide passive feeding of the fish. Often this can only be done temporarily as too much manure can imbalance a pond.

Temperatures and Salinity

Every species has a range of temperature and salinity that it prefers and usually a slightly wider range that the species can tolerate. There are species that prefer freshwater, some that prefer saltwater, and some that prefer a brackish mix. These conditions exist in all climates. All designs must take these two factors into account for any diverse, self-managed system to emerge.

---

[82] Lawton, Geoff. *The Geoff Lawton Online Permaculture Design Course*. 2014.
[83] Holzer, Sepp. _Desert or Paradise._ 2012. p. 84.

We can choose the most resilient fish while making the highest quality microclimate for our site—this increases our margin for success and reduces the margin for error, but in the end it is trial and error, with observation and reflection, that will ultimately guide any successful system. Start with researching the systems of your area, the fish that thrive there, their needs, and what you can mimic on your site.

Islands, Decks, and Floating Gardens

Floating islands with plant species thriving on them are great to feed fish with since they can be edible vegetation in the middle of the pond, adding even more edge. A traditional bamboo or PVC piping-framed raft will hold the mulch, soil, and plants just above the water's surface.

Lake Inlé in Myanmar still boasts a vibrant aquaculture system to this day that supports the Intha people with garden vegetables grown on floating gardens. While practiced for thousands of years, the recent introduction of the water hyacinth with its buoyancy and size has greatly expanded their ability to build floating gardens.[84] Compost is placed on top of the water hyacinth leaves and then covered with dried peanut branches. The peanut branches provide shade and reflect light with their high albedo while they wick moisture up from the water up onto and over the compost where the vegetables, mostly tomatoes for Rangoon, are planted directly in the compost (as pictured on the previous page).[85]

In areas with milder temperate winters, decks can be made for duck houses to give them easy access to water and us easy access to eggs and sleeping ducks with no access for predators. No doors are needed on the water side because predators will not enter the ice-cold water. To avoid predators visiting in winter by walking over ice, position the duck deck so the opening is just beyond where the water enters the system. It will not freeze there because the water is

Peanut branches cover the compost-rich soils growing tomatoes in floating rows.

---

[84] Mainguy, Pierre. *Floating Gardens of SE Asia*. 2016.
[85] Ibid.

Setting up floating rows at Phnom Penh in Cambodia.

constantly moving.

Physical islands can be created when ponds are initially built for ducks or other water fowl, so that they are housed safe from foxes or other predators. Strategically and securely placed boulders or stones a stride's length apart can make an invisible path to the island for people but not predators.

Rice

A temperate and tropical staple food, rice is an aquatic plant eaten all over the world. North American wild rice is a tall annual grass that was grown by Native Americans from Florida though Canada. In Asia, rice paddies maintained for centuries host several different rich polycultural arrangements of plants and animals, often ducks, tilapia, and crayfish. There are issues with fields that go anaerobic releasing methane gas; Masanobu Fukuoka's method of strategically and temporarily flooding fields is now being used in a variety of ways and places (including Asia) to adapt these wet systems to minimize anaerobic tendencies of permanent paddies. (Read more on Masanobu Fukuoka in the Permaculture in Action chapter). Commercially-available rice is high in arsenic likely due to arsenic in the fertilizers being used or already being present in the water, runoff, precipitation, or soil—the rice is cleaning the water of arsenic but unfortunately is passing the arsenic on to us when we consume it. It is not recommended any longer to consume commercially available rice daily because the arsenic levels are too high. If you grow rice in a system, test it for toxins it may be sequestering, and know for certain if your rice is arsenic-free.

Rice and fish farming go hand in hand in Asia especially in China's rice paddy terrace systems which are in some cases over a thousand years old. The fish manure feeds the native rice plants and the native carp feed on native plants leading to less pesticide and fertilizer usage, and many argue that no pesticides and fertilizers are needed.

Natural Swimming Pool

Natural swimming pools are pools or ponds that use biofilters to clean the water. There are numerous do-it-yourself models online as well as professional services for natural swimming pool installations. Hot tubs are being filled with plants and pebble beds to filter the water for the main pool area. Aeration is needed, but can be attained in several ways—especially if converting a standard, in-the-ground pool. Without a conversion to make, the biofiltration area can be directly next to the swimming area, so the entire pool looks natural. The pool water can stack functions as a place to raise fish and store water for irrigation, human consumption, and fire fighting.

Chinampas, Channels, and Canals

Documented by the invading Spaniards, *chinampas* are plant-growing systems originally used by the Aztec people. They are used in wetlands or in areas with high water tables. The soil is dug out and piled up above the water line until there is a channel of water and a strip of land. You can continue this and transform an entire area into small channels and

Some *chinampas* still remain today in Xochimilco, Mexico.

raised beds. The anaerobic soils from below the water are very rich but take some time to become aerobic. Once they do, they can grow rich gardens that can be harvested by paddle boat or canoe. It's an effective solution to a high water table.

Channels are narrow bodies of water that connect two larger bodies of water; they are an edge effect multiplier. Channels connect separate aquatic habitats as well as connecting water to areas previously separated from water. This exponentially increases interactions between plant and animals species.

Canals are large diversion drains for irrigation as well as narrow bodies of water that connect boats to larger bodies of water. Often in ancient cities, canals with a steady flow served as their municipal water and sanitation services: water coming in was clean water, water going out was dirty but soon gone. This seemed to work until populations grew too large or some other complication would reveal how untenable that practice was.

Veta La Palma - Spain

Once drained for raising cattle, Veta La Palma returned the waters, restored the wetlands, and is now a fish farm that is 27,000 acres (11,000 hectares) of marshland and canals.[86] It is a biologically-rich ecosystem that is nearly all edge. It is one of the largest private bird sanctuaries in Europe as well as one of the few fish farms that do not feed their fish. The shallow, sun-drenched canals frequently host algae blooms which leads to vibrant, healthy, and abundant shrimp and fish populations which in turn attracts incredible numbers of birds in great diversity. This is all happening as the farmers are harvesting fish to sell, yet they do not feel like they are in competition with the birds. The health

and abundance of the birds themselves indicates the health of their ecosystem and the fish they are harvesting, and the best part is: there is always enough fish for all.

Their system is often referred to as Algae-Culture which is similar to Joel Salatin's Grass-Farming. By focusing on the photosynthetic point of contact as the support of their enterprise, they've organized systems that are both regenerative and sustainable for as long as the sun shines and the seasons turn.

Learn more about Veta La Palma here: VetaLaPalma.es

[86] Browne Trading Company. *Veta La Palma Seafood*. 2016.

Hawaiian Aquaculture

Ancient Hawaiian aquaculture was an incredible system that started catchment further inland and upslope using swales, that often became *chinampas*, all the way down to the shore where artificial lagoons held fish for the Hawaiian chieftains. The *chinampas*/swales would catch, slow water, and overflow into each other down the slope, and fish, along with plants like taro, would be cultivated in them. Their shoreline aquaculture was most impressive—they used stones to create artificial lagoons that allowed small fish in but not larger predator fish. As the small fish grew in size in the safety of the lagoon, they were no longer able to leave the artificial lagoon and were then harvested when needed. Today anyone who visits the fishponds of Molokai can see these artificial lagoons for themselves. This design and concept can be locally replicated anywhere to create sheltered habitat.

Riparian and River Restoration

The restoration of riparian areas and rivers have not been given the same amount of attention as backyard gardening in permaculture, but it is imperative that we rewild the rivers, lakes, streams, marshlands, and all wetlands. Many historically enduring rivers and lakes have dried up or are rapidly drying up. Peatlands, which account for an enormous amount of sequestered carbon, are being mined, burned, and drained—all of which releases carbon into the atmosphere. To bring these ecologies back into balance, we have to foster and accelerate the natural processes.

Bill Zeedky's Five Ways to Build Resilience in Riparian Areas[87]

- ***Reconnect stream with the historic stream channel segments*** - access to floodplain increases infiltration and bank storage.
- ***Remove "floodplain clutter"***[88] - improve stream access to land surfaces [by removing things like old roads, earthworks, culverts, and anything else that impedes flow]
- ***Raise channel bed elevation over the length of the stream reach to access floodplain*** - apply induced meandering concepts.
- ***Reestablish dispersed flow across alluvial fan surfaces*** - enhances water infiltration and storage.
- ***Prevent gully proliferation*** - save the floodplain/alluvial fan

---

[87] Zeedyk, Bill. *2013 Quivira Conference, Bill Zeedyk*. 2013.
[88] Ibid.

Floodplains

While this concept readily brings to mind the fan of water spilling over the banks of the river into the lower valley area, there's also another floodplain in the soil. Consisting of larger soil particles like gravel and sand, floodplains allow water to move through the landscape below the topsoil. It is the flood's own water table. It is above the impermeable layers of silt and clay and can travel across large areas.

Alluvial Fans

When runoff from an erosive, steep slope comes into contact with a flatter area, the change in speed causing the water to drop the silts and clays it was carrying. Over time this creates a great fan of sediment building up into a rounded mound that holds water and leaks it often as a perennial spring. Cutting or damaging these alluvial fans creates erosion, poor pasture, aerobic water, and displaces the natural movement of water.

Gullies and Head Cuts

A gully happens when a channel of water has incised deep into the landscape, well below the floodplain. Water fails to spread out via the floodplain during flood events, making

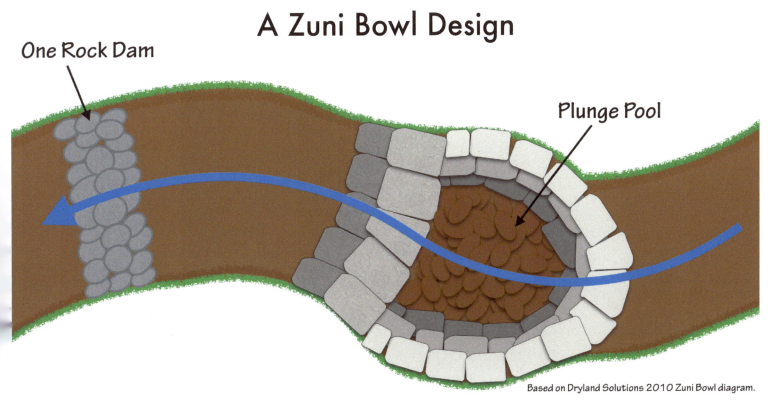

Based on Dryland Solutions 2010 Zuni Bowl diagram.

the area drier, which in turn leads to more erosion. Gullies form along the fastest, easiest path for water to travel downhill, which causes increasingly more erosion over time.

Head cuts are erosion features in the land that look like small cliffs or bluffs. They cause a score below them that erodes the land and can lead to channel incision. A head cut can be restored with techniques such as a Zuni rock bowl which features a stone-covered plunge pool to disperse the water's force without causing erosion (see diagram on the previous page) which repairs the cut and prevents further erosion. Following that is the lip of the zuni bowl that raises the water level back up while still allowing the drop to occur. There are numerous techniques using stones, logs, geotextile, sod, and even heavy machinery to prevent erosion or restore an area damaged by erosion.

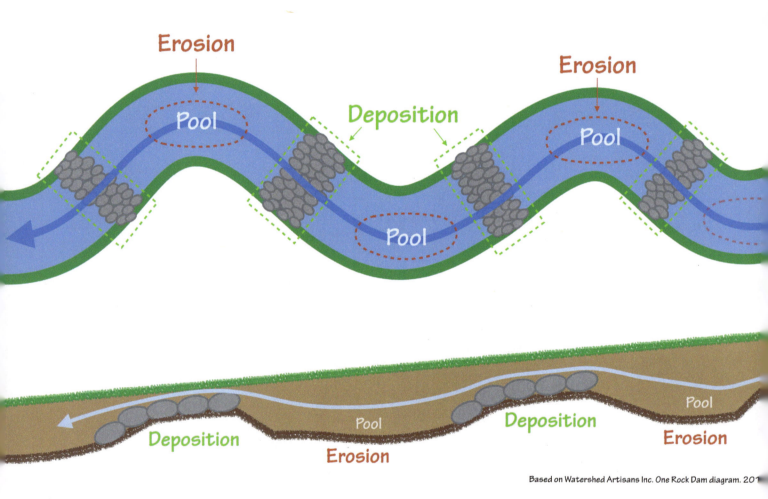

Based on Watershed Artisans Inc. One Rock Dam diagram. 201

One Rock Dams

A one rock dam is as simple as it sounds: it is a dam that is only one rock high, but it can have enormous effects. By positioning these dams in the natural deposition area, it enhances and speeds up this natural effect and raises the water level in the landscape uphill and upstream. The elbows of the channel always are areas of erosion where pooling develops. The rock dam should extend up the walls of the channel, so they prevent overflow from scouring away the vegetation on the banks.

Induce Meandering Concepts

The more meander there is to a channel of water, the more flood control it naturally has. The back and forth also develops shallow and deeper areas which increases biodiversity. Straight channels tend to incise into the subsoils, disconnecting the flood waters from the flood plain. To slow, soak, and spread the water for long-term storage, rivers and streams must be able to meander. Using baffles, erosion can be encouraged with weirs or rock dams between the baffles to imitate the natural deposition area. Over time the channel will shift and begin to meander. Extending the rock dams as the water slowly shifts will speed up restoration.

Bill Zeedyk's _Let Water Do The Work_ is a superb resource for more detailed information on riparian and wetland restoration methodologies and techniques. Craig Sponholtz of Watershed Artisans is another excellent resource.

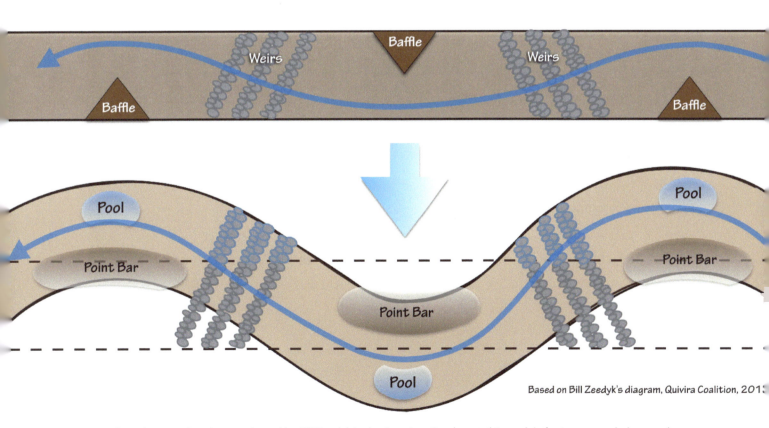

Based an actual project conducted by Bill Zeedyk in the American Southwest, this model of using one rock dams and baffles to induce meandering and erosion purposefully will restore the natural shape of the channel. As the channel shifts, new rocks need to be added to the dams. Read more about Bill's work in his book: _Let Water Do The Work_ (2009).

Urbanized Stream ReWilding

Often streams, rivers, and creeks are channelized or straightened and lined with concrete. This prevents the water from infiltrating into the ground and prevents vegetation from growing around the channel. It also speeds up the flow rate and prevents settling of sediments. This unnatural design prevents sinuous curves that slow water and lacks a flood plain. Creating a 4-level floodplain system around the channel is critical: 1 - base or low-flow

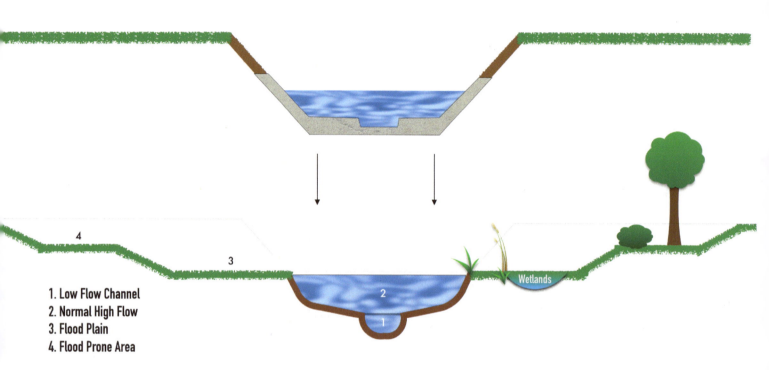

1. Low Flow Channel
2. Normal High Flow
3. Flood Plain
4. Flood Prone Area

Based on the work of Dave Rosgen Phd River Hydrologist. Matt Powers, 2016.

channel, 2 - normal high-flow channel, 3 - floodplain, and 4 - flood-prone areas. The floodplain can be planted with riparian species and a wetland can even be dug out—water will seep through the ground into that area naturally or be filled by rain or flood event. The flood-prone area can be planted with trees, shrubs, and other selected perennials adapted to an area frequented by flooding. Excavation opens up more area lower in the land profile to accommodate more water, making flooding events more manageable and safe. It also increases wildlife and deep infiltration of water. The sinuous meander of a natural stream is needed to control flood waters, slow and infiltrate water, and to support the development of biological systems which clean, slow the water more, and grow more life.

Large-Scale Ocean Repair

Rarely addressed in permaculture are the oceans and seas, where approximately half our excess atmospheric carbon has been absorbed – which has acidified the waters – and changed living conditions fundamentally for all marine life. Calcium-carbonates, the structures of coral and the shells of shellfish, are dissolved in carbonic acids which are formed when carbon dioxide is absorbed into water. There are cases of soft, nearly translucent shells and deformed shells on shellfish, coral bleaching, and reef die-offs—all linked directly to the rising acidity. (It is important to note that the ocean water has not become acidic, just less alkaline.) Without suitable habitat, these keystone species, and all the trophic levels dependent upon them, could easily go extinct; they rely upon systems we cannot even hope to replicate in a man-made setting like an aquarium. Whatever changes are needed to

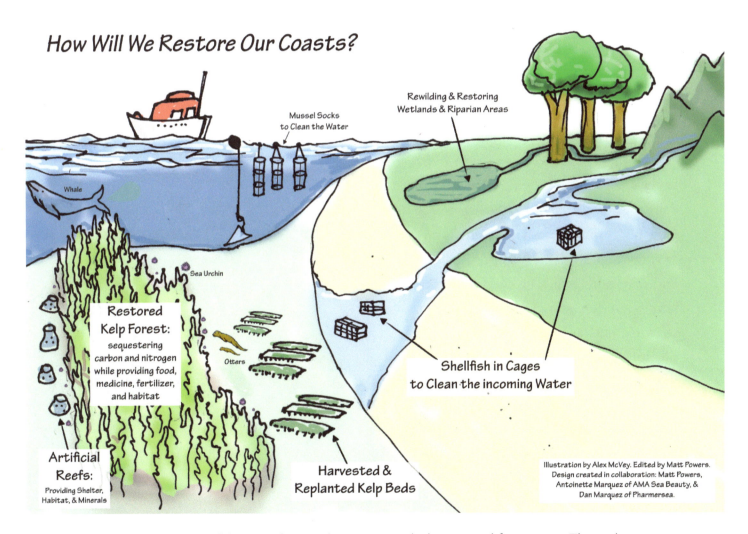

improve the oceans will have to be made on-site with the ocean life present. The only regenerative solution is one that sources nature itself. With seven times the volume of fish living along our coastlines (per cubic meter) compared to the rest of the ocean's open water, it is not outside our ability to affect the situation; in fact, it is well within our reach. There are several strategies already in development.

Keys to Ocean Restoration

- Clean and Filter Water as It Approaches and Enters the Oceans Using Shellfish
- Clean and Filter the Coastal Waters with Kelp and other Biology
- Farm Seaweed to Sequester Carbon, Nitrogen, and Other Excess Nutrients in the Water (drones can be used to know exactly when to harvest)
- Farm Shellfish in the Coastal Waters to Sequester Carbon, Nitrogen, and Other Nutrients as They Clean the Water
- Restore the Kelp Forests (recent warming killed over 90% of US west coast kelp)
- Build Artificial Reefs (reefs provide shelter and act like underwater windbreak as do kelp)
- Return the Harvested Kelp and Shellfish to the Soil to Return the Lost Nutrients and Sequester Carbon

Vertical Farming

Thimble Island Ocean farm in the Long Island Sound is doing what Veta La Palma is doing onshore, only they are doing it offshore. The Long Island Sound is an inlet of water that borders New York State, New York City, and Connecticut. Using vertical farming underwater, they are growing seaweed and shellfish. Nitrogen, phosphorous, and carbon found in excess in the water are consumed by the marine life. The shellfish and seaweed filter the water as they develop. The shellfish waste is rich in nutrients and feeds into the Long Island Sound's food web and settles into the soil at the Sound's bottom. The shellfish and seaweed are sold to restaurants and to people in the surrounding area, which

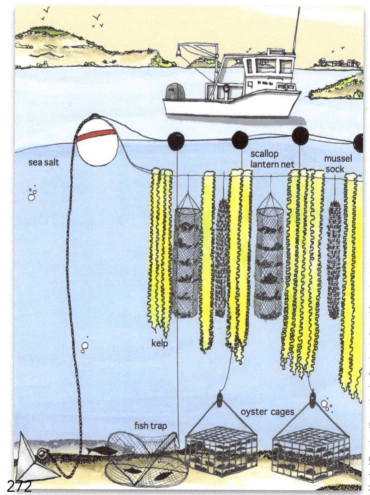

Vertical Farming illustration by Stephanie Stroud from GreenWave.org

removes the excess nutrients from the waters. Some of the seaweed is sold as a soil amendment as well—trapping the carbon, nitrogen, and phosphorous in the soil.

To fully restore the Long Island Sound and similar bodies of water, this vertical farming model must be adapted to be more regenerative and adopted widely. We can sequester the excess nutrients in the plants and animals, and in turn their waste and decomposition will sequester these excess nutrients and toxins into aquatic and wetland soils. Limiting fish takes, fishing moratoria, reducing by-catch (the accidental catching of fish other than the target fish), and protecting nursery and spawning habitats are all pieces of a greater picture of large-scale ocean repair that will change with time, but it starts with biological regeneration, sequestration of carbon and excess nutrients in aquatic soils, and abstaining from seafood consumption.

Artificial Reefs

Though not a new concept, the creation of artificial reefs needs to become more sophisticated. The basic idea behind an artificial reef is that a habitat, such as a decommissioned NYC subway car or warship, is provided for ocean life to inhabit. These man-made structures work much like a windbreak of trees does: they provide shelter in which a microclimate can develop while netting passing nutrients and organic matter. Any foothold in nature and life will take hold, but understanding flow, patterns, keystone species, their habitat, and how to install them is critical for these areas to be successful. We need an explosion of marine life to take all the excess carbon back into the biological cycles and sequester it in the soils.

We need to create microclimates that favor a biological solution to ocean repair. A large scale effort needs to be launched to create structures that are seeded with coral and the minerals for them to feed upon. On a small-scale, reef buds or Biri Buds are doing just that. Developed by the BIRI initiative, they are heavy, porous shelters for fish and marine life to develop on and within. We need

A Biri Bud is the seed of a future reef.

to create artificial reefs everywhere we can with innovations like the BIRI Bud that give the biology extra support. We also need to regenerate and protect the already bleaching reefs if we can.

Coastal Microclimates

Looking closely at Veta La Palma's example we can extrapolate certain principles: shallow waters in sunlight create algae blooms, algae is the photosynthetic point of contact for the wetland food web, and if the growth of the foundational species of an ecosystem like algae or grass is improved, the health of the entire ecosystem will improve. If we draw sea water inland in shallow channels, we can recreate the natural wetlands that used to dominate our coasts; these areas act as filters and carbon sinks.

Aeration or maintenance of oxygen levels through water movement is a critical part of the planning. Tides can serve to aerate the systems if planned properly. While many homes and much development has been focused on the beaches and shorelines, these areas are better served and serve us better when they are put back into natural landscapes that filter water, sequester carbon and toxins, provide wildlife habitat, and grow food for people and the ecosystem itself.

The rivers and the streams that lead to the ocean also need to act as filters and sinks; they cannot be clad in concrete, barren, drawn-down, or consumed. Instead they need to be vibrant, biologically-rich, and sequestering carbon. The shores are the anchors, the nurseries, the sanctuaries, and the mating grounds for most marine life.

Using the creation of artificial reefs to shelter the coastlines, with vertical farming inside and outside the reefs (to encourage and support the new ecosystems), we can create a nesting of microclimates. All of it can be allowed to become more and more abundant each season which will speed up cycling and create increasingly more diverse niches. If we allow and monitor predator populations in these areas, we can start to recreate natural systems and learn from them. Later on, we can use the information we've gathered to make our shelters predator-proof as the Hawaiians did, or perhaps we can design it to allow controlled entry to predators when appropriate to the season and status of the system.

Without predator/prey relationships, we cannot have a normally functioning ecosystem. Once we have an abundance of soils, healthy marine life, and clean waters, we can carefully begin to take the place of natural predators though we should always include them in our final systems. Careful harvesting and testing should always be partnered with holistic management of the site. Natural predation ends up feeding many different layers of the maritime food web, so we can never rule it out completely. The same philosophy should be adopted in wild landscapes everywhere as well. Predators are keystone species in all their habitats. Rewilding the shores is how we will save them.

Pollution Clean Up

Plastics ranging in size from bottles to tiny fragments are killing marine life and choking our waters, so much so that the gyrations of all the ocean are collecting islands of trash in their centers—the Great Pacific Garbage Patch being one of these. These plastics need to be removed and decomposed using fungi or cleanly burned. Oil spills can also be treated with fungi though it's easier to do on land than in place. Absorbent organic materials like straw and hay are used to soak up oil and then brought unto land for composting. Solar powered skimmers designed to gather plastic from the ocean surface are being devised and tested currently. If nothing happens, the marine life will continue to tie it up through consumption or incorporation, and it will eventually be taken out of the food web that way as well.

XVI. Renewable Energy

This book, in many ways, is about wise management of resources in relation to one's particular bioregion and site conditions. There is also an emphasis on the capture, retention, recycling, upcycling, and re-use of all resources. In the case of energy, all these things apply, but there is an additional layer to consider: our urgent need to stop using fossil fuels and shrink our energy consumption for the sake of all life on earth.

If we go back in time, we can see that for most of human history we have actually been working on different ways to harness the sun's energy stored in the wood we charred, in the fossil fuels we burned, in plants we grew, in the foods we ate, and in innumerable other ways. Fossil fuels too are ancient sources of captured solar energy that formed from ancient deposits of decomposed matter.

The last 50 to 60 years have seen the greatest acceleration of carbon dioxide released into the atmosphere, and this period coincides with the rise of consumerism in the late 1950s. Fashion and fad rather than need drove demand, and these consumer avenues were driven by novelties such as television programming and omnipresent advertising. These fashions and fads, not based on natural needs or patterns, have in turn generated abnormal and unnatural behaviors individually and collectively.

Examples of these abnormalities might be the cult of the tidy lawn and the tidy garden to garnish the homes of the American Dream, which in the Western hemisphere went on to drive consumption of fuel-powered lawn mowers and toxic herbicides, then tractor-like lawnmowers, and finally to eutrophication and topsoil loss and contamination. The constant removal of biomass to the detriment of our soils, leading to the battery-operated plastic leaf blowers of today, the bags of leaves on the curb, seasonal burning, and the seasonal bare fields of the Midwest. Companies also taught us to believe that it is "old-fashioned" to use hand tools, to repair things ourselves, to grow food ourselves, and to be self-reliant because they wanted to sell us "modern electrified" tools, homes, and lifestyles – automation and leisure became the focus. The list of "old-fashioned" equipment replaced by systems designed to be costly and to cost electricity is endless. Science became servant to consumerism as planned obsolescence took over, and we lost the thread of connection between our needs and what a natural response to our needs was. We need to draw a veil over this phase of our history and enter into a new phase.

In this chapter we discuss some of the many technologies – ancient, current, and upcoming –that pair up with natural cycles without drawing down on fossil fuel-originated power. Every bioregion will have a different set of possibilities. Some areas are dominated by the sun year-round while others are dominated by forests and long, dark winters. Some areas have moving waters while others have strong winds. Some areas have lots of people. Some areas have constant rain and others seasonal floods. It is our job as designers and responsible earth-dwellers to observe and interact ethically with what our bioregion has to offer us.

Human Power

Used longer than any other technology, humans have used hand tools and their bodies to work with animals, plants, and the soil since before we had written or oral historical records. Hand tools can be extraordinarily powerful and effective. Teams of hundreds and thousands of strong young men and women with shovels can install rainwater catchment in city landscapes across the world within a few months' time. Many machines can be replaced with smart designs that increase yields and are dependent on human power.

A European Scythe from Scythe Works

If everyone gardened locally with hand tools, we would save the energy and resources spent producing, packaging, and shipping the food as well as the energy spent earning money to buy the food. If we cut back the vegetation with scythes instead of gas-powered mowers and "weed-whackers", we would avoid fossil fuel consumption and improve our health as we work outdoors. The jobs market and the energy market could both use more human power. It will proliferate local jobs as it saves energy and money for people locally. Money that would be leaving the community would instead remain within that community and enrich it.

Bioregionality

The alternative energy systems used on any given site are determined by what is available in that bioregion and, ideally, on-site. Usually this means a collection of different energy systems are used seasonally with varying yields. Having multiple ways to store and generate power is always wise, as is having a back-up system.

Potential sources of energy can be consistent, strong winds, water pressure, moving water, falling water, solar radiation, fish oil, or even woody biomass—the possibilities are limitless. Anywhere solar energy is captured, it can be stored and released. Wind and water move in predictable patterns and are affected by the sun, trees, earthworks, and the greater terrain. All can be manipulated to focus their effect. We can speed water up, slow it down, increase its pressure, etc. These raw elements and effects of nature give us resources that never run out, though often they are either always running, or seasonally affected. We have to recognize what energies we can capture where we are currently or orient ourselves to where those energies are.

The tropics bask in persistent vertical sunlight and, therefore, experience high evaporation. The amount of biological cycling tends to make large dams economically unfeasible over the long term. The best source of energy in this region is the sun since it is constant and powering the rapid pace of all other cycles. Micro-hydraulic power, small wind systems, and many other methods can all work, depending on what is available. In hot arid regions like the Sahara, large solar farms are currently starting up.

In the cold temperate climates where each year there is significant seasonal growth, stick fires for clean heat and energy may be the best option—especially in areas where the sun is too dim in the winters for greenhouses to be effective at warming a home or growing food. In these areas, rocket mass heaters can provide heat and help grow foods not dependent on winter sunlight.

As we explore these options, feel free to combine and tinker with them since they are all based on simple principles that can be applied to almost any site or system.

The Energy Audit or EROEI

We must always consider the complete energy audit or EROEI (Energy Returned Over Energy Invested). A sustainable EROEI would be 1:1 ratio, but a regenerative (and profitable) EROEI would have more energy returned than invested. If we do not count all inputs of energy, time, or resources, we cannot know the actual energy it takes to get the desired output and whether the process is worthwhile at all. Ethanol is the classic example: it takes more energy to grow and process corn into a gallon of ethanol than is contained in a gallon of ethanol. While corn biomass can make a good fuel in a number of ways, the way we were doing it is energy intensive.

> *"Overlay [the EROEI] with your biome, climate, and context, and the choices become clear on what will work and what won't"*
> —Troy Martz, Alternative Energy Expert, 2016.

Making Things That Last

Planned Obsolescence is a design principle unique to the Age of Industrialism. It's the idea that you sell things that are designed to break, wear out, or become obsolete within a short period of time to guarantee continued company sales. This can be seen throughout consumer culture today, as in the case of constantly updated iPhones. This all came about because of the indestructible suitcase that was actually indestructible–that's how it was advertised and sold. The company very quickly realized they were going to put themselves out of business once everyone had purchased their new suitcases, so they changed the design, so the handle unravelled after a few years. The idea being: the suitcase was still indestructible, but the handle was not. This trend has led to lower quality consumer products and incredible amounts of waste.

Luckily, we can make things that last generations, and in the end, ideally compost or at least recycle them. The following sections have several examples of long-lasting technologies, even generational ones.

The Edison Iron-Nickel Battery

Thomas Edison's iron-nickel battery, invented in 1901, can last 25 years or more–some batteries have even lasted generations! This technology was bought out in 1975 by Exide, the world's second largest manufacturer of lead acid batteries, and then taken off the market. Though cheap to buy, lead acid batteries need maintenance and only last 7-10 years. Luckily, iron-nickel batteries are making a comeback as one of the best solutions to offgrid solar and wind energy storage along with the increasingly popular Li-Ion batteries. These batteries slowly take in and slowly release energy; they are DC or deep cycle batteries which are ideal for solar and wind power storage since they generate energy in trickles. Iron-nickel batteries do need regular additions of water, but with a clear container, it is easy to monitor. These batteries can also take great abuse, pay for themselves over time, and improve over time. Companies like IronEdison are currently selling these batteries as well as sealed LithiumIron batteries which are projected to last even longer with no maintenance needed similar to the now popular Li-Ion batteries which come with up to twenty year warranties.

Lights that Last

There was a time when lightbulb filaments were made by hand, and because of this, they varied in quality, with some lasting decades. The Livermore Centennial Lightbulb of Fire Station 6 of Livermore, California has been shining for over 114 years, over a million hours. The lightbulb's filament operates at a very low wattage, is complex, and has been continuously on —all of which contribute to its longevity.

We can develop products that last, but we have to revive craftsmanship by shifting our support from industrially manufactured products to locally produced ones, or we have to produce these products ourselves. We also have to use less energy in general, just like the lightbulb, to last longer. This may mean that we rely upon LEDs, bioluminescence, or candlelight at night instead of standard light bulbs. We either need lights that last and are worth the energy spent in their creation like LEDs, or, ideally, we need to use compostable, regenerative solutions that we can source locally.

The Livermore Centennial Lightbulb

Passive Systems

Passively powered systems have infinite potential but rely upon certain conditions. Whether it's the heat or light from the sun, the behavior of hot, cool, dry, or moist air, or the pressure or cooling temperature of water, we can harness the power of nature in a system that creates reliable and even constant energy. Passive systems work without or with minimal maintenance and are usually always on (in a constantly running system). An example of this could be a watermill's wheel turning with the flow of water—trying to stop its turning could very well destroy it. A rainwater catchment cistern uphill from the home can provide water for the home that is pressurized, held, and accumulated passively.

Passive Solar Heating

We can heat all our homes using the sun, even in the dead of winter. We can insulate our homes well with something natural like straw bales, earth, or cob. We can use windows facing the sun path to let the sunlight in to heat the home (the direction depends on your hemisphere). Greenhouses or even just thermosiphons designed to make hot air can be used as well. Heat rises, so any way that a designer can sink heat below the home allows this heat to slowly rise and warm everything through conduction, and then through radiation, as the objects in the home release that heat. This can be done with solar collectors that heat up rods that connect to metal running below or inside the foundation, or it can be done by having a concrete pad that connects the greenhouse to the house. There is limitless room for innovation in passive solar heating.

Passive Cooling

Using air from a well, creating microclimates, timing watering for evaporative cooling (which can be done passively), and cooling air by pulling it through the earth are all feasible options for creating passive cooling. In tropical areas or in many work areas, encouraging the movement of air is critical. Homes can be designed to pull in cool air and to release hot air as well since hot air rises and cool air falls.

Solar ovens

Using the same principle as a greenhouse that isn't ventilated on a hot day, food can be cooked efficiently with solar energy. Durable metal and glass solar ovens are commercially available though they can be built out of cardboard, a casserole dish, and a piece of glass and still be effective. Additional mirrors or reflectors can be added to increase the solar energy focused on the food. Similar in concept, food that is brought up to

Abri La Roux, 2013. Creative Commons.

a high temperature can be insulated and sealed in a container to keep the heat bottled up and continue the cooking process. Using the sun to cook food keeps the heat out of our homes, and in the hot, arid regions, this is a win-win situation.

Hydraulic Ram or Ram Pump

Using the constant flow of water and a ram pump, we can pump that water to a higher elevation or magnify the pressure of that water without using any extra energy aside from the kinetic energy of the initial flow. The hammer effect creates pressure that develops as the flow is bottlenecked and then forced out a different way; this greatly magnifies the pressure and increases the psi (pounds per square inch) of the water in that hose or irrigation system as well. For a full energy audit, the mining, refining, and creation of the ram pump would be included, so it is critical that a long-lasting ram pump be used or built.

Water Wheels

These can be used in a series of different ways, but water wheels turn with the power of water and use that passively created energy to do a variety of tasks: compressing air, harvesting fish, grinding grain, running a sawmill, generating electricity, pumping water, and many more creative applications.

The Barsha Pump being tested in Nepal

The Barsha Pump can move water incredible distances

Wheel Pump

Water can also be pushed up a hose or across a landscape using the kinetic energy of its passing to turn a wheel. Some wheels power a pump, and in some cases the wheel is itself the pump. The pressure created can easily take water uphill or into a containment area with an overspill that rejoins the flow of water. The continuous flow creates passive energy that can be used in a variety of applications. The Barsha water pump or rain pump uses these principles and is currently being trialed and tested.[89] It works continuously with the flow of water, constantly taking in more water as it turns, and the pressure of the current magnifies inside the wheel and sends that water up the hose (see pictures next page).

The Pelton Wheel

Considered the most efficient water wheel ever devised, the Pelton Wheel has a concave split bucket design that prevents back spray and absorbs the water's force efficiently. The largest one ever used was at the Northstar Mine and Power generation facility in

---
[89] Aqysta.com. *The Barsha Pump*. 2016.

California to provide compressed air for a now-closed gold mine. With 77%+ efficiency,[90] the Pelton Wheel is much more efficient than current "modern" electricity generation methods like coal and nuclear power, and that is why it is still used today in hydroelectric generation.

Pelton Wheel systems rely upon water and create numerous outputs: geysers, compressed air, electricity, hot water, and water pressure. When the 18 ft-wide Pelton wheel (refer to the picture) was in operation it would get up to 70 mph and was driven by two jets of water pressurized by a 700 ft (213.3m) drop in elevation. The nearby town could redesign and restore this Pelton Wheel power generation plant and start creating energy for the town passively and locally.

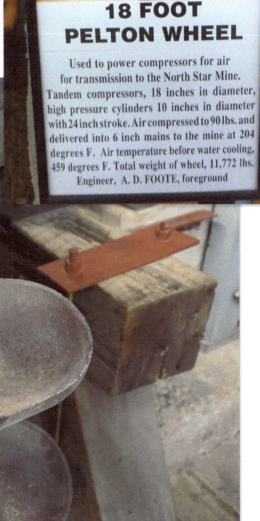

[90] Bjørn W Solemslie and Ole G Dahlhaug 2014 *IOP Conf. Ser.: Earth Environ. Sci.* **22** 012004

Bioluminescent Algae and Fungi Lighting

Many plants and animals express bioluminescence, algae and fungi being the most adaptable and malleable examples. Lamps of glowing green algae-thick water are being developed to reduce atmospheric carbon levels.[91] Easy to maintain, these oxygen-releasing algae lamps can give us nightlights that are powered by the sun and atmospheric CO_2. Bioluminescent fungi, both mycelium and mushroom fruit, also exist, and mycologists are currently working on developing a way to source them in cities and homes as well.

Making Things that Compost

Almost everything we use needs to be compostable, be renewable, or easily and safely recyclable. Making things that compost is the ideal. If our clothing turns into mulch that can be composted with our kitchen scraps, we will prevent an immense amount of ecological destruction and contamination on top of saving energy. If the fiber for our clothes can be grown and raised in the backyard or in the back field, the amount of energy saved will be even greater. From the shed, to the hand tool, to the garbage can, to the shopping bag, all these choices can be regenerative.

Fewest Moving Parts Principle

As with the Barsha pump, the ram pump, and the rocket mass heater, having few or preferably no moving parts in a system is critical to longevity, to the overall energy audit, and to smooth functioning. As with the principle of the least change for maximum effect, the few-as-possible moving parts principle in an engine or energy-capturing system helps maintain the system, reduces work, and keeps things simple.

---

[91] Nguyen, Tuan C. *Can an Algae-Powered Lamp Quench Our Thirst For Energy?* 2013.

Old and New Technology

Gasification

Over the past 200 years, gasification has appeared and disappeared at different times when fossil fuels became scarce for various reasons. During WWII, wood gasifiers ran engines in millions of vehicles because fuel was hard to come by. This is an old technology that is now making a comeback again as scarcity and an ethical obligation to the environment has led many to search for other options. It should be noted that gasification happens all the time in nature in a way as part of releasing gases in decomposition though this process in specific is about capturing and harnessing those gases.

Gasification is the process of turning organic matter into synthetic gas (called syngas or producer gas) which works like regular fuel in a gas engine. Steam is combined with charred biomass to create carbon monoxide, hydrogen, and methane. The char can be made with coal (char-coal), peat, wood, walnut hulls, peanut hulls, and numerous other sources of carbonaceous biomass. While years of research still lie ahead, there are promising innovations surfacing for homesteaders that can be managed and fueled easily. Waste streams from farms and cities can easily turn into sources for reliable electricity, heat, and fuel. Most gasification systems are currently not carbon neutral, but designs can be adapted to sequester carbon if redesigned to capture all exhaust. This process creates carbon monoxide and can be extremely dangerous if done indoors if there are any leaks.

An excellent example is the Enki Stove which is a portable pyrolytic stove that burns the gases from biomass rather than the biomass itself uses gasification in real-time in a contained system that burns clean with no smoke. It uses a battery or solar panel system and creates biochar as a byproduct! It is a precursor to a new way of cooking and gasification.

BioFuel

The original fuel that Professor Rudolf Diesel, father of the diesel engine, used was biofuel: peanut and vegetable oils. Today, coconut oil in the tropics is used in vehicles in places that eschew outside imports, such as **Bougainville**.[92] Hemp, recently legalized in many US states, is currently being scaled up for industrial and commercial uses again, biofuel included. Biofuels themselves cover a spectrum that includes different processes and products—some of which could be regenerative and others which need to be redesigned to be carbon neutral or carbon-sequestering. It is imperative that all combustion engines capture all their exhaust before it enters the atmosphere, and it is also critical that they burn

---

[92] Rotheroe, Dom. *The Coconut Revolution*. 2001.

cleanly enough to only release carbon dioxide and water in that exhaust.

Alcohol is a biofuel that is being used worldwide. Alcohol can even be used to run an engine as has been done in Brazil with sugar cane biomass. Alcohol is a reliable fuel source that can be created easily and regeneratively from numerous biological sources and waste streams on a home-scale or even a citywide scale.

Biogas is short for biological gases released when decomposition occurs. When wood breaks down in a large quantity without any turning it begins to release CO_2 and methane (CH_4), both of which can be captured and used. Biogas can also be sourced from already partially broken-down wastes like human and animal manures or mixed sources of rotting biomass. The sewage, the kitchen scraps, and all the other organic wastes can be composted, and the gases can be captured from that event. Latrines

A small HomeBiogas unit turns kitchen waste into biogas.

can even be designed to do all the work passively. That gas can be used in numerous ways, and its CO_2 accompaniment and the resultant CO_2 from burning the CH_4 can also be sequestered if systems are properly designed. Biogas backpacks are being used to transport methane for cooking fuel to decentralized locations, spreading self-reliance and trade, in places like Ethiopia, Africa.[93] Septic tanks could be redesigned to capture the biogas they generate.

A **biodigester** for biogas can be made on a small scale easily but offers only a modest amount of gas while a larger neighborhood or village-sized biodigester can easily generate more biological activity and therefore more gas. It should be noted that smaller size biodigesters are easier to manage. Transporting and storing the gas will always be a dangerous process, but if systems that generate the gas are built with extra capacity for high production or high-heat time periods and generally designed to provide the amount of gas regularly needed, then there will never be a build up of gas for that to become a problem.

---

[93] Teutsch, Betsy. *100 under $100: One Hundred Tools for Empowering Global Women*. 2015. p. 41.

Additionally, containment vessels need to be double-chambered to withstand high pressure times. The waste product at the end is a fungal food or a soil amendment once it has been aerobically composted—the same is true of anaerobic fermentations like "bio-fertilizers" which are different from biofertilizers (biological fertilizers) like rhizobia, AMF, etc.[94] It should be noted that most of the valuable nutrients are gassed off during the digestion process—compost whenever possible instead of biodigesting.

Algae-culture farms are the fastest way to sequester carbon on either land or water. Algae can also be grown in a closed loop system. Microalgae sequester carbon at 10 times the rate of trees and perennial grasses; they've even been proven effective at sequestering CO_2 emissions from coal combustion.

> **"Aquatic microalgae are among the fastest growing photosynthetic organisms, having carbon fixation rates an order of magnitude higher than those of land plants"**
>
> *Microalgal Removal of CO_2 from Flue Gases: CO_2 Capture from a Coal Combustor, US Department of Energy, 2002.*

Though they currently involve a costly process, algae fuels could replace the fuel needs of the US on a fraction of the land currently being used for growing corn in the US—though it remains an expensive solution.[95] Algae-fuel generation systems are often called carbon neutral because algae pulls CO_2 out of the atmosphere to make the lipids, or oils, that can be refined into biodiesel; however, biodiesel and biogas release CO_2 and water when efficiently burned. Much like trees, both these carbon neutral sources are sustainable, but algae fuel is not regenerative on its own—a way to sequester the carbon released from the fuel combustion is needed.

Thinking about this fuel differently for a moment, if we have algae biofuel plants that only use the biogas coming off the algae as it breaks down to generate electricity, we can sink and sequester all the biofuel and the carbon it holds. All the carbon dioxide from burning the biogas for electricity can be returned back to feed the algae, making it a closed loop system. This is a perfect permaculture power plant because the by-products are used to continuously feed it, and the surplus (which is the main product of the reaction) is quantifiably taking carbon out of the atmosphere and returning it to the original source of the fossil fuels.

The pumps needed to run such a system could be powered by electricity generated from burning the biogas, with no outside inputs needed to run the plant. These carbon

---

[94] Sawada, Kozue, and Koki, Toyota. *Effects of the Application of Digestates from Wet and Dry Anaerobic Fermentation to Japanese Paddy and Upland Soils on Short-Term Nitrification.* 2015.

[95] Cornell University. *Marine microalgae, a new sustainable food and fuel source.* 2016.

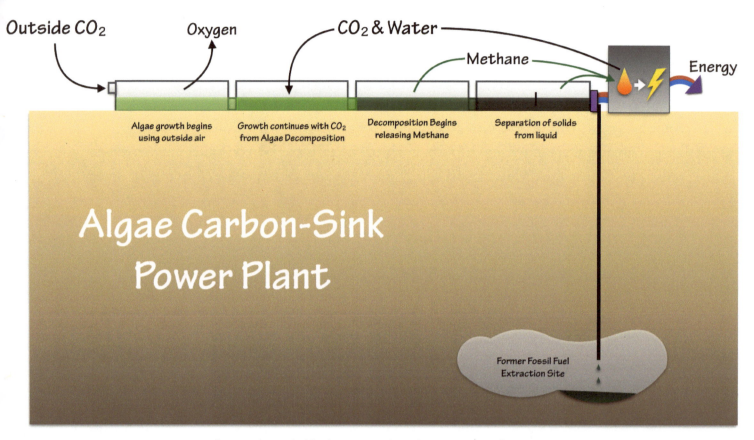

sequestering power plants, though likely minimal in their return of energy to people, could be made to care for the earth and our collective future by sinking enormous amounts of carbon into the earth. They could be made anywhere that there has been drilling in the past to return carbon back to where it came from in a form nearly identical to the original fossil fuels.

Wind

Though wind power companies are currently building gigantic windmill farms for centralized grids, decentralized wind power will be even more important to communities that cannot afford large investment and infrastructure. Small-scale wind turbines are relatively simple to build, repair, and maintain on a home or community scale, but should be positioned high above the tree line to catch winds above the air closer to the ground which experiences drag from the trees, buildings, and plants there. An entire local industry maintaining and supplying parts within each bioregion, adapted to that area, is possible where wind is a viable option in terms of either dependability or seasonality. Wind can be used to cool homes, objects, or areas, to passively spread seed or spores, and to move objects like windmills, sailboats, and so many other things.

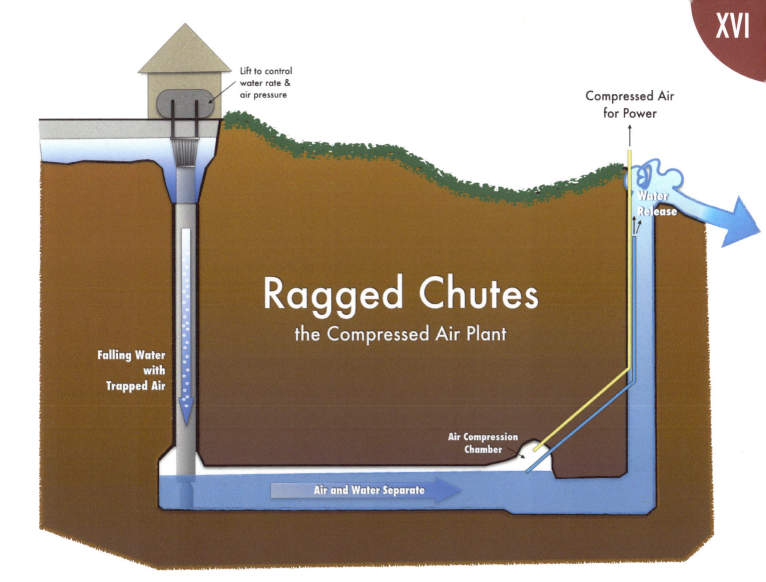

Falling or Running Water

It has long been recognized that running or falling water could be harnessed for hydropower. From the Roman aqueducts to the irrigation methods of the Mesopotamians to modern times, we've used moving water to clean, irrigate, flood, fill, push, cool, and power everything we could possibly conceive of.

"The Ragged Chutes Air Plant opened in 1910, and operated exactly as Taylor had predicted. The plant operated at 82% efficiency, required no fuel, and cost almost nothing to operate. The compressed air from the system had the added advantage of actually being drier and cleaner than air from conventional compressors which reduced the need for maintenance for the drills."
- Charles Dumaresq, Cobalt Mining Legacy: Power to the Mines, 2009.

Trompes utilize the air captured in falling water to create compressed air continuously. This can be done on a small scale or a large scale as it was done with Ragged Chutes, the Compressed Air Plant in Ontario, Canada, which generated compressed air for over 70 years with falling water, using an unused cobalt mine shaft and some ingenious piping.[96] In recent decades Ragged Chutes has been repurposed by the TransAlta company into a hydropower plant after an insurance company forced the destruction of the air plant, but locally and globally the air plant still intrigues and inspires.[97]

Miniature and adapted trompes are being made using the principle of the ram pump but with check valves to build up even more pressure.[98] Some designs source the compressed air while others use the water pressure. A combination system that allows for both increased water flow, like a ram pump, and compressed air, like a trompe, would be ideal for a homestead near running water or to use with water falling off a roof in the monsoon tropics. Places where it rains nearly continuously for a season like in Seattle, Washington could have entire housing tracts or city blocks harvesting and channeling water for energy. In areas with seasonal flooding, or a constantly running river, trompes can be employed to do almost anything.

The Sun

The cornerstone of all available energy on earth is the consistent daily visitation of solar energy. This off/on effect, caused by the spinning and tilting of the planet, created all the variation and rhythm to life here on earth. Without it we wouldn't have the weather patterns we have, nor the rest and growth **phases that occur** both daily and seasonally. From our beating hearts to the binary code of computers, we focus on off/on pulsations for expression, control, and growth. All life is a reaction to and in alignment with this diurnal/nocturnal, dual reality.

All life relies upon the energy from the sun, and in that way, all life is like a battery capturing solar energy. Even our skin, bones, and hair represent solar energy recombined with the digested elements of our diets.

We can use the sun's energy biologically or mechanically. When we transfer the energy into biology, we are growing plants or living organisms. Transferring it physically uses the heat or the light of the sun to power a process like drying fruit or creating steam. Whether it's to passively heat a home, power a turbine, or cook a meal, solar energy can be leveraged in innumerable ways to save on electricity generation and fossil fuel consumption.

While solar panels still rely upon fossil fuels for their creation and maintenance, they can offset their carbon footprint within one year of usage, and they can operate for over 20

---

[96] The Corporation of the Town of Cobalt. *Ragged Chutes: A Modern Wonder.* 2016.
[97] Lennox, James. *Phone Conversation with Matt Powers.* 2016.
[98] Water Powered "Air Compressor and Water Pump". The "Trompe Hammer", Trompe and Water Ram, 2015.

years. The carbon and the energy needed to make solar panels can come from regenerative sources. We can recycle waste from landfills for all we need.

In nature, plants, animals, and fungi are solar-powered and store energy in themselves like a solar panel and its battery. Recent innovations using houseplants to generate electricity using photosynthesis are promising. If forests can be used as solar networks by tapping into the mycelium that connects the forest, there's no limit to how regenerative our power generation and communication systems can get. Research is still pending as is binary coding and mapping of the communications of fungi mycelium in a native forest, but the possibilities are hopeful and promising.

XVII. Urban Permaculture

Cities generate and trap heat and seal off bare ground with impermeable surfaces; they mismanage their resources of water, waste streams, and potential energy. Cities are however attracting an ever-greater proportion of our human population as their preferred habitat. This is clustering human populations like never before, and this focusing of human populations may help take the stress off rural landscapes. Cities offer the convenience of clustered services, resources, employment, education, and cross-cultural diversity and venues, and cities also provide nexus points for trade and exchange. Applying permacultural thinking to urban systems is generating tremendous new market opportunities for effective solutions, but widespread and large-scale adoption has yet to occur. When it does, the effects will be dramatic.

Cities from their inception were microclimates interspersed with smaller microclimates inside them — often parkland, wind tunnels, still areas, or cooling water surfaces — and our megacities of today still offer many features that can be directed to work towards sustainability.

Concrete planters can be built on top of the ever-present concrete. The concrete and blacktop can be painted white and shaded to retain the cooler temperatures of the night instead of the hotter temperatures of the day. More easily incorporated into the design of new builds, green roofs are becoming important heat absorption and water-retention elements. All the organic matter in the waste streams of the city can be repurposed to create energy, soil, and more food. There are also numerous sheltered areas in these urban spaces that can be used for growing daylight-sensitive plants or fungi. Their windbreak can shelter tender plants or seedlings. These sheltered areas can provide habitat for ponds and other

Food can be grown in pots on balconies or in small spaces.

ecosystems that extend and magnify the cooling effects. Greening the city with biological solutions to mechanically created problems is possible, and it is already occurring with the highest yields-per-square-foot found in urban gardens where soil is often scarce but resources many—though more than the food systems of the cities must change.

Urban Waste Water

The way toilets are constructed and waste is handled must change drastically. The toilet must only be used for urine and manure, and both need to be separate from each other —toilets designed for this are available (and there are historic examples as well). Urine can be used at low concentrations on plants as a fertilizer because it contains many important nutrients. Urine is also used as a fungicide at a higher concentration because of the ammonia it contains. Humanure from healthy individuals can be processed in a hot compost; humanure can also be processed for a minimum of three months in a biodigester, biogas system, composting system, or even outhouse pit, but examine and test the final product always as it has many variables.[99] Human waste is an end result of food grown in soil. All of these could be resources used by city administrations in green spaces, street planters, community food gardens, and more.

---

[99] Ingham, Elaine. *Celebration Farm Tour*. 2016.

Natural cycle principles hold that animals that consume a plant will pass the waste from digesting that plant along to the soil food web—either close to that plant to nourish it, or further from that plant, to spread that plant's genetics. If our soils are to be sustainable and regenerative, they must be built on and be part of the whole cycle of soil—from food, to digestion, to waste, to the soil, and back again to food, restoring the cycle that never should have been broken in Western agriculture—it wasn't in most Asian agriculture. In China it is called *nightsoil*. In Britain, they have used the biogas from sewage for cooking and heating.[100] It is not just human waste needs to be put back into the proper cycle, but also all waste water.

Gardens can grow vertically

Mitigation Ideas

Graywater from showers, sinks, and washing machines needs to be used wisely. It can't be drained into the ocean after one usage as has been the practice in many areas such as Los Angeles.[101] Water falling in and on skyscrapers has to be leveraged to generate energy, filtered in place with gravel, ceramic/clay, carbon, and sand filters, and it has to be used to water vegetation or reused on site. Reed beds at the bases of buildings can filter water further, allowing it to be recycled for watering or even human use if clean enough. Energy generated from falling water within buildings can be used to pump water back up to the top of the system. Rainwater and graywater can be focused into larger catchment in lower buildings for power generation. Any excess energy generated can charge the building's main batteries or be put to some other use. Adding in air as the water falls at any point can increase both water and air pressure. Depending on how buildings are

Portable Pallet Gardens by Stuart Muir Wilson.

---
[100] BBC.com. *First UK homes heated with 'poo power' gas from sewage.* 2014.
[101] Powers, Matthew. *Where Did the Water Go in Central Valley California? PowerTalk.* 2014.

Raised gardens grown on top of concrete in Los Angeles, CA.

designed, they can save energy and store it.

Vines that shade the roof can lower the temperatures in top floors of buildings in summers. Climbing vines, like hardy kiwi fruit growing from the bottom story up, allow folks to vacation without having to worry about watering. They also provide shade, privacy, food, cooling, air filtration, and oxygen. Vertical gardens can be installed on balconies as well as bag gardens hanging inside and outside the apartment wherever there is space. Rooftop gardens with light soil mixes (to keep pressure off the roofs) and trellises above them (to further shade the roof from the sun's direct rays) help further insulate buildings from high summer temperatures. These areas also clean and filter the rainfall and air while mitigating overall consumption of resources: people stay at home more, grow their own food, heat and cool their home less, and enjoy better health and, therefore, consume fewer medicines.

Sidewalks, many roadways, and any pointless concrete need to be removed and perennial, edible, shading and ground cover plants need to be put into place. Regardless of where you live, greening the urban landscape is the only way to make cities sustainable and, ideally, regenerative. Any space like curb strips or front lawns can be productive and yield food and other products.

Rainwater catchment has to be encouraged from the home scale to the municipal; if we funnel all the rainwater to a municipal site for it to be cleaned and returned, that represents an immense amount of energy. To prevent water intake from being too great at any one time for the city system, all home systems need to catch rainwater and store it wherever they can. From a rain barrel, to a 5,000 gallon tank, to a buried concrete cistern, all the spaces we can get are needed to catch all the water that passes over the urban hardscape. City streets, sidewalks, and parks can all be places of water catchment. Swales, earthworks, and careful catchment can happen everywhere. Many US cities like Phoenix and Los Angeles thought to be parched actually receive enough rainfall annually to match most if not all their water needs. If water is recycled, reused, and sunk into the landscape, there will soon be an abundance that translates into improvements in other areas: socioeconomically and health-wise. The desertification of our lands and urban areas keenly affects the local economies as well. All the water that falls on concrete should be slowed, spread, soaked, or stored for later use.

Empty underground car garages can be cleaned, sealed, and used to raise fish and aquatic plants in an aerated system, or they could even be turned into giant aquaponics systems with the fish in a lower lot and the plants on the lot above in stacked shelves of flooded plant trays like Will Allen's greenhouse farms in Milwaukee—on a giant scale. His system is creating 10,000 lbs (4,500Kg) of food on a quarter of an acre (1000m²), so imagine what a system the size of a city block could generate, or even imagine, if the water were routed through sidewalk hydroponic growing beds that double as street corner farm stands for fresh produce.[102] You could buy your dinner's greens fresh off the street fed by rainwater and fish manure. You could likely even buy a delectable carp for dinner that fed your swiss chard and in turn the swiss chard cleaned the water for that fish. It could be all part of your own apartment complex's design, and as part of your rent, you get a certain amount of fresh produce and fish for free or at a discount.

Chickens, ducks, rabbits, or any animal that can develop fully while in confinement are good choices for urban environments. Chickens can be noisy, but good design can muffle a rooster in a straw bale coop and bell dome. Rabbits are always quiet and create over 100 lbs (45Kg) of meat per mature rabbit doe per year. Ducks, quail, and chickens can even feed on rabbit manure as well as the worms and soil life that develops in the composting rabbit manure. Birds can be kept below rabbits with wire cages that allow their manure to fall through.

---

[102] *Growing Power - A Model for Urban Agriculture*, 2010.

All organic matter is composted either as part of a home, apartment, neighborhood block, or municipal system. All paper, woody waste, and cardboard is first inoculated with mushroom mycelium then vermicomposted to worm castings for parks, gardens, and compost teas—these can be free or sold cheaply to keep the nutrients and waste cycling. Food in turn would become continuously cheaper and more widely available with more choices as soils build and plant offerings increase—all of which sequesters carbon dioxide in the landscape itself.

Energy Solutions

Falling water, trapped heat, the sun's light, wind tunnels, and decomposition are all available sources for energy solutions in an urban area.

Sun, Water, & Wind

All urban surfaces facing the sunpath can be leveraged to capture the sun's rays in solar panels and stored in batteries (hydrogen, lithium ion, nickel iron) that power buildings, streetlights, the local communication grid, and local municipal transportation.

Many large international cities are positioned near coast lines with tidal activity that can be leveraged for consistent power generation. Large strategic wind turbines are used to power many parts of the European Union. Renewable energies of this kind are key to replacing the fossil fuel-based energy economy. They are continuously lowering the costs and increasing efficiency, and soon the renewable energy sources will be unstoppable on the market. Widespread adoption of clean energy technologies are around the corner - how fast and how deeply we adopt renewable energy is up to us.

Apartment and Municipal-Scale Trompes

Built for Peter the Great, Peterhof Palace outside St. Petersburg boasts elaborate fountains that perform sophisticated water performances perpetually with nothing but water pressure, very similar to a trompe. In Canada, a municipal trompe, the "Air Plant" mentioned in the Alternative Energy chapter, was in operation for over 70 years. The engineering to create trompes has been with us for centuries but not implemented widely. Trompe systems that run large turbines with only compressed air and, with little if any maintenance, are the future of passively harvested energy for any city with a considerable body of moving water nearby. Apartment, homestead, and municipal trompes can all generate and store electricity passively as well as use that cooled air and collected water for other purposes.

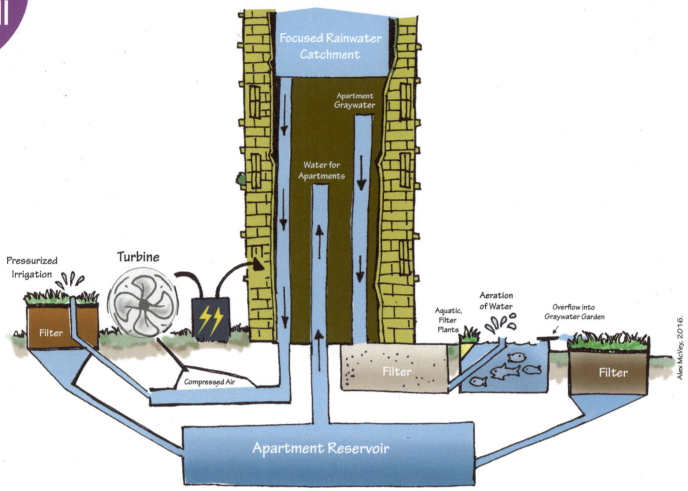

Rainwater catchment from larger buildings could be routed and focused into smaller building systems for energy generation, water pressure, or graywater usage.

Rocket Stove Turbines

Since stick fires in a rocket stoves J-tube easily give us clean burning heat, we can apply rocket stove technology to the classic turbine system: steam or even a stirling engine. Commonly a serious concern in creating hot-water systems with a rocket stove, steam can be deadly in almost any situation. Putting it under pressure is even more deadly, so precision and caution are needed in design. In spite of the associated danger, rocket stove turbines offer the cleanest, most regenerative, and sustainable renewable energy source available for generating electricity where woody or combustable biomass is available.

Municipal steam turbines already exist and are used in nuclear, waste-burning, and coal-burning power plants, among others. We just have to design a rocket stove system that can replace these other heat sources in place, making the least amount of change for the greatest effect. Rocket stoves have also been recently found by researchers Erica and Ernie Wisner to be able to burn plastic cleanly. In addition, the Wisners have been able to burn wood at nearly the full BTU of wood itself which is 4000°F (2204°C)![103] That means that we could burn our trash at home inside and have it burn cleanly, power the lights in the kitchen, warm the floor or walls, and then feed in a closed loop to an algae pool that feeds into a biofuel system, becomes a soil amendment, or feeds fish. The possibilities are endless.

Solar & Thermal Walls & Windows

Using the greenhouse effect we can trap heat in chambers attached to the outside of buildings like a facade or siding and let that air flow into the building for heat. Similarly, we can cover the sunpath facing walls and building faces to have solar panels embedded in them. There are numerous ways we can draw in heat and capture solar energy in our architecture.

Transportation

Mass transit is possible in a city that can leverage its waste to power it; it can save energy by helping large amounts of people move all at once. Otherwise all vehicles can be pedal-powered with battery back-ups like the Rahtracer[104] or the PodRide[105] which is strictly pedal-powered—both are engineered to enhance the pedal power of the driver to make it easy to pedal up hills and over long distances. Pedal-powered vehicles are lighter, smaller, and slower than gasoline- and diesel-powered vehicles. These vehicles are safer, healthier, and have zero emissions. They also encourage local, healthy living. Longer trips can rely upon mass transit or enhanced pedal vehicles. In time these innovative designs will improve even more as they are adopted widely, and new complexities will also arise to further adapt their designs to

A crowdfunded idea, the PodRide makes pedal-power look like a lot of fun!

[103] Wisner, Ernie. *Email*. 2016.
[104] *RAHT RACER: Cycling vehicle - pedal as fast as a car*. 2015.
[105] Kjellman, Mikael. *Podride a practical and fun bicycle-car*. 2016.

We can convert existing bicycles to electrically powered ones with conversion kits like the BionX which consists of a lithium battery, controller, and motor that can help us use bikes over longer distances, more consistently in our daily lives, and uphill for longer periods of time.

the new needs.

Flight will likewise be adapted to use batteries, solar power, and gliding aircraft where possible. Gliders can fly for five hours after being towed to altitude and speed by a biofuel- or solar-powered plane.[106] Many current efforts in flight are focused around a solar-powered drone designed to stay in flight perpetually. The first day/night solar-powered flight successfully occurred in 2005 with more recently the "AtlantikSolar UAV [Unmanned Ariel Vehicle] in 2015… [breaking] the flight endurance (world) record for all aircrafts < 50kg total mass."[107]

Nearly 1000 km (621 miles) of highway in France are being turned into large solar panels themselves.[108] Sweden is creating an electric highway using the technology used for electric trains, only with trucks, and sourcing the electricity from alternative energy sources.[109]

---

[106] Yeomans, Alan. *Priority One*. 2005. p. 53.
[107] Oettershagen, Philipp - Representative of AtlantikSolar. *Re: Photo Request for Alternative Energy chapter in the first permaculture high school textbook*. 2016.
[108] Markham, Justin. *France to pave 1000km of roads with solar panels*. 2016.
[109] Cooper, Daniel. *Sweden debuts the world's first 'electric highway'*. 2016.

Electric vehicles have taken off all over the world but not everywhere is using clean electricity - many are using fossil fuels or nuclear power to recharge their electric cars, trucks, and buses. As gas prices rise these vehicles will begin to dominate the market - already diesel vehicles and repair shops are becoming a rare sight as clean energy technology and engineering become more prominent in the highest levels of academia and prevalent in the market.

On top of all this innovation, we have growing culture of sharing: we'd rather share bikes and cars than own them in many cases. Uber is a decentralized peer-to-peer car service where users can ride Uber vehicles when needed or drive Uber vehicles to earn money. Bike share programs are popping up in cities all over the world, and even further, driverless technologies are being adopted and tested everywhere. Self-driving cars that plug into decentralized solar, wind, and micro-hydro power grids are the future in many places.

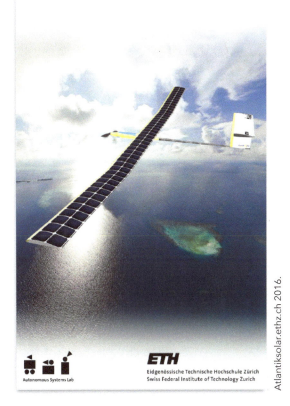

Atlantik-Solar
The first-ever crossing of the Atlantic Ocean using a solar-powered Unmanned Aerial Vehicle (UAV)

Atlantiksolar.ethz.ch 2016.

The Edo Period

At the start of the Edo Period in Japan (1600s–1800s), there was deforestation, overpopulation with densities greater than modern Tokyo, and a growing waste problem, but Japan overcame and reversed these trends using permaculture thinking and techniques.[110] Within a generation the people adopted what they called "multiform solutions" that helped people, the environment, and the future; this concept mirrors stacking functions.

By using less, reusing, and recycling, the Japanese were able to live on their islands in isolation and refuse trade with Westerners. They built small homes with small rooms. They had tinkerers, pottery healers, and buyers as well as purveyors of nearly every kind of waste from paper to ash to manure. They used latrines or earth pit toilets that were used to compost the humanure in an advanced fertilization program where individuals sold their humanure on the free market, and the compost was later sold to farmers.[111] Vegetarian homes sold their

---

[110] Japan for Sustainability Staff. *Japan's Sustainable Society in the Edo Period (1603-1867)*. 2005.
[111] Ibid.

manure at a higher price point—it had less nitrogen and so needed less carbon to compost, so it meant less work and input of resources to fully compost.

In order to use less fuel, clay pots of water were warmed in the sun before being heated to make tea.[112] Considered a plant-based or agricultural country, Japan in the Edo period exemplified a culture staying within its allotted annual solar energy budget. They didn't use anything outside their system, and they avoided anything degenerative.

Malmö, Sweden

The third largest city in Sweden and home to more than 300,000 people, Malmö is leading its country in renewable energy generation. The Western Harbor section has an ingenious heating and cooling system that uses summer-warmed water to heat the area in winter and winter-cooled water to cool the area in summer. They use cisterns 90m (295ft) below the city to hold the water, and they use wind-powered electricity to pump it in and out of the cisterns. "Over the course of a year over 5 million kWh of heat and 3 million kWh of cooling is produced."[113]

25% of all travel in Malmö is done by bicycle.[114] The city fosters a bike culture by showing preference for bikes in the road design with raised bike paths and providing free bike parking. By encouraging carpooling and car-sharing by keeping parking to the edges of the community, cars roles are minimized and, therefore, they are used minimally.

Green roofs are widely used in the Augustenborg section of town. Over 9,000m^2 (29,500ft^2) were installed after an initiative was begun to prevent flooding in the area.[115] Before green roofs, they initially diverted the stormwater into ponds and canals. The roofs clean the air and water as well.

While all government buildings are powered by renewable hydroelectric power, the largest offshore wind farm in Sweden is just outside Malmö. The 48 turbines provide enough electricity for 60,000 homes, a large portion of the population. Fast becoming a place to visit and study, Malmö is researching and testing out many different new designs and technologies to reach their goal of being 100% renewable by 2030.[116]

---

[112] Ibid.
[113] Guevara- Stone, Leslie. *How a decaying Swedish city became an eco-friendly hub*. 2014.
[114] Ibid.
[115] Ibid.
[116] Ibid.

The Brooklyn Grange Farm

Using marginal spaces in cities is how urban farming is making fast inroads into city markets: rooftops are a natural choice in places like New York City, with little green space. Occupying two rooftops covering 2.5 acres (1 hectare), the Grange has taken urban gardening to new heights. Using a green roof design that is waterproof and lightweight, soil and moisture are held in place.[117] This new system of growing food has interesting advantages: there are no large browsers like deer to worry about, and there is less air pollution—the particulates are too heavy to reach them![118] The farm operates a modest CSA program providing members with seasonal, fresh, clean food though most of their produce is sold wholesale directly to restaurants, similar to Curtis Stone's business model featured in the Permaculture in Action chapter. They adhere to organic principles but eschew the USDA's certification. The Grange was also the first and currently is the largest commercial bee farm in New York City. They also teach urban beekeeping, sell honey, offer design and installation services, and put on public events. In a city of eight million people, they are a leading example of growing food in the concrete jungle.

---

[117] McFarland, Kathy. *The Brooklyn Grange*. 2013.
[118] Plakias, Anastasia Cole. *Re: Can I get a picture of the grange to feature in my book?* 2016.

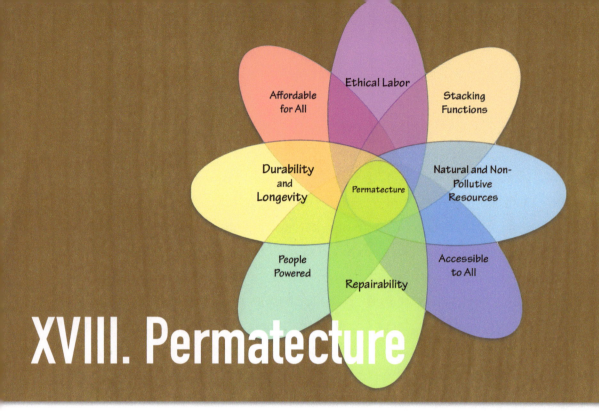

XVIII. Permatecture

The term "Permatecture" was coined by Stuart Muir Wilson, the grandson of Bill Mollison (the co-creator of permaculture), as a component of his Master of Architecture program. Currently Stuart consults and raises funds in Australia and online to build regenerative housing for the homeless of Mexico.

> "Permatecture (regenerative shelter) is a system of design that integrates [and] stems from site specific sustainability and client contextual ideas. Implemented in a practical way through geometry edge, design, integration, multi-functional use and a regenerative impact on the landscape"
> —Stuart Muir Wilson, DesignForHumanity.org.au

Permatecture Principles

While there have been examples of permaculture applied to the design of buildings and construction in other areas of the book, here instead the focus is on principles we can use to help guide us in planning, building, and maintaining regenerative structures.

Durability and Longevity

The idea of long-lasting, permanent structures made from natural resources is not new. In fact, we have many famous examples of ancient structures made from stone and earth including Petra (300 BC) and the Pantheon (126 AD); there are even a few timber structures still standing, such as the Great Buddha Hall (782 AD) in the Nanchan Temple, in Shanxi province, China.

These natural structures are the oldest buildings still standing, and they use natural resources in a way that can be cycled and repaired naturally—though much of the methodology has been lost. Their construction didn't create persistent environmental toxins. Construction today is an energy intensive, highly polluting, and wasteful short-term solution to a constant human need: safe shelter.

Natural and Non-Polluting Resources

Since we are all part of nature, almost anything can be portrayed as "natural", but our specific definition of "natural" here means something that is easily generated and cycled in a non-harmful, ecologically sound manner. Natural resources are those that are found in nature readily and can be easily cycled back to their natural state. These can be living and non-living resources like wood, clay, sand, stone, cob, manures, and animal and plant fibers. Resources we can find on-site or locally are ideal as long as their usage does not harm the ecology or the people. Ideally, the resources should be regenerative and, at the very least, sustainable.

Straw is commonly used in natural building.

Trees can be regrown regeneratively. Earthen structures like cob homes weather and decompose without pollution or loss of integrity of that earthen resource; they simply return to the soil.

This concept goes beyond just the initial sourcing. How was the material shipped? How was it treated, refined, or processed? Sourcing natural building materials takes research, perseverance, a minimalist approach, and sometimes legal challenges if you are the first to use this type of building in your area.

Sometimes these natural resources have a greater financial cost than their toxic competitors, but the health, environmental, and social costs of using toxic building materials are too high. Taking full responsibility for our choices costs more in money terms unless sourced from our site, and then can often cost more physically and in terms of time. There is always a cost. This principle is always in conflict with accessibility for all.

One of the best compromises for this conflict is to recycle waste into building materials. Bottle bricks, plastic bottles packed tight with dirt or trash, can be both a great building material and responsible way to be accountable for one's waste. Earthships are made primarily out of old car tires sealed in cob or adobe. In recycling waste into building materials, we turn a problem into a solution and lessen the building costs at the same time.

Are our current "modern" homes even safe? If you have vinyl siding on your home, and it catches on fire, the fumes from the burning vinyl siding will be more dangerous than the smoke or the flames because the fumes from the combusting siding can kill you faster.[119] Many remember asbestos being ubiquitously used in everything from car brake resurfacing equipment, to the insides of electric guitars, to insulation—today it is removed from homes by workers in hazmat suits. Toxic sheetrock from China ruined approximately 100,000 US homes it was installed in.[120] The list of toxins found in our homes is seemingly endless—from glues, to plastics, to heavy metals, and more.

This raises the question: do we even know what we are doing currently in drafting of construction regulations? We are not building to last. We are not building for the future. We are not even building for our health, our own longevity. This has to change for

Phragmite Bundles
Bundles of Reeds serve as the core of a chicken coop wall. The next step is to apply cob, sealing and covering the bundles.

---
[119] *Blue Vinyl*, 2002.
[120] Allen, Greg. *Toxic Chinese Drywall Creates A Housing Disaster*. 2009.

A straw and wood house being built at Laughing Oak Village, California.

housing to become ethical again.

People-Powered

Construction methods have to be as people-powered as possible—with the least amount of reliance on fossil fuel-powered machines and manufacturing as possible. Building methods like using cob that only use hands and feet invite everyone of all ages and ranges of ability to participate.

People-powered does not necessarily mean no machines—there are numerous, creative ways that common, people-powered machines can be used regeneratively. Bicycles have been transformed into numerous creative forms to generate power or to run machines like blenders, generators, threshers, water pumps, and more. Pedal-powered machines can also be used—they are similar to bikes but can often require less materials and can be built on-site with resources like bamboo. It also shouldn't rule out fuels we can make locally and on site as long as they are regenerative and we can sequester the carbon we release.

When we participate in making life better for each other together, we stack functions socially: building trust, friendship, interdependence, and resiliency.

Ethical Labor

Much of the financial prowess and consumer choice we have today has been built on a legacy of deplorable activities ranging from historical slavery, to warfare, to imperialism, to colonialism, to child labor, to modern wage-slavery—they are all based on exploitation of human labor focusing on the weakest of any given society (usually minorities, women, and children). Historically, lower classes, labor classes, and unions fought to achieve ethical labor and social standards which are present today to certain degrees and in certain places, but in many places and at many jobs, many of our working rights have been erased or redefined to remove or restrict access to those rights—it should be noted that some places have never seen these struggles and have never achieved ethical standards.

This trend can be reversed by closing the distance between the producers and the consumers through local markets. Just as gratitude is directly correlated to the proximity of giver to the receiver, being transparent and local builds gratitude because it fosters a closer relationship between the consumer and the producer, and, therefore, a greater appreciation for the work put in. Because the factories of today are out of sight and out of mind, most people are buying products they would never buy if they had to visit the actual factories, refineries, mines, and sweatshops where the products were made for pennies on the dollar; they would be appalled, disgusted, and outraged. Luckily, we all are able wherever we are to make a positive change and start producing products locally and making ethical choices.

The working environment must be free of toxins. Standards must be observed for safety. Safety equipment and training must be providing for those working with dangerous equipment or materials. The risk must be accepted by both parties, workers and hirers, for responsibility over the health and safety of the workers. Mutual agreements and mutually designed systems for the highest level of safety are needed. This extends to emotional and mental safety of workers.

Fair Compensation

Fair Compensation is in many ways the foundation of ethical labor; workers must be compensated fairly for their time and work. If workers cannot support themselves and their families, they will not be a productive and stable part of that community—or company—for very long. Instead they will need financial assistance and perhaps medical and housing assistance, or in countries where these social supports do not exist, they will become criminalized by their circumstances and/or suffer from deficiency. Stable societies are based on workers getting fair wages that allow them to live comfortably and afford all their necessities.

Compensation doesn't even have to be monetary! There are many types of capital: social, intellectual, spiritual, cultural, material, financial, experiential, and perhaps more. We can be compensated in many different ways as long as it honors and respects all involved.

Accessible to All, Affordable by All

When we have people-powered systems, many hands can be included and, often, many are needed. The easier it is physically to do (like Cob), the more people can be included. The more difficult it is physically, the fewer people can participate. The difficulty could be cost-related, or it can be related to the complexity of the process; both block access. The best designs are open to all walks of life for adaptation and utilization in as many situations as possible. Designs can be open source and made with common materials or waste streams. This does not rule out craftsmanship—which is open to all willing to take the time to work and achieve.

Not every design or strategy will work for everyone everywhere, nor will they be affordable or practical everywhere. Locally derived designs are often the most powerful because they cater to the regional costs and complexities, often implicitly addressing these issues.

Cob is often mixed by stomping.

Repairability

Repairing a tool or appliance instead of replacing it is often not an option in many areas despite repair shops being a long, human tradition. Globally, repair shops are found nearly everywhere though in the Western hemisphere, the widespread availability and cheapness of factory-made goods combined with planned obsolescence's effectiveness proved too much for consumers to resist, and repair shops went out of business all over the United States and other "developed" nations. The human desire to fix machines and electronics has never ceased, and repair shops persist in high density population centers in the Western world and could return if more emphasis was put on the ethical imperative of repairing our broken electronics, tools, and machines instead of replacing them—in fact, repair cafes, freely associating local collectives, are forming to address the need to repair electronics. If enough market demand manifests, universal compatibility of parts and modular models could be next. Computers, for instance, in the 1990s were modular enough that home computer users would upgrade, repair, and modify them on their own. Modern iPhones and laptops encourage the opposite—stay out or risk your warranty! Repairability is also an access-for-all and affordability issue. Many of these ideas and concepts dovetail, work together, or overlap.

Homes, too, must be repairable. Cob and natural plaster can be patched easily at little to no cost. Homes built using people-power usually involve the people who will live there. When people are involved in the design and construction of their own home, they learn how to build, properly maintain, and repair the home as well. When designs are made using local materials, sourcing repair materials is also easy and often can be done on site.

Stacking Functions

The home, each room, wall, and surface must serve as many functions as possible. Tiny homes are often filled with creative responses to limited space. Tables fold away, chairs collapse, and beds emerge from couches or walls. Every space can be re-imagined to improve function and form.

XIX: Invisible Structures: Community, Commerce, & Governance

Compassion is the recognition of another's suffering and the desire to do something to stop that suffering. Empathy is the ability to deeply understand other people's suffering and point of view. Hopefully our compassion will lead us to seek to be empathetic. From this place we should lead, guide, dispute, debate, and govern.

Introduction

We are living in times of exponential change. We are in the midst of the one of the largest extinction events in known history as well as a transition from one stable ecology to another on a global scale, and no one knows quite where we are heading or what tipping points we will reach and their consequences.[121] There is also a perceptible breakdown in social order, governmental efficacy and integrity, the environment, biodiversity, and our economies. Ultimately, all these problems are related, as are the people systems involved and the peoples' assumption that they are somehow immune from the natural consequences of ecological collapse. We cannot burn fossil fuels, or even biofuels, and release their fumes into the atmosphere indefinitely. We cannot destroy all the rainforest without consequence. We

---
[121] Drake, Nadia. *Will Humans Survive the Sixth Mass Extinction?* 2015.

cannot poison or drain all the aquifers and reservoirs without unintended negative consequences being incurred. We are not separate from nature and its systems. Human systems are subject to natural laws as well: we cannot inflate money or tolerate a dictator forever; eventually a realization spreads and change occurs.

Fortunately, with the advent of the internet and international communication, people everywhere from all walks of life and systems of belief are realizing that many of us are focusing on the same concerns. Many people from diverse cultures are seeking new patterns to live by. We need to adapt not just our gardens, local farms, and building codes, but the very systems we live within— government, community, and economic systems. We must begin to live lives of integration, not separation, from the natural world. We can then apply those new insights and patterns to adapt our economies, governments, and communities.

While it would take many books to fully address each area of this chapter, we instead explore some examples of social permaculture in hopes that it will encourage further study into those examples listed and the generation of many more books and resources on social permaculture. In many ways, food production and landscape repair are easy compared to the reform of people systems.

Principles for People Systems

These are distilled from my own thinking mixed with all that I've ever read or learned from everyone I've ever encountered.

- **Treat Others Better than They Expect to be Treated** - If we want people to be willing to dig swales, grow their own food, etc., we have to make it a pleasant experience. If we treat others better than they expect, they will be more trusting and open.
- **Show Trust and Be Trustworthy** - The only way to gain trust is to show that we are worthy of it by demonstrating trust first and then being trustworthy.
- **Be Clear** - When we are clear, our intentions are understood, and others can meet expectations—which also builds trust.
- **Set Clear Boundaries** - Like all edges, boundaries are areas for productivity. When we set clear boundaries, it shapes expectations and guides behaviors towards mutually beneficial ends. It also protects us and allows for growth, reflection, privacy, individuality, autonomy, healthy relationships, and much more.
- **Educate by Example** - The only way to spread good examples of design or behavior is through living those behaviors or thriving inside those designs. People have to see it, touch it, taste it, experience it, and know the story of it to adopt a significant change in the way they live their lives.

- **Share as much as you can** - This may be difficult as we are starting out, but everything we design can turn quickly into abundance and enable us to return surplus and prepare for the future. It is also a main component of the Third ethic.
- **Be Self-Reliant (Be Prepared)** - While seemingly in juxtaposition to the last principle, the first step to helping others is being able to help ourselves especially in relation to unforeseen complications. Planning and preparation are vital components of self-reliance.
- **Be Patient** - A tree does not grow in a day nor a forest in a year. The best things take time and patience—especially with people.
- **Be Local** - When we focus our time, energy, spending, and production locally and regeneratively, we improve the local area and reap all the associated holistic benefits.
- **Be Open** - Trust in you or in your business can only happen if you are transparent and open about how you conduct business.
- **Be Timely** - Whether it's plant, animal, or people systems, timing is everything.
- **Solutions, Not Complaints** - Though critical thinking and pointing out places for improvement is vital for reflection, complaining and finding solutions are worlds apart. When we speak in the language of solutions, we are more likely to arrive at solutions.
- **Smile First** - Enthusiasm is the energy of will made manifest. We all want to be around someone who is enthusiastic and ready to work.
- **Family First** - Families are the most basic units of all communities, the foundation of all civilization, and the purpose for our lives.
- **Work First on What Matters Now** - Always ask yourself: What is important now? Always keep your priorities focused on what is needed now to accomplish your holistic goals.
- **Always Innovate and Adapt** - Nature is always innovating and adapting to the constantly changing environment. If we are to work with nature, we must change with it.
- **Don't take Offense, be Better** - Often it is the most difficult challenge, but, ultimately, the ability to consider criticism without taking offense brings the greatest set of socio-environmental rewards both individually and collectively.
- **Look to Elders** - Experience is priceless, and we all can learn from earlier generations' perspectives even if we choose to do something different. All perspectives play a part in holistic management.
- **Celebrate Common Interests** - When we meet someone who shares a passion we have, we cannot help but feel like we've gained a new friend and ally. A common bond can be found between almost any two people because we all have universal needs.
- **Listen to and Make Space for Children and Youth** - Children, youth, and young adults also tend to be marginalized, despite families having done so much in order to raise

children; making room for the younger people to be heard and to sense themselves as autonomous beings is important.
- **Include Everyone When Possible** - Permanent cultures rely upon holistic representation and celebration of everyone, regardless of culture, gender, or beliefs. The cultural edges can bring a spectrum of experience and perspective that would be impossible otherwise. This might not always be possible - some issues might be personal, trivial, inappropriate, outside their concern, and a myriad of other possibilities.

Patterns for Permanency

Care of the Future

We need a culture that is permanent and adaptive. Culture is what determines how we interact with the land and with each other. Contemplating the effects of our actions up to the 7th generation, the approach taken by the Iroquois First Nation, or making a multi-generational business plan can be forms of caring for the future. A culture that is using its surplus time, food, energy, and biomass to invest in the future is a stable one. Even as we focus on what can be done now, we must split that lens to also envision what it will be like in the future, based on our current actions. A great number of polluting and destructive practices would end overnight if all decisions passed through this split lens.

Decentralization

Permaculture encourages decentralization and localization in design out of necessity. Decentralization is the only way to manage complex systems. When most of our food, fuel, fiber, water, and electricity are sourced locally, at home ideally, and regeneratively, all the money supporting the large, corporate distribution chain will shrink dramatically and shift to the local economy. Transporting electricity from a distant power plant to homes is inefficient. The same is true of shipping fresh fruit and vegetables in winter from the opposite hemisphere's summer. Homegrown or bioregional food, electricity, fuel, and fiber with roof rainwater storage saves the most energy, carbon, and money over time.

Autonomy

Autonomy is the concept of self-governing or self-management. Parents want their children to grow up to be autonomous, and designers want large landscape restorations to be completely autonomous. Autonomy is also the essence of freedom, the ability to make choices without external controls or pressures. As is being found in schooling, teachers must be facilitators, not dictators. If we want to access the highest levels of understanding, application, creativity, and performance, students must maintain their autonomy. The same is true for the environment—it cannot be monocultured into submission and get sustainable results. People and the natural world must be allowed autonomy and the resources to self-manage.

Community Patterns

Family

The needs of families constitutes the majority of what defines communal life or community: education, food/water, housing, and healthcare. Without those elements, most families would not thrive. It is up to our collective culture to provide the means for all people to have easy access to all of these services whether it be through self-reliance, interdependence, or charity. Those core elements sustain our culture; families are the gardens in which we grow the seeds of the future culture's caretakers. How we treat families is how we treat the future.

Neighborhoods

Consisting of anywhere between a few families in a rural area to hundreds in urban settings, neighborhoods of families are a geographic grouping of people. They are incentivized by this proximity to work together harmoniously and regeneratively.

Groups

Local groups can be formed based around common ethical interests such as permaculture, homeschooling, pasture-raised pigs, rocket mass heaters, silviculture, community enrichment, or other topics that draw people together. These groups, both physical and online, have immense power to make change happen without confrontation or even criticism. They can provide positive solutions and examples that invite others to embrace a new way of living. With the power of the internet, new examples for better living are being uploaded daily and shared publicly from the morning breakfast routine, to growing tomatoes, to building a cob oven, to repairing washing machines—the possibilities are endless!

Ethical interests cross all boundaries of culture, ethnicity, and the physical bioregion. This is most visible on the internet where we form groups purely based on interests, values, and beliefs. These groups have the ability to foster interactions outside our traditional circles of influence (family, friends, locals, and local officials). The internet and social media have opened the door to the global community and spread a realization that fast-growing numbers of us desire a more ethical, regenerative, and abundant world. This global online phenomenon of inclusivity and outreach is also driving the formation of bioregional groups that attract people from a vast array of cultures and ethnicities, a convergence perhaps unprecedented in human history.

There is a commonly held view that the maximum size for any functional group is 150 people, a concept known as Dunbar's Number. Any larger, and it starts to fracture off into separate groups within that group—which isn't a bad thing in itself: it's just a matter of forming a second group. While Dunbar's Number is an imperfect approximation, it can give communities a rule of thumb by which to preserve functionality of organizational units, communities, schools; at least it can make communities conscious of their size and relative functionality. In addition, within a group of that size, you can fit all the roles needed for a thriving local economy and vibrant community. In a city this could be the members of one apartment building, an entire block, or the entire neighborhood in the suburbs.

Rural, Suburban, and Urban Villages

Traditional villages still persist throughout the world—primarily in rural areas. These traditional models cover all aspects of community needs: healthcare, food, shelter, security, community, family, spirituality, and work. Villages historically have been bioregional and small (between several hundred and several thousand people) because their size, vitality, and mobility were determined by the bioregion's available resources.

Permaculture-based villages can form in any context: suburban, rural, or urban. The smaller the village, the more closely-knit the community, the lower the crime rate, and, arguably, the better the relationships. To build resilient local communities, cities and towns need to foster the development of villages within their town or city and seek to make each as autonomous and self-reliant as possible. Decentralization of town resources and services allows for quicker deployment and more informed interactions.

Shared Community Structures

Shared community structures are communal buildings and locations that are multipurpose spaces. These buildings can be hyper-local, serving a neighborhood's needs, or they can serve the needs of any city-based institution. They can be used for weddings, meetings, classes, government, commerce, and more. They can include multimedia libraries, parks, playgrounds, tool libraries, seed libraries, record halls, town halls, etc. The structures needed depend on the community's bioregion and specific needs. They can be collectively managed through election, through rotation, or by those who use the facilities.

The more service the people who use the facilities give to maintain it, the more appreciation they will have for it. Gratitude and community trust are linked directly to how close recipients are to the source of their food, welfare, or any community service. Including everyone, especially those who receive benefit from any shared structure or service, in managing that structure or service is critical to maintaining that service or structure's efficacy in serving those people, and those people valuing that service.

Intentional Communities

These are groups of people who choose to live near each other for a common purpose. Though many different organizations of intentional communities exist, there are several types commonly attempted. Some pool their income and resources but maintain separate lives, others share ownership of land and select community structures and patterns, and still others are based on a business that all are participating in collectively on-site. Each of these models has a different level of communalism that reflects that particular community's mission statement and values. Intentional communities called EcoVillages have a focus on ecology and living regeneratively as a group. Historically, Christian abbeys in the Middle Ages qualify as intentional communities focused on their religion and providing for themselves bioregionally.

Land Trusts

Though many definitions persist and have been held throughout history, land trusts are generally legal contracts that grant rights of stewardship over a piece of land for a specific conservation or sustainable use, ideally, for a very long time. Originally land trusts were plots of land being held for others temporarily, usually by relatives, while they grew up or for other purposes (like avoiding being enlisted in the army or tax evasion). Rising real estate prices are making it harder and harder to purchase and own land globally, but with community land trusts, we can stabilize the price of the land itself and prevent it from inflating with the market. The real estate market by design raises property values. Taking land off the market and putting it into a trust prevents it from appreciating on the market and keeps land taxation from increasing.

Led by an ethical mission statement and guided by any series of decision-making processes and governing structures, the community land trust, through its operation, protects the land and fulfills the community's mission statement, its holistic goals. These communities or families have to obey the laws of the county, state, province, territory, or country they are in, but they also follow their own by-laws. Methodologies of management and governance are stipulated as part of the mission statement or constitution. These land trusts can be managed by a caretaker, a single family, a group of relatives, a small village, or, conceivably, a town.

Population Size, Bioregional Capacity, & Empowering Women

While overpopulation in urban areas of developing countries has long been an area of concern, areas of depopulation are also growing steadily, especially in rural areas, as cities draw in more people. Developed countries, including Japan and many Eastern European countries, are suffering from rural depopulation while thousands of small towns and villages across America and Europe are suffering from negative population growth even as population explodes in India, China, and mega-cities across the globe.[122]

Population sizes find their ethical equilibrium by living within the limitations of what their bioregion can sustain. Stable populations linked to bioregions are observable in remaining indigenous populations today. It is only when a temporary resource is exploited that populations soar; the rise in population spanning the last 150 years is directly correlated to the rise in fossil fuel usage. The combustion engine, synthetic fertilizer, and pesticides all were sourced from fossil fuels. Fossil fuel-powered machinery for fertilizing, weeding, harvesting, and planting with fossil fuel-based fertilizers and biocides as the only inputs besides seeds became the backbone of industrial agriculture and the poorly named "green revolution". Fossil fuels are the concentrated mineral and oil remains of organic matter that was originally grown with the sun. Using oil is literally like using more sunlight than you get in a day to power whatever you are using that oil to do. The moment we turn off the carbonaceous diet and become aware of our bioregions carrying capacities, the population boom will end. Carbon constitutes all the structure in all life on earth, so the less we are inputting into our systems, the less products we will get. Phasing out fossil fuels will force populations to align to bioregions which will normalize population densities.

While it may not seem related, educating women is the other critical component in maintaining population equilibrium within a bioregion. Women that are educated will understand and use contraception wisely and choose when they want to have a child. Women that are educated are much less likely to have children that later on get involved in violent extremism, crime, or terrorism. The work of Greg Mortenson, author of *Three Cups of Tea*, has shown that educating women in Pakistan and Afghanistan is the most powerful way

---

[122] Kassam, Ashifa, Scammel, Rosie, Connolly, Kate, Orange, Richard, Willsher, Kim, and Ratcliffe, Rebecca. *Europe needs many more babies to avert a population disaster.* 2015.

to prevent the spread of terrorism and conflict, and his work is but one among many examples. Women are the key and often, along with children, the most vulnerable. If we are to stop overpopulation, we have to do more than see it as a numbers game: it's a series of imbalances - each must be addressed.

Governance

Function

Governments set the behavioral boundaries for society like the game-maker sets the game's rules. To be precise, governing includes management of shared resources and information, dissemination of information, managing and directing labor, creation and enforcement of law, and the management, collection, and disbursement of money. Governments are also caretakers of culture—they manage the culture's collective story and work to pass it on to the next generation.

The traditional hierarchical systems of government tend to create a parent/child relationship between the leadership and the citizenry which often becomes problematic. However, government does not have to be hierarchical. It can be ecologically patterned and based on nature, where autonomy is the basis for all complexity, fertility, and stability.

Governments bring balance when they are based on the three ethics. When they do no harm and, instead, encourage regeneration culturally and environmentally, governments can present boundaries that enhance life in the same way that edge effect does. Getting past hierarchy and bureaucracy is possible, but it requires the empowerment and engagement of the people currently affected by those systems—it takes a vibrant culture. This chapter provides real solutions to obstacles found in government, politics, commerce, and community by introducing new decision-making, conflict resolution, and governing systems, but each requires the participation of passionate people.

Sociocracy

Often called a deeper democracy, Sociocracy is a governance system that uses the self-organizing principles determined by members of a common-interest group to allow all the members to hear and speak to proposals in order to shape it until all members feel able to give their consent to a proposal.[123] In this way rich inputs can be obtained and thorough understanding of issues reached by the members. This is very similar to holistic management in decision-making but has an organizational component added. All proposals and

---

[123] Buck, John and Villines, Sharon. *We the People: Consenting to a Deeper Democracy*. 2007. p. 27.

appointments are made with the participants' consent. Each group is organized around common interests, and each group manages itself. These groups work on projects and policies and seek out objections to improve the projects and policies from the people that those projects and policies affect. If it was inside a business, it would be the customers' objections informing the business' next product improvement. Instead of a board or an individual making decisions for all members, small groups or circles (whose members have been nominated by the members of the whole group) hold the power of decision-making in a prescribed area of concern. The decisions of such groups are the result of discussing exhaustively and amending proposals until every member of the group feels able to give their consent.

Sociocracy can be sourced for a business, a single event, a long-term project, or even a local or federal government; it can be added or overlaid onto any system. By seeking out objections, decisions are arrived upon that everyone recognizes as being holistically beneficial for the time being since later things may change, and the agreement can be adapted at that time; adaptability is key.

Often we see schools of fish or flocks of birds in flight as a whole moving as one, but it is not so. Instead they are all constantly responding to each other and expressing themselves even as they work together as a whole or holon. It is a natural example of sociocracy in action.

Holacracy

Holacracy is a trademarked, self-organizing government system for self-management based upon a constitution where all involved work towards an agreed set of common rules and mutually understood goals. This enables companies and organizations to behave and evolve in the way ecologies, organisms, and cities naturally do.

> **"Like Sociocracy, Holacracy is also organized into circles, groups of roles that work together for a common purpose (such as 'marketing') within the organization. Individuals act as the 'sensors' for the organization, taking action on behalf of their roles within the company, and channeling feedback back into the company to improve the way it works. Individuals who fill roles in the marketing circle may sense 'tensions' (opportunities for improvement) that impact the work of that circle. Any individual filling a role within the circle may make a proposal for a way to resolve a tension that he or she senses. Others may object if they see a reason, based on known information, that the proposal will cause harm to the organization. Using a special integrative process, these objections can then be addressed. The goal of the process is to allow the 'tension-sensor' to author a**

> **change that will address his or her tension without harming the work of others in the circle. Through these incremental changes, the structure of the organization evolves and improves"**
>
> –Tara Everhart, HolacracyOne, 2016

Zappos.com and many other businesses are already using Holacracy to be more adaptive, creative, active, empowering, and competitive. Where adoptions of Holacracy have been successful, participants have noted shorter, more effective meetings, clearer interactions, more creativity, more engagement, more productivity, and less stress. Seemingly without explicit ethical guidelines, Holacracy cares for people and the earth by imitating natural evolutionary patterns as the focus of the method. It fosters the voice and involvement of the individual as it accomplishes the goals of the collective in an regenerative, ethical, and holistic way.

> **"Evolution seems to favor processes that allow peer to peer emergent order to show up in response to real tensions"**
>
> –Brian Robertson, *Holacracy: The New Management System for a Rapidly Changing World*, 2015.

NonViolent Communication

Created by Marshall Rosenberg, NonViolent Communication (NVC) is an empathic language technique that recognizes universal human needs and their communication as the key to unlocking conflict.[124] Honesty, compassion, and empathy typify NVC's non-judgmental communication process. NVC is being used in businesses, schools, families, prisons, war zones, and more places to improve communications and resolve conflict.[125] When both sides of a conflict can recognize the needs of each other without hearing judgment from the other side, the process for seeing how we can help each other can begin.

Practitioners focus on observations, feelings, needs, and requests. When we can observe or share an observation without judgement, we can also say how it makes us feel in relation to what our needs are, and then they can make a request. NVC can be used to develop self-empathy, find empathy for others, and to communicate honestly.

---

[124] Rosenberg, Marshall. *Nonviolent Communication Training Course Marshall Rosenberg CNVC org*. 2014.
[125] Ibid.

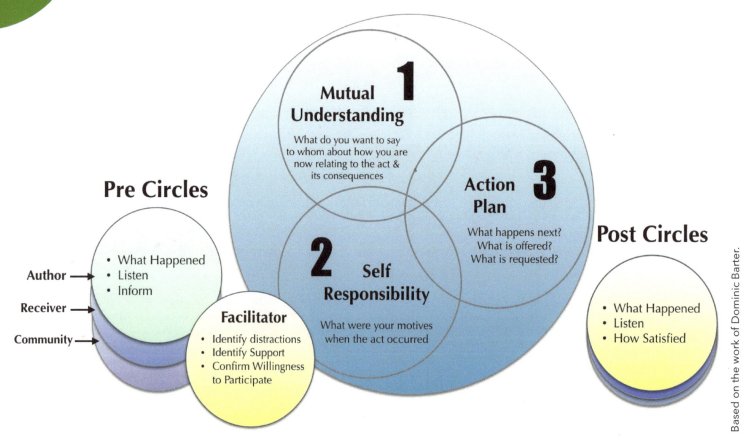

Restorative Circles

Much like a concentrated formula for NVC, this restorative justice method was pioneered by Dominic Barter of Rio de Janeiro in the mid-1990s.[126] It is a holistic method that includes all voices that are involved with a conflict. Instead of punishment or judgement, it seeks restoration. The circles open and close with a reflective sharing process that allows everyone to have an initial and final say (see diagram).[127]

Community Commerce, Alternative Economies, and Other Emergent Models

The Gift Economy

The gift economy is a system of sharing and freely giving resources to one another with no expectation of a return - while this may seem out of touch as a concept, the internet is the primary example of the gift economy where information, services, and social capital are exchanged in a vibrant economy of free exchange. This book itself is offered for free digital download on my site for example, and I offer daily posts, videos, teaching, and more for free

---

[126] RestorativeCircles.org. *Restorative Circles*. 2014.
[127] Aurovilleradio.org *Restorative Circles*. 2014.

online as well. In many areas, the gift economy is outcompeting the market and barter economy: the best example is the music industry. Music is now free through online services with advertisements embedded or nearly free through streaming services that offer astonishingly large libraries of music for negligible costs. MIT and Harvard courses are now being streamed online for public access - information is becoming free as our communication networks decentralize, interconnect, and collaborate exponentially.

The Third Industrial Revolution

A book and concept put forth by Jeremy Rifkin, the Third Industrial Revolution (TIR) is already upon us in many ways and being adopted as policy moving forward in China, the EU, parts of South America, and many other parts of the world. It is a vision of post-carbon world, decentralized, and laterally composed in contrast to today's top down hierarchical design powered by fossil fuels. The basic premise is that the internet is the first step in the Third Industrial Revolution: it is decentralized, people powered, and social capital-based. Once energy and transportation join communication in the TIR, the end of the petrochemical Second Industrial Revolution will be complete: our cultures and economies will be plugged into a decentralized alternative energy grid with laterally shared energy, communication, and transportation. Once we do that, we will be able to pay off the infrastructural investments within decades if not years, it is a road to a debt-free society: the sun is always shining, the wind is always blowing, and the tides are always changing. Currently since the world markets and economies are fundamentally linked to oil, the price of oil affects everything, and since it is running out and the price will rise out of reach in time despite current consumer prices, all the largest economies and markets of the world are stagnating and in decline because of this clear fact: they will be outcompeted by businesses and countries with businesses plugged into a decentralized and alternative energy grid. The efforts of Rifkin's organizations currently are to create masterplans for select international cities focused on the TIR mission, to generate ways to bring the decentralized grids online as fast as possible in as many areas as possible, and to influence all nations to adopt these concepts. Despite it sounding like a political battle, the TIR concepts are being widely adopted all over the world: they are a sure path out of debt and into a regenerative economy. The Third Industrial Revolution concept is detailed and holistic - it cannot be covered here in detail.

Companies

When people join together for a common goal or vision, they literally form a company. That is the root of all business: to fulfill a goal of common purpose or need. Today, companies fall into several groups: sole proprietorship (one owner), general proprietorship (two or more partner owners), corporations (shareholders own the company), and limited liability or limited partnership (these combine different aspects of the other types).[128] Each type has its own advantages and disadvantages, with many of the most successful businesses, present and past, receiving criticism for abusing both people and the environment. In general, our economy relies upon the extraction of natural resources and their transformation into marketable products. Industrialization over time has exponentially increased the liquidation of the natural world. Despite this trend and proclivity, companies <u>can</u> be ethical because in the end they are simply a group of like-minded people who, if ethically aligning themselves with their bioregion socially and ecologically, can be regenerative. There are several examples of companies that are regenerative. Ethics can overlay any company type and make it a regenerative part of any community.

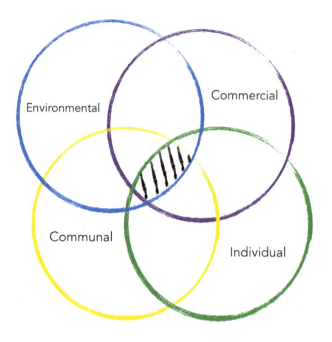

Entropy vs Syntropy

Entropy is the thermodynamic concept that states that energy is always lost in each exchange, leading to the eventual diffusion of that energy until it is no longer identifiable. It is the idea that all order is continuously breaking down into disorder. Syntropy, from Greek meaning 'cross-feeding', is the idea that disorder turns into order infinitely in a properly designed natural system. Natural systems (or syntropic systems) will always expand upon the amount of life and complexity given even the smallest amount of opportunity. Life begets life. The current economic model, a scarcity model, is based on entropy, and not on an

---

[128] Spadaccini, Michael. *The Basics of Business Structure*. 2009.

abundance model. The more businesses that begin to operate syntropically, the faster our current system will transition into a regenerative economy.

Return on Investment (ROI) through a Permaculture Lens

Much like an energy audit, the ROI determines how profitable a venture is. Knowing how much was invested in contrast to how much profit was made exposes clearly the strength and success of a business with a simple metric. In nature, you must make a return of surplus back to the system to minerals, energy, organic matter, nutrients, and water cycling. In a business based on natural systems, the surplus is divided and returned to the business itself, the workers of that business, the environment and soils the business relies upon, and the customers that continue to support the business financially.

Local = Bioregional

The local economy should be the primary or foundational economy—the more local, the closer to the home, the better. Food, fuel, fiber, and water should all be sourced as locally as possible. The decentralization of basic utilities and needs strengthens community resilience while it boosts the local economy. Entrepreneurism and innovation within a stable larger economy depend on the existence of vibrant local markets. Face-to-face interactions between the business owners and customers, transparency of business practices, and a sense of community, all these can happen at the local scale, but are more difficult to maintain at the industrial level. Because of this, properly managed small businesses are more effective at serving their communities and maintaining integrity over the long term.

We need the local butcher, baker, shoemaker, repairman, compost expert, seed saver, and more specialized vocations per region. The old crafts and trades of our ancestors, what defined their lives and passions, are returning as ethical people seek out ways to avoid consuming fossil fuels.

The online communities we form can work with our online businesses in much the same way as local communities do with local businesses. They are leveraging technology through webinars, video chats, regular pictures, blogs, or podcasts to maintain an intimate connection with their customer base.

Commuting

Commuting must end. The average daily US commute to work is 25 minutes long which is a huge waste of time and natural resources transporting people all over to do work they should be doing in their own community.[129] The future is going to have less and less meaningless travel. It will instead be meaningful and occur less often. Telecommuting will continue to spread as well. This cuts down on costs for everyone and supports communities where they are.

Currency

Local currencies are fast becoming a new American phenomenon with the Ithaca Hour dollar, The Coastal Marin Fund, and the Portland bartering economy, but their roots are ancient and multicultural. When local goods are purchased with local currencies, it keeps funds in the community and can lower taxation. Tying local production to a local currency also protects that local industry from industrialized competition, allowing for entrepreneurism and craftsmanship to flourish locally.

The first local currency, the Ithaca Hour, was started by Paul Glover in 1991. Equivalent to $10 US dollars or an hour of work, the Hour is based on historical accounts Glover read about where a British Industrialist would let workers earn "hours" to spend at the company store. Though the Hour went into decline with Glover leaving the town of Ithaca, the currency still endures, and the model remains a strong one. LETS (Local Exchange Trading Systems) are systems that create a secondary economy, or exchange of goods and services, independent of the monetary economy; the LETS operate as a book-keeping, non-profit organization to assist the members. The Brixton Pound was launched in 2009 as a local currency in the United Kingdom as part of the Transition Town movement.[130]

Bitcoin is a digital currency system that is peer-to-peer and decentralized. It is run using an encrypting software that no one controls. It has no intermediaries and optional transaction fees. It is a free market but requires money initially to buy bitcoin credits. It behaves like a local currency would, only it exists internationally through the internet.

Micro-loans, Micro-Insurance and Micro-Franchises

Micro-lending is the practice of loaning very small amounts of money to low-income individuals using adapted risk management methodologies instead of collateral. These are amounts that no commercial bank has any interest in dealing with but are critical to impoverished individuals, businesses, and families.

Starting out as an experiment in Bangladesh, Muhammad Yunus created the Grameen Bank, the first micro-credit bank, with a focus on lending to impoverished women because, as

---

[129] US Census Bureau. *Commuting in the United States: 2009.* 2011.
[130] Hickman, Leo. *Will the Brixton pound buy a brighter future?* 2009.

Yunus says in a PBS interview, they "give immediate attention to children... [and] have longer vision."[131] He discovered that over 98% of the loans were paid back while for his customers, there was an increase in savings and profits.[132] His work especially benefited women, children, and women's enterprises. In 2006, Muhammad won the Nobel Peace Prize for his work.

Micro-Insurance follows the same principles of adaptation to the conditions of low-income households, and micro-insurance is most often applied automatically as added security for lenders. In order to make financial sense, micro-insurance policies tend to be contracted village-by-village in Asia. Communities elsewhere can pool resources or money to protect themselves against crop failures, illnesses, or the expense related to family members passing away. The small contributions add up and mitigate risk faced by the members of each group.

Applying the same idea to franchise businesses, micro-franchises help impoverished individuals, especially women, become literate in business models, marketing, going door-to-door selling merchandise, managing a kiosk or stall, and more.[133] For people from small villages with limited experience and education, having a model to follow like a franchise can be a powerful guide and learning tool.

This concept of scaling down to include and care for all peoples is powerful. We can apply this in all communities to reach and serve all peoples with our designs, services, and businesses.

Formal and Informal Economies

The formal economy of paper money has always had an informal counterpart: bartering, the gift economy, public food forests and gardens, seed and food swaps, free food stands, neighborhood libraries, etc. The informal economy provides stability, so that we are not dependent on the formal economy alone. The informal economy is the safety net for our cultures. During the US Depression of the 1930s, the informal economy helped many families survive. When money was scarce, food was still plentiful on a farm or homestead.

NPOs

Non-profit organizations (NPOs) can be charities, cooperatives, or mission-driven businesses designed for a public or social/environmental good. Guided by a mission

---

[131] The New Heroes. *PBS New Heroes Ep3 03 Muhammad Yunus Microcredit Bangladesh*. 2005.
[132] Ibid.
[133] Teutsch, Betsy. *100 under $100: One Hundred Tools for Empowering Global Women*. 2015. p. 126.

statement and a board of directors, these organizations rely upon donations, grants, volunteers, and tax exemption. They can be easier or harder to set up depending on where you are—in some countries like the US and UK, they are easy to register, but they can take time in other areas, especially if the objectives include co-owning land. They may or may not make a profit; most NPOs return any surplus back to their operations in order to develop additional programming and actions mandated by their mission. NPOs in the fields of regenerative agriculture/permaculture and restoration projects will typically share surplus production that arises from abundance.

Cooperatives/Co-Ops/Buying Clubs

When groups get together, form a new organization, and pool responsibility for the management for a communal benefit like lower prices or organic foods in an urban food desert, they are forming a cooperative organization. These range in their purpose and their actions, but they all involve people cooperatively owning and doing the work to run the business or pooling their resources to get better products for lower prices.

Crowdfunding

Kickstarter, GoFundMe, IndieGogo, Barnraiser, and many other crowdfunding platforms offer entrepreneurs, artists, authors, and individuals a place online to legally raise money. Connecting demand directly to innovation and production, startups get the feedback and the funding they need to have a successful start. While not entirely a local tool, it behaves like a local one in terms of connecting the consumer to the producer, or the funder to the project.

CSAs

Community Supported Agriculture (CSA) programs are where customers directly support farmers and ranchers by buying shares of the farmer's produce for that season or year at the beginning of the season and sometimes throughout. In return, the farmers then provide weekly/biweekly/monthly fruit, vegetables, dairy, meat, or any combination thereof. In the future this could also include energy and fuel. Membership in these community-supported programs usually starts with a downpayment early in the growing season to support the farmer while most of the food is still in the ground (and sometimes this is a deposit you can get back). Payments can happen monthly, weekly, or annually upfront depending on the farm. CSAs are like local crowdfunding programs for farmers!

FarmShare Farms

FarmShare farms apply the concept of the CSA and crowdfunding to the land, facilities, and startup capital needed to run a farm. Participants invest several thousand dollars for a share in the farm in exchange for below-wholesale pricing on all the farm produce (which ideally would be all the food needs of their community of supporters). These can be organized with a representative board of shareholders that manages and directs the farm as well as providing reports to the supporters and including them in the decision-making process. Shareholders can form businesses around the products they receive if these are abundant enough, or they can buy multiple shares in the farmshare to facilitate the development of a business.

The contents of a CSA box from La Ferme de Quatre Temps (the 4 Time Farm)

Permaculture teacher, designer, and humanitarian aid worker Geoff Lawton has been working on setting these up in Australia, Japan, and the United States.[134] These farms have animals and plants, and they have very little waste as they feed and fertilize with what they generate on-site.

Stacked Careers or Products

By stacking functions in a system, we can increase the yield of the original unit of land; the same can be done with our time, careers, and even our company's products. We can do two things at once like visiting our grandmother and pruning her fruit trees for cuttings. We can work jobs that work together like seed saving and farming or physical therapy and yoga, or we can make products that work together in some way like seeds with a gardening book.

---

[134] *Episode 8 Geoff Lawton on the future of Permaculture & Food Production, Children & Permaculture*, 2015.

Leading Examples

Guayaki Yerba Maté Company

Yerba maté is a rainforest holly vine found in South America and used to brew a hot tea that is unique and energizing. It is a traditional cultural drink that represents the bioregion from which it comes from. It is a symbiotic plant which aptly captures the spirit of Guayaki 's business practices.

Guayaki is an example of a company trying to do everything to the highest ethical standards possible. The company partners with Atlantic rainforest communities in South America to sustainably harvest yerba maté and actively restore the rainforest. They are Certified Organic and Fair Trade. Their metallic packaging is compostable. Their offices use solar power; they offset their carbon footprint–sequestering over 55 tons of carbon.[135] Their marketing is viral and largely carried out by fans of the company and their drinks. By leveraging market demand for ethical, energizing drinks, Guayaki has created a powerfully regenerative business and community. It is a model that is scalable and replicable.

> "Our mission is to steward and restore 200,000 acres [81,000 hectares] of South American Atlantic rainforest and create over 1,000 living wage jobs by 2020 by leveraging our Market Driven Restoration business model"
> –*Guayaki.com*, 2016.

Fibershed

With a focus on carbon sequestration and local textile networks, Fibershed, the nonprofit organization, was founded by Rebecca Burgess to educate "the public on the environmental, economic and social benefits of de-centralizing the textile supply chain."[136]

Shortly after Burgess' project began in 2010, it inspired a worldwide movement based around the new term "fibershed" which sounded bioregional like a watershed but clearly focused on the entire garment lifecycle.

---

[135] Guayaki.com *Sustainability*. 2016.
[136] Fibershed.com. *About.*, 2016.

If all our garments and fibers were grown and processed locally, an enormous amount of pollution and unethical labor would be erased from the world.

> "Fibershed develops regional and regenerative fiber systems on behalf of independent working producers, by expanding opportunities to implement carbon farming, forming catalytic foundations to rebuild regional manufacturing, and through connecting end-users to farms and ranches through public education."
> —Our Mission and Vision, FiberShed.com

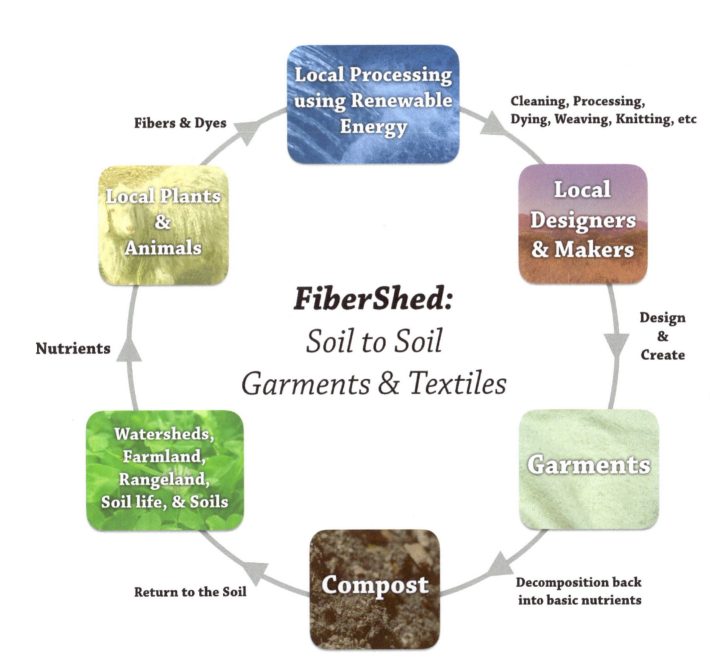

The Permaculture Skills Center

Erik and Lauren Ohlsen co-founded and together run The Permaculture Skills Center in Sebastopol, California. There they have been developing and helping others develop collaborative businesses for years in addition to teaching classes on specific permaculture applications such as graywater harvesting, landscape design, and social permaculture. Collaboration between autonomous agents leads to a pattern of resiliency similar to the growth of a city, forest, or meadow. The Permaculture Skills Center hosts several businesses on-site, fosters fledgling businesses, and forms relationships with other businesses. It works much like an edge ecosystem—it contains its own autonomous systems yet overlaps with outside cycles and systems, making it extremely resilient and valuable to the community as a whole.

The Center has a demonstration food forest that is open to the public and an extensive water harvesting system. A multitude of educational programs are held regularly on-site. Each program is managed as a separate business entity—the Farm School, Ecological Landscaper Immersion, Advanced Water Harvesting Workshops, and more. Red H farm rents a section of the center grounds to run a no-till CSA. Permaculture Artisans offers landscaping and design consultation and installation work; they work with local governments, businesses, and homeowners. They also host guest speakers and events open to the public. Each entity is financially separate, but they all work with each other. It is an ecosystem of businesses occupying one space with many people playing many roles. Rather than competition, the Permaculture Skills Center is fostering a regenerative economy that uses collaboration, permaculture principles, and synergistic niches to develop the way that ecosystems do. Skill centers like this are needed everywhere.

> "[The Regenerative Economy] focuses on relationships rather than outcomes"
> –Erik Ohlsen, Permaculture Voices: PV3, 2016.

Business Guilds for Community

While governments guide, businesses build, feed, and power communities. Businesses form guilds that cooperatively support each other in an ecosystem of support that, if stable enough, a town or village will form around. Unstable economies and business models, like the California Gold Rush and high-pressure hydraulic fracking for natural gas, always lead to a boom and bust pattern. When many businesses are involved and invested in the local area, communities are born. A business guild can be as simple as the components of the local food chain: farmers, ranchers, value-adding businesses, and consumers. A business guild can both represent a cycle and a series of interrelated businesses.

XX. Regenerative Agriculture

What is Regenerative Agriculture?

Regenerative agriculture implies different practices in different scenarios, but for our purposes we can say that regenerative farming uses processes that continuously restore, maintain, and enrich the ecology of the area being farmed. This ecology would include the soils, soil life, plants, animals, people, and even non-living elements—for example, check dams made of stone are very important non-living regenerative tools. In practical terms, "enriching ecology" means the farmed area contains more biodiversity, more water is available on-site (or wiser management of water is in evidence), and more soil is retained on-site. Meanwhile, farmers are making a profit with a marketable yield that is safe, nutritious, and improves and maintains the health of those who consume it. While all these methodologies are similar in their goals and to varying degrees in their application, they are each distinct and represent different branches of study.

Biological Farming

> **"Biological Agriculture or Biological Farming: Production of healthy plants and animals using management promoting a healthy soil microbiome, using natural processes to grow crops and animals that promote healthy animal and human microbiomes. Biologically-based production methods select for the desired crop and against weeds that might compete with the crop, promote natural processes of nutrient cycling so the soil is not mined of nutrients, leaching is prevented, soil structure is built, and plants/animals are protected from diseases and pests"**
>
> –Dr. Elaine Ingham, 2016.

Biological farming is a way of growing food and raising animals without amendments, as nature has done for at least a billion years.[137] Biological farming respects the whole ecosystem while making a profit for the farmer and generating high yields of nutritionally-dense foods. At the same time, biological farming cares for the future by building soil and increasing biodiversity—which leads to economic stability for the farmer and ecological stability for the ecosystem their farm interacts with. An example of a farm using biological farming techniques is Celebration Farm in California, USA, where Dr Elaine Ingham is running biological farming trials using compost, compost tea, compost extract, mulch, no-till or light tillage to test their effects on both garden-sized and farm-sized plots. For instance, "weeds" get sprayed with a compost tea that changes the soil food web constituents at the weed's root level. This makes the plant so weak that when chopped it either rebounds weakly or dies back— meanwhile, the same compost tea strengthens the garden crops growing among the weeds.

Dr. Ingham uses microscopes and sophisticated, modern analysis and testing methodologies (some of those that she pioneered are now seen globally) to calibrate her compost to her exact garden needs down to the exact soil food web population densities and ratios in her compost, compost tea, or compost extract. Her methods are being adopted by commercial enterprises and small farmers alike the world over.

Carbon Farming

Carbon sequestration is an important aspect of regenerative agriculture which increases the capacity of soils to sequester carbon in the soil. Carbon farming makes carbon sequestration and the tracking thereof the farmer's main focus, even as he/she strives for

---

[137] Ingham, Elaine. *Email*. 2016.

profitability and competitiveness. Regenerative agriculture processes most importantly feature returning as much organic matter or humus to soils as possible, where it can improve soil quality and foster plant and fungal growth, sparking a beneficial cycle of carbon sequestration from the overburdened atmosphere. Joel Salatin's family farm (featured in the next chapter) is an example of a multi-generation carbon-farming operation.

While carbon farming is often associated with land-based farming systems, our definition here also includes carbon farming in the oceans, seas, and other bodies of water through the growing of algae, fish, shrimp, plants, birds, and their deposit in sludgy, carbon-rich soils below. Veta La Palma (featured in the Aquaculture chapter) is an example of an ocean-based carbon-sequestering farming operation.

It may seem to some that the claims that carbon farming can take atmospheric levels of carbon back to pre-industrial levels within 10 years or more of full adoption are exaggerated, but it is actually mathematically possible. Half the atmospheric carbon has already been absorbed by the oceans (leading to their well-documented acidification). The remaining half in the atmosphere is well within the capacity of our soils to take back.[138]

The missing organic matter or humus from our agricultural soils can also be returned at the same time since humus is mostly carbon. All the soil depletion can be remedied through carbon sequestered from the atmosphere through plant and fungal growth. Industrial agriculture and aquaculture have over time diminished the capacity of our soils and seas to sequester carbon; carbon farming using regenerative processes can bring back that capacity – all our "problems" are linked but so are their solutions.

The caveat to Carbon Farming is that we need everyone to be involved, and we need fossil fuel usage to be phased out as quickly as possible and left behind. It is a one-time sequestration event using the soils and agricultural lands, and it has a limit. In all reality, this plan has to also include the oceans since they are holding the rest of the excess carbon which also can and must be biologically sequestered.

Natural Farming

Developed by Masanobu Fukuoka, natural farming is often called the Fukuoka method or "do-nothing farming". His no-till, no machines, no fertilizers, and no chemicals stance seemed idealistic at the time, but by mimicking nature, Fukuoka's methods accomplished surprising successes like growing rice without permanent paddies, naturalized annuals, and productive fruit trees without pruning. He not only ran an economically successful and productive farm in a counter-cultural way but promoted his methodology worldwide.

---

[138] Pickerell, John. *Oceans Found to Absorb Half of All Man-Made Carbon Dioxide*. 2004.

High in the Austrian alps, Sepp Holzer's Krameterhof operates in much the same way but with one diesel twist. Plastic, greenhouses, transplanting, fertilizers, and tillage are not used, but Sepp does use heavy machinery such as bulldozers to build ponds, terraces, and other earthworks which burst into life shortly after the disturbance.[139] Sepp grew up on his family farm and was given great autonomy over himself and a section of land the family gave to him. He eventually took over the farm, and using the intuition and experience from a lifetime of observing and working with nature, he began to turn heads quickly with his own natural farming methods. His seemingly strange marriage of heavy machinery, minimal inputs, and natural management led to marketable outputs, scrutiny, excitement, and endless attention.

Almost all his designs, after the initial installation, can operate without human intervention. Several of the sites he has worked on have been abandoned for years—only to be found thriving later on. In fact, Sepp even took over one of these abandoned sites and dubbed it Holzerhof. It is where he lives and farms currently when he is not traveling the world speaking, designing, and installing sites.

While there are several types of natural farming (like KNF) and it may seem the territory of the generational farmer or the rural guru, the methods of natural farming are understandable and replicable in any climate working with permacultural methods, scientific research, local knowledge, and minimalism as your guides. It is a philosophy in practice more than any set of protocols. Interestingly, it naturally sequesters carbon by design but has no specific protocol for carbon sequestering.

Perennial Farming

In many ways, perennial farms hark back to the way humans gathered and foraged for food in the wild, seasonally. Established orchards, food forests, wild forests, and permaculture sites can potentially go on for centuries and millennia without human caretakers if the conditions are right. Perennial farming is a stable farming system when it is closely patterned after natural ecosystems.

A component of carbon farming, perennial staple crop farming is what large groups of people used to rely upon for their staple foods before fossil fuels shifted our reliance onto a handful of annuals primarily—and this predates the concept of carbon farming. Perennial crops such as chestnut, baobab, and coconut trees each provide a wide spectrum of foods, fibers, fuels, and medicines. In the 1800s, fallen chestnuts in the eastern United States were so

---

[139] Holzer, Sepp. *Sepp Holzer's Permaculture: A Practical Guide to Small-Scale, Integrative Farming and Gardening*. 2011. p. 4.

plentiful that shovels could be used to gather them.[140] Baobab trees can live for over a 1000 years, are gigantic trees, and bear fruit that tastes like sherbet![141] Coconuts provide a cooking oil that can be refined into biofuel! Wisely, humans traditionally paired with long-lived, climax species.

Long-lived perennial crops need a polyculture around them for strong and resilient growth. An example of perennial monoculture cropping that is visibly failing is the large-scale strawberry production of California, which has led to widespread soil fumigation practices using methyl bromide—which makes for toxic fruit sold in stores across the US. However, due to popular outrage, methyl bromide is being phased out in stages at the time of writing.[142] Only a polyculture will support a strong perennial system as seen in nature. Mark Shepard's oak savanna system (described in the next chapter) is a perfect example of imitating nature to create a perennial farming operation. Its resilience comes from imitating what naturally occurred in that bioregion before it was brought into dysfunction.

Perennial farming can come in many iterations and under many names, from polycultural orchards to agroforestry to alley cropping to edge cropping. They all use trees, shrubs, or bushes to support and enrich the annual production and provide a second or third yield on top of the annual one—meanwhile saving soil, stopping erosion, preventing flooding, soaking in more water seasonally, and enriching the soil, the other plants, and themselves in the process.

Holistic Management of Cattle

Using the holistic management framework but focusing on cattle, this framework was created by Allan Savory and helps farmers use observation to inform management decisions, and imitate prehistoric grazing patterns. Joel Salatin, a widely known HM farmer, allows his cattle to mob graze a small patch of pasture, approximately 100 cattle on 1-2 acres (4000-8000m²), for 24 hrs, and then shifts the cattle onto a new patch of land.[143] They will only return two to three times a season to that same patch. The daily cattle shift happens without coercion, only invitation. Using portable electric fencing, cattle can be kept safe and easily shifted daily. They eagerly await the chance to chew on fresh pasture each morning, so all the managers have to do is peel back the electric fence and let them through.

Portable fencing allows for creative adaptation to terrain, area, and timing. It is also cheaper than permanent fencing which doesn't accomplish the same goals at all. Joel says

---

[140] *Woody Agriculture and Breeding Trees with Phil Rutter. (Part 1 of 2) (PVP057).* 2015.
[141] Phytotrade Africa. *The Baobab Tree and its Fruit.* 2012.
[142] De Witte, Melissa. *Pesticide predicament for California's strawberry growers.* 2016.
[143] *The Salatin Semester,* 2016.

that if you don't move a fence for three years, you can make that permanent.[144] You may have a series of small, permanently fenced paddocks that you shift cattle through, but then must rely upon observation and timing for when to shift the animals and at what densities. It is imperative to keep the right amount of pressure on the herd, so they graze indiscriminately, so the plants can be grazed thoroughly, and so they can fully rebound. It is also important to know the history of your grazing site: was it an area of intense or light grazing in the past? What herbivores thrived there? How can we imitate what was natural? How can we prepare for climate change with all this in mind? For some areas, an extended rest may be called for before grazing can resume as was the case with the Loess Plateau Restoration project.[145]

Holistic management above all embraces the complexity of holism in an agricultural setting with careful planning and management principles and techniques. Understanding your animals, their digestion, their waste products, their inputs, their outputs, and anything else particular to your operation can improve your understanding and management of the business. Doing your best to work with complexity requires reflection, note-taking, developing metrics, collecting and analyzing data, and adapting in response to feedback.

---

[144] Ibid.

[145] Wuerthner, George. *Climate Change and Livestock Grazing*. 2015.

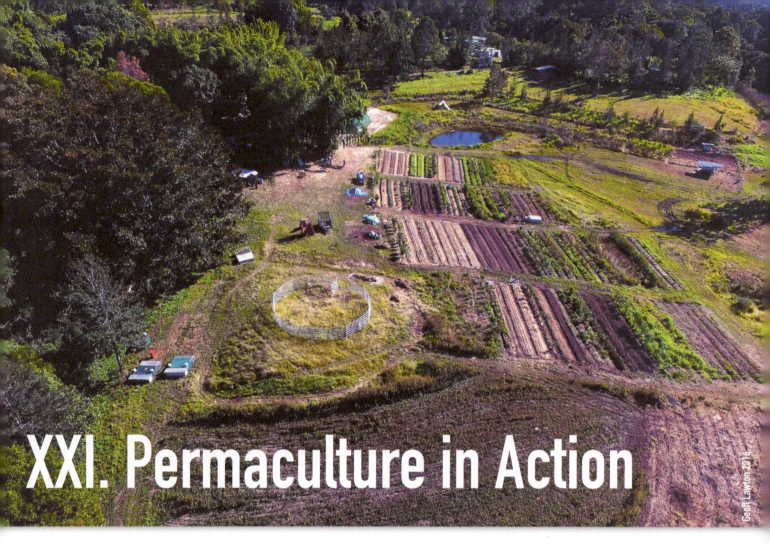

XXI. Permaculture in Action

This list of applied-permaculture examples shares a spectrum of examples where permaculture is applied. They do not represent all that is possible, nor do they represent all who are actively using permaculture to be more regenerative and profitable in their business or service. They are simply a selection of regenerative examples to share the many ways permaculture is being applied.

Zaytuna Farm reviewed by Geoff Lawton

This permaculture demonstration site in northern New South Wales, Australia, was once an ecologically-degraded dairy farm, and is now 66 acres (27 hectares) of detailed and vibrant subtropical food forests, gardens, ponds, and more, designed to educate and inspire.[146] It is an off-grid permaculture paradise! In development since 2001, world-renowned permaculture educators Geoff and Nadia Lawton regularly host groups of students, interns, homeschoolers, and tours.

Geoff's farm has seemingly everything; permaculture techniques and methods can be found in practice all over.[147] They are off the electrical grid with solar power and their own generator. Their hot water comes from a rocket stove water heater. The toilets are all

---

[146] *PRI Zaytuna Farm, NSW, Australia.* 2011.
[147] Lawton, Geoff. *The Geoff Lawton Online Permaculture Design Course.* 2014.

composting toilets, and the graywater reed bed system is inspected and approved by the local state officials. There is a metal shop, student kitchen, a commercial kitchen, and housing for their family, workers, and interns. Sustainable materials like bamboo (which is on-site), straw bales, and natural plaster are sourced for building projects, big and small, whenever possible.

Zaytuna is a full-cycle farm that includes animals, people, and plants working together in harmonious cycles to produce an abundance of food that supports 45 people on-site.[148]

---
[148] *PRI Zaytuna Farm, NSW, Australia.* 2011.

XXI

Permaculture techniques like remineralization of soils using animals, compost, insect hotels, cell-grazing of cattle, food forestry, biointensive gardening, and more are demonstrated and taught to students on-site and through Geoff's online courses.

Animal systems using ducks, chickens, cattle, and more are also constantly in operation.[149] Using permaculture as the lens, interns learn about gardening, farming, dairy operations, omnivores, and cattle in context using permaculture techniques hands-on. Zaytuna farm yields topnotch students and educators who continue to spread permaculture to people all over the world.

---

[149] Lawton, Geoff. *The Geoff Lawton Online Permaculture Design Course*. 2014.

Polyface Farm reviewed by Joel Salatin

Joel Salatin's 550-acre (222-hectare) family farm, Polyface, began in 1961 as an eroded, gullied, and worn-out farm. Joel and his father before him stewarded the land through an amazing recovery using a sophisticated holistic management system involving chickens, turkeys, cows, pigs, earthworms, and soil microbiology.[150] Joel happily calls himself a "grass farmer" because that is the basis for his entire ecosystem: the annual solar energy trapped by grasses.

Grass-based farming focuses on increasing the perennial grasses' potential to photosynthesize, store carbon, and rebound from grazing while turning a profit. Specifically, that means rotational grazing timed perfectly where animals are held at beneficial ratios of grazing space to animal density for a day or less. Cattle grazing is followed by chickens a few days after to spread out the cow manure—ending pest cycles while feeding the birds high protein fly larvae. This imitates the phenomenon observed in nature where birds travel and live in symbiosis with herbivores, spreading their manures, eating larvae out of manures, and keeping parasites in check.

---

[150] *The Salatin Semester*, 2016.

The Millennium Feathernet (above) has an electric fence and is used as a portable egg production facility. It does not follow the cows in rotation like an egg mobile though the cows can come through and graze the grass down for them routinely. It is a way to raise chickens in a pasture and shift them regularly.

In terms of animal succession, turkeys can follow the chickens. Goats and sheep can clear brush and grasses to prepare the way for cows initially. Joel's operation is mature and focuses primarily on cows, chickens (layers and broilers), turkeys, and pigs. He integrates their systems ingeniously, following the animal's instincts and desires as his guiding ethic.

Salatin's approximately 100 cows graze 100 acres (40 hectares) about one acre (4000m^2) at a time using portable electric fencing. The paddocks are roughly all at the same elevation since different elevations have grasses germinating and developing at different times and rates. They return to the same area two to three times a season at most. They only spend 24 hrs in any one space to avoid the second bite effect.[151]

Perennial pasture grasses have evolved to thrive when eaten to the ground in one grazing event by herbivores and then allowed to regrow as the tightly packed herd of grazers moved on, pressured by their natural predators. This process, unhindered in prehistory, created the deep, rich grassland soils that humans naturally were drawn to for practicing agriculture. If we allow a "second bite" by grazing the following day, we interrupt the plant recovery and regrowth which will begin to push the pasture holistically (soil, animals, and plants) into decline. This is why grazers have been blamed for creating desertification—while more accurately speaking, it was grazing management that was at fault.

---

[151] Ibid.

Following after the cows are the laying hens in an egg mobile, a simple, portable shelter (featured earlier in *Convection*, p. 22) that lacks an electric fence. They have feed, pasture, and the freedom to range, but they primarily focus on spreading the cow manure patties and eating the insect and parasite larvae inside.[152] They spread them out and sanitize them by increasing surface area and exposing them to the air, sun, and wind—this means that the cows rarely have flies on them and never need any special treatments for parasites. The Salatins don't use vaccines or antibiotics; they cull sick animals quickly, monitor closely, and mimic nature which makes for strong genetics and functional immune systems that don't need vaccines or antibiotics.[153]

Turkeys often follow the chickens, and then the pasture rests until the return of the cows, or they grow it out for hay. This system makes for delectable eggs, pasture-raised beef, and pasture-raised soup birds in a sanitized and carbon-sequestering pasture system. Joel's system sources lumber from his property that is sustainably cut and milled on-site. He does supplement his laying hens with local non-GMO grain which Joel admits is their weakest point.[154]

---

[152] *Polyfaces*, 2015.
[153] *The Salatin Semester*, 2016.
[154] Ibid.

In winter, the Salatins rely on a clever deep bedding method where the cattle feed on bailed hay from the pastures they visited lightly that previous season.[155] They eat from troughs on pulleys that raise and lower to unload the hay from the top of the stacked hay bales as well as keep the feeding troughs off the rising ground. This all occurs inside a hay barn that is half stacked hay bales and the other half deep bedding. A layer of straw and wood chips with some grain mixed in gets laid down after each feeding to combine with the manure from the visiting cows. By mixing in grain with this new layer of bedding, it creates a compost heap littered with fermented grains which for a pig is a cross between a buffet and a treasure hunt. In their enthusiastic searching, they turn the matted layers into a fluffy mix making it easy for the farm interns to remove and relocate to where it can be fully composted. Most of the compost gets spread out onto the fields to support the foundation of the entire system, the pasture, but some goes into the Salatins' vegetable garden.

The deep bedding method of adding carbon throughout a season can be applied to many different animal systems. Their chickens overwinter living below rabbits inside a *Raken* (Rabbit+Chicken) house where they live in a deep bedding situation. The rabbit manure and chicken manure are mixed with a carbon source like wood chips or straw regularly to keep the ratios composting and sanitary; this leads to a rising floor. By using pigs to loosen the bedding initially and using a tractor to transport the bedding to a large compost area where it can be turned further with the tractor, Joel has created a system where labor and time are saved, and animals are utilized as much as possible in synchronization with their instincts.

Joel hosts regular farm tours, has an internship program, and has a farm store that is open regularly while his son, Daniel Salatin, runs the daily business of the farm. It truly is a farm of many faces but all pointed towards regeneration.

New Forest Farm reviewed by Mark Shepard

Designed to mimic the oak savanna that once dominated the Midwestern United States, Mark Shepard's New Forest Farm in Wisconsin, established in 1995 at 106 acres (43 hectares) is the oldest commercial-sized permaculture site in America. Versaland and New Forest Farm together are the leading examples of permaculture production farming in the Midwest.

Growing up, Mark recognized a dissonance between the freely given abundance of the forest and the hard sought-after products of the garden where serious work and constant management were required.[156] His insight led him on a quest to solve the greatest holistic

---

[155] Ibid.
[156] Shepard, Mark. *Restoration Agriculture: Real-World Permaculture for Farmers*. 2013. p. 19.

problem of our day: how to live regeneratively. Inspired by Fukuoka, J. Russell Smith, Bill Mollison, and Henry David Thoreau, Mark, with his family, transformed former dairy farmland into an oak savanna polyculture using a network of keyline swales. It is managed with silvopasture operations using chickens, pigs, and cattle as well as alley cropping of annual vegetables.[157]

Ecological restoration is a foundational component of Shepard's operation. His work is undoing the damage done by overgrazing and poor management even while producing food crops. His farm is a member of Organic Valley which is a farmer-owned consortium with a product line recognized nationwide in the US grocery stores.

While establishing his system, Mark understood that there were no known successful chestnut varieties for his area and realized that getting the right genetics would take a variety of trials. Mark has developed a system for getting the hardiest, most resilient, and vigorous plants for his area—with vigor being sourced from their genetics, not water or fertilizer. The STUN method stands for Strategic Total Utter Neglect.[158] Mark plants his trees close together which makes it obvious which trees perform better than others and then aggressively culls out the weak plants as Joel Salatin does with his sick animals. This has led to selecting and propagating the best of the best of the best of the best and so on for over 20 consecutive

---

[157] Ibid. p. 210.
[158] Shepard, Mark. *Farming. It's Damn Hard. An interview with Mark Shepard.* (PVP091), 2014.

NewForest.farm, 2016

seasons with all his varieties. His genetics are now highly sought after and support a profitable plant nursery business.

New Forest Farm has oak and chestnut canopies with cherry and apple trees in the understory, hazelnuts at the shrub layer with cane berries, currants below them, and grapes climbing throughout.[159] It is a thriving oak savanna wonderland of food, timber, and fiber where the focus is not on NPK but on ecological health.

> "Nature has never spent a dime on pest or disease control or fertility"
> *Mark Shepard, Permaculture Voices, 2014.*

Krameterhof reviewed by Zach Weiss

Nearly 5,000 ft (1,500m) above sea level, nestled high in the Austrian Alps in a place known as Austria's Siberia, lies the Krameterhof, Sepp Holzer's family farm where he grew up and took over management in 1962.[160] Since that time it has been a place of natural farming experimentation. His experimentations and stalwart determination have earned Sepp the title "Agro Rebel".[161]

Krameterhof is a 111-acre (45-hectare) site with 72 ponds and a more than 1000ft (400m) difference in elevations, with 116 days of frost and an average temperature of 41°F (5°C).[162] The landscape is a polycultural myriad of microclimates on large terraces that prevent the slopes from eroding and allow for the water they soak up to slowly release during the summers. It is covered with ponds stocked with European crawfish, carp, trout, and pike. Sepp

---

[159] Shepard, Mark. *Restoration Agriculture: Real-World Permaculture for Farmers*. 2013. p. 66.
[160] Holzer Permaculture. *The Krameterhof*. 2016.
[161] *Sepp Holzer The Agro Rebel*, 2016.
[162] Holzer Permaculture. *The Krameterhof*. 2016.

does not irrigate; his area receives enough precipitation.[163] Selecting seeds from the best plants in the poorest soils, and spreading those seeds, has over time built a collection of extremely hardy and exuberant plants. His microclimates at one point even supported citrus which is unheard of in his area. He uses no soil amendments but focuses instead on the whole system's health.

Sepp uses animals to assist in managing the farm, especially pigs.[164] Pigs will turn the soil, make ponds, and remove stubborn plants like a blackberry patch just by following their natural inclinations. Housing mixed species cattle and pigs inside earth shelters, Holzer keeps them in large habitat paddocks held in place with an assortment of fencing types: living, metal, wood, electric, and more. Sepp also uses pigs with portable electric fences to reforest areas.

Sepp Holzer at times, like Fukuoka, does not prune, especially in a broad acre setting. He allows the branches to grow and fall below the horizon line which stimulates fruiting rather than vegetative growth. He also grafts his trees with sometimes very interesting combinations. He protects his trees from wildlife with a homemade bone sauce that's flicked or dabbed onto tree trunks (see the picture to make your own). The sticky, stinky substance stays on the trunk's bark for several seasons, though some, like Zach Weiss, make it annually and simply

---

[163] Weiss, Zach. *Email*, 2016.
[164] Holzer, Sepp. Sepp Holzer's Permaculture: A Practical Guide to Small-Scale, Integrative Farming and Gardening. 2011. p. 83.

Zach Weiss www.ElementalEcosystems.com, 2016.

flick a dab on each tree in early spring.[165] It is primarily animal fat so it hardens when cool and is non-soluble, so watering or rain won't remove it.

Sepp is also known within the US for popularizing *hugelkultur* gardening, or mound-culture gardening, where wood and woody biomass is buried under soil in a mound.[166] As the wood breaks down it releases nutrients, retains water, and can generate heat at times, but usually does not. Sepp grows annuals and perennials on his *hugelkulturs*. He also uses them to generate topsoil. Once the wood is fully broken down, the mounds shrink in size but are piles of soil rich in organic matter—perfect for growing annuals and perennials.

Throughout his systems, Sepp

[165] Weiss, Zach. *Email*. 2016.
[166] Holzer, Sepp. <u>Sepp Holzer's Permaculture: A Practical Guide to Small-Scale, Integrative Farming and Gardening</u>. 2011. p. 21.

also uses rocks and boulders.[167] Boulders serve as thermal masses in his microclimates, keeping areas warm overnight and lengthening their growing season. Rock piles are also used as moisture condensers and animal habitat. Some herbs are even grown in rocky areas on the Krameterhof because they provide the best medicine—many wild herbs prefer naturalized soils to rich garden soils. Sepp came to this realization through observation, trial, and error.

Sepp's childhood was immersed in nature and led to his deep understanding of it. Today, Sepp teaches and consults all over the world and lives at the Holzerhof. His son Josef manages the Krameterhof, where there are tours, classes, and new discoveries regularly occurring—such as Josef's successful grafting of both apple and pear onto mountain ash for an extremely hardy rootstock and to keep fruit above the natural browsers.[168]

Paradise Lot

With over 200 different perennial plant varieties on a tenth of an acre (400m^2), Paradise Lot would have been impressive in any climate, but to be in the cold temperate climate of Massachusetts, USA with banana plants for mulch in the front yard and a subtropical plant-filled greenhouse in the backyard for year-round food from the garden is beyond impressive.[169]

When best friends Eric Toensmeier and Jonathan Bates purchased the duplex, a house split into two separate apartments, in 2001, the backyard was a barren lot, devoid of vegetation. With this blank slate, Eric and Jonathan designed an abundant vision and put it into reality with hard work, persistence, and good design.

From the plants surrounding the pavement of their parking spaces (which hold thermal mass) to the dense growth ringing the edge of the property like a green wall, there is no bare earth; every space is used. Chickens are tucked onto the side of the backyard which is only 90x45 ft (15x30m).

There's even aquaculture inside and outside the greenhouse. The water inside the greenhouse serves as a thermal mass, radiating heat overnight in cold winters to keep the plants warm. The side yards and small backyard are a rich perennial food forest of pigeon peas, plums, mulberries, hardy kiwi, grapes, asian pears, blueberries, pawpaws, Chinese yam, watercress, arugula, and much, much more.[170] Jonathan and Eric sell the new plants the

---

[167] Ibid.
[168] Weiss, Zach. *Email*. 2016.
[169] Permaculture Research Institute. *OVER 200 FOOD PLANTS ON JUST A TINY 1/10TH ACRE OF COLD CLIMATE URBAN LAND*. 2014.
[170] Ibid.

perennials generate and give workshops teaching locals how to make their own lots abundant. Paradise Lot is a superb cold climate urban permaculture example.

The Urban Homestead reviewed by Jordanne Dervaes

Since 1990, the Dervaes family has been working on being self-sufficient in the middle of Pasadena, California, 15 minutes away from downtown Los Angeles and 100 feet from a major freeway with only a tenth of an acre of land to grow on (3,900 ft²/1189 m²).[171] When the Dervaes learned about how pervasive GMOs were in commercial foods, the family mulched over the lawns and installed a tightly packed garden and homestead system, providing for 75% of their food needs.[172] They are a perfect example of hyperlocal seasonality in the city.

The Dervaes use biointensive planting densities to save water. The family raises pygmy goats, ducks, dwarf rabbits, and chickens for milk, eggs, and manure, but no meat; they are vegetarians. They sell their extra eggs, edible flowers, and organic heirloom produce to local restaurants and run a front porch farmstand earning an annual average of $65,000.[173] Their kitchen waste is either processed by the animals or composted in various ways, and then it is put back into the garden or used in compost tea. Since 2003, the Dervaes have grown 5,000-6,000 lbs (2,200-2,700Kg) of food every year.[174][175]

Using a graywater system, olla pots, rainwater harvesting, and mulch, their annual water bill is only $600.[176] They have a solar panel array, but no back-up battery, so they are still hooked up to the grid which is less expensive and easier for their situation—though they've expressed that outside of a city they would like to be completely off the grid. Using less than 6 kWh a day, the Dervaes use less than a quarter of the electricity used by the average Pasadena citizen, which is 25 kWh a day.[177]

The Dervaes lifestyle is the glue that binds all the systems together. They use less, save more. From home-brewed biodiesel from the restaurants they work with, to the assortment of hand-crank kitchenware, to the solar and cob ovens, to the pedal-powered wheat grinder, to seed saving, to homeschooling, to staycations, to upcycling crafts, the Urban Homestead is a holistic system that is not just about food or business; it is a way of life that they find ethical.

---

[171] The Urban Homestead. *By the Numbers*. 2016.
[172] *Homegrown Revolution (Award winning short-film 2009)- The Urban Homestead, Dervaes*. 2011.
[173] Ibid.
[174] Dervaes, Jordanne. *Email*. 2016.
[175] The Urban Homestead. *By the Numbers*. 2016.
[176] Ibid.
[177] *Homegrown Revolution (Award winning short-film 2009)- The Urban Homestead, Dervaes*, 2011.

Growing Power

In Milwaukee, Wisconsin, former professional basketball player Will Allen started Growing Power, a program to help teens grow food for their community. It would become a new kind of urban farm that would feed 10,000 people on 2 acres (8000m²) year-round.[178] In a frigid climate full of fisherman, aquaculture is a logical solution to the contamination of waterways and fish stock in the North American Great Lakes, yet before Will Allen, it was scarcely seen in the Midwest.

Using a vertical stacking system of growing beds, Growing Power has nearly doubled the growing area of their greenhouses. The plants feed on the fish manure and filter the water for the fish. Plant-eating fish like Tilapia can be fed vegetation that is grown by their waste in systems like these. They raise Tilapia and Lake Perch as well as grow various greens, edible

---
[178] *Growing Power - A Model for Urban Agriculture*, 2010.

XXI

By stacking the growing shelves over the fish cultivation area, they can multiply their yields by at least three to four times per square meter (or foot).

flowers, tomatoes, and peppers.[179]

Vermicomposting is a critical component to Growing Power with entire greenhouses filled with composting material. Through a CSA program, they sell produce, fish, oyster mushrooms, and compost, with their primary focus being on health. They feed over 10,000 people with local foods, free of biocides and synthetic fertilizers in a cold climate in an urban food desert using greenhouses, solar power, water heaters with a computer monitoring temperature, vermicompost, aquaculture, and local community support.

Will Allen's systems are always being improved upon, and innovations are constantly being tested, such as old washing machines being used to sift compost. Experimentation, observation, and adaptation led Will to his successful systems. Winner of the MacArthur Genius Award, Will Allen works currently in both Chicago and Milwaukee as the founder and leader of Growing Power.[180] He teaches biointensive indoor wetland development and management, and shows us all how to fight hunger and build strong, healthy communities anywhere.

Ernst Gotsch's Family Farm

In 1984, Ernst Gotsch, a Swiss immigrant, acquired a parcel of dryland in Brazil that was completely denuded of trees and began to establish a homestead and farm for his family.[181] Through careful observation of the pioneer species that appeared first on other recovering clear-cut properties with similarly poor soils, he selected commercially valuable species out of that mix and conceived a polyculture of bananas, pigeon peas, cassava, and native trees like Eretrinas and Ingás.[182] He planted them all over his land. Within a few months, the new growth was large enough to prune, so he chopped and dropped their leaves, but he left the stalks, allowing them to regrow and prepare more mulch to be chopped and dropped. It is this regular pruning, that he does perpetually, that builds and maintains his soil fertility. With the tropical decomposition cycle being so rapid, composting is difficult, yet chop and drop mulch is broken down by the area's soil food web quickly. Within only a few months after the initial pruning, Ernst had enough soil on the ground to start planting valuable trees and crops.

Forty years have passed since he first established those trees. Where once a dryland stood, 17 streams run year-round.[183] There is substantially more rain, and the atmosphere is

---

[179] Ibid.
[180] GrowingPower.org. *Will Allen*. 2014.
[181] AgendaGotsch. *Films*. 2016.
[182] Ibid.
[183] Ibid.

cooler than it was before the establishment of the homestead.[184] His farm is a large, diverse rainforest that he manages with a chainsaw, machete, and loppers. His forest system produces some of the most expensive and sought-after cocoa beans in the world. He also grows numerous other crops in his agroforestry and alley cropping system.

The constant pruning releases gibberellic acids from the plants into the soil which hormonally stimulates a bloom in soil biology which in turn creates a burst in root and stem growth. Pruning is his fertilization method. The plant exudates and the mulch feed the soil food web and then support the recovering plants.

Ernst's systems have rows of annual crops between rows of legumes and mulch trees that are regularly pruned down to just their trunks. The pruned perennials provide timber, mulch, and an enormous amount of biological activity in the soils through decomposition. Ernst has clearings for annuals, forage for his chickens, and fruit and nut trees—all within his rainforest's system where he and his family satisfy nearly all their needs.

"During my whole life I haven't seen an invasive plant. I haven't seen any harmful plant"
–Ernst Gotsch, Agenda Gotsch, 2016.

Ernst Gotsch's farm has not gone unnoticed. Fazenda da Toca, a 2,300-hectare (5,700-acre) organic commercial farm in Brazil, worked with Ernst for several years transitioning their farm system to agroforestry using Ernst's methods and guidance.[185] Ernst had to adapt his manual management techniques to machine management techniques for the large-scale farm which led to the development of new technologies. In only two and a half years, dramatic changes and benefits were recorded in the soil's ability to hold water and provide nutrients to the plants. The farm produces fruit, eggs, and grains and has an educational outreach program teaching ecological farming methods with farm tours and through the Toca Institute. Fazenda da Toca's goal is to prove that Ernst's methods scale up and are syntropic economically, socially, and environmentally.

The Al Baydha Project reviewed by Neal Spackman

Al Baydha, in the kingdom of Saudi Arabia, is in the foothills of the Hijaz mountains south of Makkah, where nomadic herding tribes, known as *Bedou*, have grazed for thousands of years. These traditional peoples have ceased being nomads and are now cutting their forests for charcoal to sell to the city while overgrazing the land, causing both deforestation and desertification. Since 2010, the Al Baydha Project, headed by The King Faisal Foundation

---
[184] Ibid.
[185] Ibid.

and run by Neal Spackman and a crew of local tribesman, has been greening 90 acres (36.5 hectares) of desert.[186] This research site is a permaculture prototype that aims to create a new rural economic model that enables the Bedou to continue to practice their animal husbandry, which is intrinsic to their culture, in a regenerative and sustainable way. The project also has long-term goals related to infrastructure, public health, education, business development, and housing.[187] It is nothing less than an attempt to save a dying culture and land.

Neal Spackman, 2013.

    With an average annual rainfall of 2-3" (5-7.5cm) a year, the site experiences droughts that can last anywhere from two to four years. It is so dry that biological decomposition is not possible; dead plants and animals left exposed are weathered and oxidized—they do not incorporate into the soil as decomposed organic matter.

    When rain events do happen, they cause dangerous and erosive flooding. Moisture is fleeting in this climate and region. The Al Baydha Project puts more water in the ground through flood management than it takes out in drip irrigation.[188] This guarantees that they are always recharging the aquifers more than watering their trees.

    Starting with check dams and weirs far above the site followed by 5 km of swale, the flood waters are slowed, spread, soaked, and saved. The site is watered with drip irrigation, but there are plans to wean the system off of drip irrigation to find the hardiest plants, to propagate those varieties, to replant the system with those, and then sell those genetically superior varieties. Only a few years after planting the swale berms, both the berms and beds are now completely covered with vegetation with no bare earth showing.

    The site development began in 2010. The first rainfall occurred in 2011 which allowed them to plant 1000 trees and water them for an estimated four years. The next rainfall

---

[186] Spackman, Neal. *Facing Fear and Stepping into the Unknown - The Al Baydha Project.* 2015.
[187] Ibid.
[188] Spackman, Neal. *Email.* 2015-2016.

XXI

In only a few years' time, vegetation has claimed the swales completely, leaving no bare ground in the swale beds.

wouldn't occur until 2014 when they harvested 8.5 million liters (2.25 million gallons) of water in the earth berms and 5 million liters (1.3 million gallons) in the swales—which allowed them to plant 1000 more trees and water them for an estimated six more years.[189]

During this last event, water accumulated in a large area held in by earth berms which allowed the water to slowly leak through, forming a slow flow on the opposite side of the berm wall. In time, when many of these sites are installed above and below each other, they will develop seasonal streams, vegetation to soak and retain the water longer, and forest to help precipitate rain events.

The system is designed to eventually be used for silvopasture but is being used for alley cropping currently. As Neal puts it, in his area, the bacteria they need in the soil is only found in the stomach of ruminants, but their system needs to be resilient enough to withstand grazing and browsing first. If the tribes are able to produce their own dairy, meat, forage, food, oil, honey, and other farm products from a perennial system fed by annual rain events and powered only by the sun and man-power, the Al Baydha community would sidestep an

---
[189] Spackman, Neal. *10 Keys for Greening Any Desert*. 2016.

environmental and economic disaster and instead embrace a regenerative economic and ecological model.

At this point in development, the Al Baydha model is an early success that shows unlimited promise in larger scale applications. In years to come and as larger sites are developed using Neal Spackman's model, the return on investment will be regular precipitation, water and vegetation on the landscape, wildlife, regenerative land-based incomes, and cultural preservation.

Sambajo Lestari reviewed by Dr. Willie Smits

In 2002, biologist Dr. Willie Smits founded the Borneo Orangutan Survival Foundation in Indonesia (locally known as *Masarang*) in an effort to save and restore the habitat of the orangutan and increase their dwindling population. Deforestation has orphaned and killed thousands of orangutans. Dr. Smits has rescued over a thousand, but housing them is difficult, especially if it brings no benefit to the local people who are the ones destroying the habitat.[190]

Indonesia, one of three nations that has staked claims to the right to administer parts of the island of Borneo; the Indonesian part of Borneo (which accounts for 76% of the island) is known as Indonesian Kalimantan. Indonesia is the world's leading producer of palm oil

---

[190] Smits, Willie. *How to Restore a Rainforest*. 2009.

which is a common ingredient in cosmetics and processed foods such as Doritos and is also used as biodiesel. In order to create new oil palm plantations (or any monocultural farm system in Indonesia or in Indonesian Kalimantan), you must first destroy the rainforest that is already there. The rate of destruction is very rapid with reports of 25% of their forests lost in only two years (2009-2011), and there are plans to expand by millions of hectares for more biofuel production. In addition these operations are destroying areas that are swampy peat marshes and peatlands which are carbon-rich and very sensitive environments. The primary method of destruction is fire for all these natural systems. These activities make Indonesia (including its activities in Indonesian Kalimantan) the fourth largest emitter of greenhouse gases in the world and at times a larger producer than the entire US economy.[191]

Once these plantations are installed, they are maintained with machines, chemical fertilizers, and pesticides which all lead to erosion and toxic runoff that destroys the river and stream ecologies, creating dead zones along the coastal regions where the rivers meet the sea. These dead zones destroy the coral reefs, and, soon after, the fisheries collapse, putting more Indonesians out of work.[192] On top of all of this, coal is exported and consumed in Indonesia, and coal mining has consequences like coal seam fires which restart forest fires every summer, consuming the peatlands and the rainforest alike. The more habitat that is destroyed, the more Indonesians take part in the destruction to provide for their basic needs. The entire country is a delicate, tropical island ecosystem upon which their economy rests unsteadily.

This has left the two species of orangutan, the sumatran tiger, two species of elephant, and the sumatran rhinoceros along with all their ecosystemic peers without a habitat. Alan Savory's point in *Holistic Management*, that the environment and the people are always linked, holds here as well. With the destruction of the forest comes the collapse of the local economies. Once the forests have been gone for a few years, the rivers start to dry up, making life even harder for all. Willie Smits has an economic model to save the orangutans, the forest and all its creatures, and the local people.

Dr. Smits and his foundation are creating protected habitat for rainforest flora and fauna that pairs with regenerative agriculture symbiotically. It is his belief that only in a holistic context will the rainforest and the orangutans survive, so there must be a way for people to be involved and generate a long-term yield from the forest that is competitive with the oil palm industry.

Sugar palms are the answer. They require little watering and are both fire- and drought-resistant. Sugar palm branches are manually tapped for sap daily, and it does not

---

[191] Freedman, Andrew. *Indonesia's peat fires make it the 4th-largest of carbon emitter in the world.* 2015.
[192] Ibid.

impede the plant's growth or photosynthesis. The sugary sap can be used in food or as fuel. It is more productive than sugar cane and sequesters carbon since it is a no-till cultivation operation. It produces more ethanol per acre than oil palm and doesn't require gas-powered machinery to manage. It creates jobs that are hands-on and an economy that supports local workers.

Dr. Smits' idea is this: create a nucleus of rainforest plants and animals protected by a ring of spiky palms to keep the animals inside and people out.[193] Surround the sanctuary with a sugar palm forest that people tend and have a perimeter of sugar palm and other fire-resistant trees to protect the area. Dr. Smits believes that all the rainforest in Indonesia needs to be co-owned by the people through the state to encourage local peoples defending it as their own.

Sambajo Lestari is the demonstration site for Willie Smits' proposed system. Working with locals, his foundation teaches, develops demonstrations, and encourages local farmers to teach other local farmers regenerative practices. While the site is not yet a fully mature rainforest, there are already changes occurring, with increased cloud cover and rain. Dr. Smits has even devised a business plan that follows the succession of the forest, so farmers arrive at a sugar palm forest through several seasons of various polycultures. The Sambajo Lestari, which translates to "Everlasting Conservation of Sambajo", is regenerating the rainforest, creating jobs, planting trees, cleaning waterways, fostering a culture that values the natural world, and reintroducing orangutans to their natural setting.

Currently Dr. Smits' foundation is working with almost 2,000,000 acres (809,371 hectares) in Kalimantan Indonesia using these advanced planting systems, which continue to evolve as more is learned from nature.

"In the beginning, maybe pineapples and beans and corn; in the second phase, there will be bananas and papayas; later on, there will be chocolate and chilies. And then slowly, the trees start taking over, bringing in produce from the fruits, from the timber, from the fuel wood. And finally, the sugar palm forest takes over and provides the people with permanent income"
–Willie Smits, TEDTalk, 2009

The Green Belt Movement

The late Wangari Maathai's story starts with her trying to enlist women in planting trees, and it ends with her precipitating large-scale political upheaval and environmental restoration—leaving a legacy of planting trees in her wake that is still being carried forward.

---

[193] Ibid.

An adept student, Wangari earned a Kennedy scholarship to attend college and get her masters degree in the US and returned home to a country that no longer looked the same to her eyes.[194]

GreenBeltMovement.org, 2016.

The rivers, trees, and landscape were degrading. Children were suffering from malnutrition. Since the forests were cut down, they were eating a simplified diet of mostly carbohydrates. As the trees continued to disappear, firewood became scarce. Recognizing the missing link between their health and the landscape, Wangari organized and taught women to plant trees, going against custom and locally held beliefs. As the first woman in East Africa to earn a PhD, Wangari was empowered by her education, and she encouraged other women through education to be active agents in their own lives and bioregions.[195] Through her efforts, thousands of women planted trees using seeds they themselves foraged from their local area's remaining native plants.

This effort began to draw attention and then harassment from the post-colonial government that was continuing the colonial practice of liquidation of natural resources. The more she organized to protect the forests, parks, and rivers, the more need for social restoration work she encountered. The environment was so tightly linked to the economy and the society that all their problems were one problem; they'd forgotten their culture and lost their habitat with colonization. The European agricultural methods of cleaning the land included clearing the tribes off the land and into tightly-controlled impoundments called villages. The new government, though run by Kenyans, failed to break the pattern of abuse and corruption that imperialism had visited upon them.[196]

---

[194] Merton, Lisa, and Dater, Alan. *Taking Root: The Vision of Wangari Maathai*. 2008.
[195] Ibid.
[196] Ibid.

Wangari Maathai's position as a lecturer at the University College of Nairobi, her success with organizing women, and her academic achievements gave her public credibility. It also gave her the ability to use her voice to protect land from land-grabs, deforestation, and abuse by the post-colonial Kenyan government—though the conflict was fierce at times. At one point, Wangari and many mothers of political prisoners started a public hunger strike which quickly gained immense popular support and attention. Within a few days, their protest led to a violent dispersal of the crowd by government troops. Relocating in a nearby church's basement, the women continued their protest for 11 months, until finally, the prisoners were freed. Eventually, due to the opposition inspired and facilitated largely by Wangari's passion, example, and organization skills, President Moi stepped down, and Wangari Maathai was elected to the Kenyan Parliament where she continued to use her influence to help everyone see their lives in connection with the trees around them—even the military.[197]

> **"We believe that soldiers and trees are brothers. Our role is to protect the country, and the role of the tree is to protect the environment, so we play a very complementary role"**
> –Major General Njuki Mwaniki, Taking Root (Documentary), 2008.

Wangari's education, confidence, passion for fairness, love of nature, and open mind changed Kenya. Her work with women and trees had unexpected consequences that were political and drew her up into a larger context and conflict. Ultimately, however, she had to address the entire system of ethics (people care, earth care, and care of the future) in order to solve problems involving any of them. They were all inseparable.

As with the people of the Loess Plateau, the women were incentivized with compensation. Only four US cents a tree in compensation from the Green Belt Movement was enough for tree planting to spread like wildfire overnight. It was also not a custom for Kenyan women to plant trees, so it felt like a breakthrough for women's rights, pairing social liberation with tree planting. At the same time, they were reconnecting to their past by rebuilding the bioregion that maintained and inspired their culture before colonial occupation.

The mixture of newness, empowerment, education, and environmentalism gave the people a new perspective on life and the natural world. It forever changed how they perceived right and wrong, fairness, ethics, plants, themselves, and soil. The Green Belt Movement and Wangari Maathai have inspired many thousands to have respect for themselves and their environment.

---

[197] Ibid.

Masanobu Fukuoka reviewed by Larry Korn

As much philosopher as farmer, Masanobu Fukuoka is considered the father of Natural Farming. As a young man, Fukuoka was trained as a microbiologist and agricultural scientist. His first job was as a customs inspector in the city of Yokohama. There, at the age of 25 years old, he had an epiphany that changed his life forever. He saw that nature was completely inter-related and perfect just as it is. He realized that human beings could never improve upon nature. He tried to explain this vision to his co-workers without success, so he quit his job and returned to his family farm to create a system of natural agriculture that was based on what he considered "true nature."[198]

Fukuoka began farming through a new lens. He scrutinized and studied every agricultural practice, such as plowing, flooding the rice fields, pruning, and making compost, to decide whether or not they were actually useful. In the end, he decided that almost none of them were necessary. Instead of asking what he could do, he asked what he could avoid doing. Fukuoka used no fertilizers, no pruning, no pesticides, no tillage, no compost, and no agricultural machinery. He didn't burn his barley and rice straw after the harvest; instead he

Masanobu Fukuoka planting rice in ripening barley.

---
[198] Fukuoka, Masanobu. *The One-Straw Revolution*. 2009. p. 9.

Rice sprouting through clover on Fukuoka's farm.

spread the straw over the fields as mulch. He pelletized his rice seeds with clay to prevent them from being eaten by birds and insects. Seed balls are a combination of seeds, manure, and clay which provides the seeds with soil biology and nutrients regardless of where they land.[199]

> "To loosen the soil, he scattered seeds of deep-rooted vegetable such as daikon radish, burdock, dandelion, and comfrey. To clean and enrich the soil, he added plants that have substantial, fibrous root systems, including mustard, radish, buckwheat, alfalfa, yarrow, and horseradish. He also knew he needed green manure plants that fixed nitrogen, but which ones? He tried thirty different species before concluding that white clover and vetch were ideal for his conditions. The roots of the white clover form a mat in the top few inches of the soil so they are effective at suppressing weeds. The vetch grows well in the winter, when the white clover does not grow as readily"
>
> —Masanobu Fukuoka, <u>Sowing Seeds in the Desert</u>, 2012.

---

[199] Fukuoka, Masanobu. <u>Planting Seeds in the Desert</u>. 2012. p. 161-163.

XXI

Photo provided by Larry Korn

What gained Masanobu Fukuoka the most attention both locally and internationally was his way of growing rice and barley organically without tilling the soil. He only flooded his fields for seven to ten days instead of during the entire summer–just long enough to weaken the weeds and clover and to allow the rice to sprout through. Once the rice was growing strongly, he emptied the field of water, and the clover and weeds revived. Weeds are an important habitat for insects, so preserving them is critical for the overall health of the system. Flooded rice fields, in contrast, turn anaerobic by being constantly flooded, and they leak methane.

While the roots of his neighbor's rice were rotted from sitting in standing anaerobic water all season, the roots of Fukuoka's rice plants grew down three feet (1m) or more and were a healthy white color. His rice plants were shorter than those of the neighbor's, but Fukuoka's rice had more grains in each seed head. The result was that Fukuoka's rice yields were comparable and often higher than those of his neighbors who used chemicals and fossil fuels for their tractors and transplanting and harvesting machines.[200] Plus, his method created no pollution, and the fertility of his fields increased with each passing season.

In many ways, clover was the basis of his entire system. Fukuoka planted annuals directly into the clover and then chopped and dropped the weeds and clover to cover the seeds while they germinated. He used alley-cropping in his orchards of fruit trees which were prolific and nearly disease- and pest-free. Using free-range chickens and goats, Fukuoka was able to get complimentary yields of eggs and milk in the same space as his crops of citrus, nuts, berries, and cultivated mushrooms. His entire system allowed him to work less and get high yields.

Masanobu Fukuoka's farming techniques were an extension of his philosophy. He believed that people should not impose their will on nature but should rather be sensitive to what nature is trying to express on that site and at that time. Once nature is whole and balanced again, it is imperative that we interfere as little as possible. His example has inspired many to observe, respect, and mimic natural systems.

> **"The ultimate goal of farming is not the growing of crops but the cultivation and perfection of human beings"**
> –Masanobu Fukuoka, <u>The One-Straw Revolution: An Introduction to Natural Farming</u>, 1975.

Green City Acres reviewed by Curtis Stone

Curtis Stone, within a few short years, has gone from being a professional musician struggling to save money, to a leader in sustainable urban farming with a thriving, regenerative business.[201] Curtis has redefined what an urban farm can be with his model which encompasses six separate farm plots equaling a total of a third of an acre (1,000m²). His model is efficient, effective, and adaptable to any urban region and situation. Through much trial and error, Curtis Stone has found that a blend of permaculture ethics with biointensive methods works best for farming on limited acreage in an urban setting where the waste streams are rich and diverse. Biointensive farming requires regular working of the soil–though lightly–and using compost and compost tea on the soil in order to support the constant growing of nutritious and beautiful food.

When Curtis started out in British Columbia, Canada, he didn't have land, so he started growing his food on other people's property, their front lawns or backyards, in exchange for

---

[200] Ibid. p. XIX.
[201] Stone, Curtis. *$80,000 on Half An Acre Farming Vegetables - Profitable Mini-Farming with Curtis Stone.* 2014.

TheUrbanFarmer.Co, 2016.

membership in his CSA. He used a bike with a bike trailer to make pickups and deliveries to save on fuel costs and be more sustainable.²⁰²

In Curtis Stone's system, he is using every square inch and measuring the time, money, and energy put into each step of his operation. Seeds are closely planted to create a canopy that holds moisture in and shades and covers the soil. Every bed is exactly 30" (76cm) wide to fit his tools and to make for easier harvesting. He uses soil blocks, pressed earth instead of seed trays, to save money and produce no waste. While initially he does use a rototiller to turn over plots, from then on he only forks often and cultivates the top 1-1.5" (2.5-4cm) of soil only.

Curtis focuses on smaller-sized, high-value crops with a short growth cycle (60 days or less from seed to market) that can be sold to high-end restaurants, at grocery stores, and at weekly farmers' markets.²⁰³ This allows him to have a seasonal crop rotation where he grows three to four crops in the same bed in rotation in one season.

Curtis has greenhouses that allow for year-round production; they maintain a microclimate on his property, so that he's always first and last to have something at the farmers' market because he's warmer first in the early spring and longer into the fall.

In his greenhouses, he hard prunes his **indeterminate** tomatoes, which involves removing all the branches below the lowest bunch of fruit as well as removing all suckers

---

²⁰² Ibid.

²⁰³ Stone, Curtis. *The Urban Farmer: Growing Food for Profit on Leased and Borrowed Land*. 2015.

PermacultureVoices, 2015

(growth between the branches and the trunk) to promote a straight and singular trunk. This allows in more light over time as they get taller and encourages maximum fruiting. They are trained on a string, so they don't get tangled with each other and are easy to work with.

In winter, he grows on shelves in his greenhouse because the severely low angle of the sun allows it to reach all the levels without shading anyone out. Outside the greenhouse, production can start earlier because of the heat conducting through the soil and radiating out into the air. On top of all this, the land faces south, making his farm a well-developed microclimate itself.

One of the fascinating new tools that Curtis uses Green City Acres is a Tilther, invented by Elliot Coleman, which tills the top 1-1.5" (2.5-4cm) of soil only and uses a 18-volt power drill to run.[204] It doesn't disturb the soil deeply, and it doesn't turn over new weed seeds, so it is an ideal production tool for those trying to respect the soil while growing annuals (which always requires disturbance). Another new drill-powered machine is the Quick Greens Harvester (pictured) which reduces labor exponentially—from a four-person job that would take 4-6 hrs to a one-person job that takes under an hour, harvesting at a rate of 6 lbs (2.7Kg) per minute cleanly without leftover debris (keeping things sanitary and beautiful).[205] Curtis

---

[204] Stone, Curtis. *Re: Case Study on Green City Acres.* 2016.
[205] Stone, Curtis. *Tool review: Quick Greens Harvester.* 2014.

also uses a flame weeder which, while it sounds like a flamethrower, is a way of flashing the surface of the soil with heat to kill off weed seedlings before planting. Curtis uses row covers, cold frames, and seeding machines to save time and labor.

He also spends time analyzing and re-designing his business. Careful energy auditing and fiscal analyses allow Curtis to figure out how profitable everything he is doing is and then figure out how to improve. Using the zone mapping concept, his crops on other sites are planted in relation to their tending needs; crops that don't need tending as often get planted on the properties furthest from his own farm. Intensive operations (like his greenhouse tomatoes) are as close as possible to home, often on-site.

Each site has different lighting and different soil temperatures at different times of the year. For what he doesn't grow himself, he makes trades for, in order to round out his CSA boxes; he can barter for almost anything in his community that he needs.

The fact that he is earning an incredible $75,000 annually with his farm system is challenging two of the most commonly held beliefs about farming: first, that you need own or lease a lot of land to do it, and second, that you can't make much money at it.[206] His hyperlocal food model could be used to convert the 40+ million acres (16 million hectares) of lawn in North America to food production, which would change the food system entirely within a few months' time—while substantially boosting local economies and providing lasting jobs for young people everywhere.

The Broad Fork Gardens/Les Jardins de Grelinette
reviewed by Jean-Martin Fortier

Jean Martin Fortier and Maude-Helene Desroches, his partner and wife, co-own and run a biointensive market garden located in Quebec, Canada that is only 1.5 acres (6000m^2) in size but grossed over $150,000CAD in sold vegetables in 2015 with only four people working.[207]

Each winter, they take a three-month break. They spend less than half their year at the farmers' market and running a CSA; the rest of their time is spent in prepping their soil and systems. Their profit margins are a stunning 45% which is unheard of in agriculture; they are making retirement money on a small plot of land using permanent raised beds.

They don't use a large tractor, but they do use a BCS walking tractor which many of their amazing machines attach to. They use a small amount of fuel in their tractor and delivery truck (which is biodiesel) as well as a small amount of propane in the greenhouse to jumpstart

---
[206] Stone, Curtis. *$75,000 on 1/3 acre. Profitable Urban Farm Tour. Green City Acres.* 2015.
[207] Fortier, Jean-Martin. *Profitable small-scale farming. How design sets the stage for success.* 2016.

their season by warming the soil with hot water piped beneath the soil surface.

Jean-Martin feels that they are more effective without a large tractor. He prefers using small walking-machines that adhere to the 30" (76cm) crop bed width and an ergonomic 18" (46cm) wide path.[208] Using a stale seed bed method using either flame weeding or a black impermeable tarp for several weeks of smothering, they prepare the beds to have zero weed pressure on their crops and then plant their seeds 10 to 50x closer than the seed packaging recommends.[209] The canopy this intensive planting creates holds moisture. When there is no bare soil, traveling weed seeds cannot find a place to germinate. The beds are all on a 10-year crop rotation plan alternating heavy and light feeders, using green manure, compost, and granulated chicken manure before heavy-feeder beds are planted (crop rotation chart featured earlier in the Food Forests and Gardens chapter).[210]

Loading a walking seeder machine.

They use wheelbarrows, shovels, hand tools, broadforks, wheel hoes, regular hoes, seeders, row covers, insect covers, biodegradable mulch, and more to minimize their reliance on fuel. However, they also use attachments on their walking tractor such as a harrow which horizontally mixes the top 1-2"(2.5-5cm) which is followed by a roller attachment, creating a uniform soil surface.[211] This perfectly prepares the soil for planting without turning the soil over or pulverizing it as a rototiller would. A flail mower is also used as a tractor attachment; it shreds the green manure. They always use a rotary plow to reshape the beds and dig out the paths to cover the flailed green manure, so it

---

[208] Fortier, Jean-Martin. *The Market Gardener with Jean-Martin Fortier, Six Figure Farming*. 2016.
[209] Ibid.
[210] Fortier, Jean-Martin. *The Market Gardener*. 2014. p. 69.
[211] Fortier, Jean-Martin. *The Market Gardener with Jean-Martin Fortier, Six Figure Farming*. 2016.

Harvesting at QuatreTemps Farm, the large-scale experiment sourcing permaculture and Jean-Martin Fortier's market gardening techniques.

can incorporate into the soil without tilling it in (which is commonly how green manure is used). Covering the row with a tarp will speed up the breakdown process, so that in only a few weeks, the area is ready for planting. The farm's main inputs are labor and compost.

Jean-Martin and Maude-Helene spend their winters planning their garden calendar out over tea. They try to always plant their transplants the same day they harvest, so that their most valuable real-estate, the seed bed, is always being occupied and used during the growing season. Their crop rotation plan groups crops by their botanical family; they plant the same families in the same beds every four years. They have 10 field blocks on a 10-year light-to-heavy rotation. With these two requirements, it takes a bit of planning, but once the planning is done and the calendar set, it makes for a predictable, profitable, stress-free, and efficient year. They rotate greens and root crops between heavier feeders like solanums, brassicas, cucurbits, and alliums—a total of 40 crops.[212] They have a perfectly timed out, arranged, and scheduled plan for harvesting, planting, transplanting, and any other task necessary for their season. They know nearly exactly how much they can make per bed and what quantity they will need for a 200-share CSA season.[213]

The greatest challenge at the Broadfork Gardens is insect pressure. In response, Jean-Martin and Maude-Helene currently use certified organic biopesticides strategically, insectary netting (set to the size of the insect), and daily inspections. One cucumber beetle can infect an entire row of plants with bacterial wilt. Beneficial insect and bird habitat is the permaculture solution they are testing now in a larger, experimental plot with the hopes of eliminating biopesticide usage. Though far from a Fukuokian paradise of do-nothing farming with minimal inputs, it is a superb example of profitable and sustainable biointensive market gardening, and they are increasingly applying permaculture especially on their large-scale experimental farm, QuatreTemps Farm, which will ultimately lead to marketable, regenerative biointensive farming techniques that can be used everywhere.[214]

Miracle Farm/Les Ferme de Miracle reviewed by Stefan Sobkowiak

Converting a conventional orchard on a 12-acre (5-hectare) farm in Quebec, Canada in 1992 to certified organic by 1996 was only the first step in Stefan Sobkowiak's journey to developing one of the largest permaculture orchards in North America by 2007. With 1500 fruit trees in a diverse and regenerative series of planting guilds, the orchard showcases a model that orchardists and permaculturists in cold temperate climates can replicate.[215] Through much trial and error, Stefan has created a simple yet elegant perennial system.

Stefan Sobkowiak, 2016.

The most basic building block of his orchard system is the NAP polyculture: Nitrogen Fixer, Apple, Pear (or Plum).[216] The idea is that every tree will have a nitrogen-fixer on one side. He prunes his orchard,

---

[212] Ibid.
[213] Ibid.
[214] Pineault, Jonathan, and Fortier, Jean-Martin. *Permaculture Meets Market Gardening*. 2016.
[215] Sobkowiak, Stefan. *Miracle Farm*, 2016.
[216] Asselin, Olivier. *The Permaculture Orchard: Beyond Organic*. 2014.

grafts his fruit trees, and spreads a berry shrub layer with cuttings. He uses honeysuckle bushes to draw birds away from his cherry trees instead of using netting. Sharing 5-10% of the farm's yield with the birds, animals, and insects is expected and monitored.[217]

He also has raspberries, currants, gooseberries, and other shrub-layer fruits.[218] Onion, garlic, and other alliums as well as herbs like thyme, echinacea, arugula, and oregano are present as a secondary crop and to ward off pests and provide a medicinal yield. In the first three years while the trees and shrub layer established, they even grew annuals like ground cherry, winter squash, and watermelon. The stacked yields makes for a minimum of three annual harvests from each row.

The orchard rows are laid in 10-day harvest windows, so that every week before market, Stefan can go and harvest the food that is ready from only one aisle and the rows to either side which saves time and energy. This setup takes careful grouping, investment, and research. He has 60 apple varieties to accomplish this feat as well, winnowed down from 100 that they trialed.[219] This spreads and diversifies his yields out over the season. Their CSA members harvest most of the food themselves in a U-pick operation with the rest being sold at the roadside kiosk or eaten by wildlife.[220]

He also focuses on varieties that are disease-resistant, taste incredible, and are pest-resistant if possible. Stefan grows his own rootstock from root suckers collected from fully grown rootstock trees. Once these new rootstock trees are a year old and large enough, he grafts desirable varieties onto them using scions from the pruned branches

---

[217] Ibid.
[218] ThePermacultureOrchard.com. *The Farm*. 2016.
[219] Sobkowiak, Stefan. *Miracle Farm*, 2016.
[220] Ibid.

from his own fruit trees! To minimize the need for pruning so many trees, Stefan uses wires to hold his young tree branches below the horizon line for approximately two months. When a fruit tree's branches are pointed downward or horizontal, the trees focus on bud growth (reproductive) and not upward growth (vegetative); trees allowed to develop upward pointing branches have lower fruit production. For a home orchardist, he provides a model for a permaculture orchard that is easy to manage, abundant, and regenerative.

Using misters, Stefan prevents flowers and fruit from freezing on colder nights—the constant movement of water prevents it from freezing! He also uses red dots painted on sticky yellow pieces of plexiglass as his apple fly traps, and uses non-biodegradable plastic over his rows and plants his trees, shrubs, and herbs in holes in the plastic. This traps moisture, heat, and prevents weeds from competing with his plants as well as helps prolong the lifespan of his dripline irrigation system.[221]

Stefan is constantly observing the birds and insects, knowing they are the indicators of his system's health. Stefan mows only half of every third row, a sixth of the orchard, at a time to allow the insects to migrate and remain in the immediate area of his plants, and then returns to mow the next row's half the following week, and so on, for six weeks.[222] This helps

---

[221] Asselin, Olivier. *The Permaculture Orchard: Beyond Organic.*
[222] Sobkowiak, Stefan. *Miracle Farm*, 2016.

maintain beneficial bug populations. They also have habitat for bumblebees, leaf cutter bees, and mason bees in addition to honey bee hives.

With different sets of trees and plants flowering every week or two in the orchard throughout all the growing seasons, the pollinators, especially the honeybees, are consistently feeding on pollen and consistently facilitating fruit production. The diversity of one supports the other. It leads to an abundance of both new bee colonies and honey production. Leveraging nature and using permaculture patterning, Miracle Farm truly is a miraculous model for permaculture orchardists everywhere.

Celebration Farm

The Celebration Farm was born out of Dr. Elaine Ingham's desire to prove biological farming's efficacy in a side-by-side comparison with conventional farming. Outside Oroville, CA, on the edge of the Plumas national forest, the Celebration Farm was established in 2015.[223] It is partly a restoration site and partly an experimental farm. The previous owners had sprayed glyphosate and other herbicides profusely for many years, and the soils in the meadows and the open areas around the house were all bacterial-dominant, compacted, and contaminated with biocides. Using compost tea and compost, Elaine is transforming the soil all over her farm.

Peas are grown using conventional fertilizers, compost, compost tea, and compost extract side by side to easily compare the results.

---

[223] Ingham, Elaine. *Celebration Farm Tour*. 2016.

The goal of biological farming is to grow nutrient-dense foods regeneratively using the microbiology in the soil. Without using earthworks, Elaine used soil biology to create a self-sustaining palm oasis in Saudi Arabia in only six months.[224] She is doing more using less energy and effort by leveraging the power of the soil life. Oases that have survived for thousands of years in the deserts are maintained by the same system Elaine developed in only six months.

There are several ponds at different levels on the 60-acre (24-hectare) property. They provide water for irrigation via a concrete trough installed in the 1850s and piping in the areas where the trough is damaged. It is one of the many ongoing restoration projects. There are also heirloom apple varieties, planted by gold miners who worked in the mine at the top of her hill, that she is grafting onto rootstock to preserve their genetics. In addition to experimentation and research, Elaine teaches composting courses and is developing a large composting facility on-site.

The side-by-side comparisons of the effects of conventional, compost, compost extract, and compost tea on plant performance are invaluable, and it's surprising that it has not been done previously. The site was initially rototilled to break up the layers of compaction and trapped pesticides (which create deposition layers and impermeable layers as they are watered). Following a fungal-dominant regimen of compost tea applications, the garden plots were built on contour for flood irrigation.

Dr. Elaine Ingham is setting out to prove that if the pH and the soil biology is correct, weed seeds will not germinate even if they are present in the soil and receiving plenty of water and light. She is convinced and determined to demonstrate unequivocally that recipes of compost, compost extracts, and compost teas that focus on feeding particular components of the soil food web produce specific beneficial effects. There is no standard recipe for soil remediation; it is all site- and goal-dependent.

At the George W. Bush Presidential Library site in Dallas, Texas, the 13-acre (5-hectare) grounds were transformed from lifeless, compacted soils into authentic native prairie using compost as the only soil amendment.[225] Dr. Elaine Ingham was an integral part of the team that made this happen. It required her soil expertise to determine the exact fungal:bacterial ratios needed. Once the 300 tons of compost were acquired, the installation only took a single year to fully establish despite it having been flattened parking lot in many areas.[226]

Fully-grown live trees were installed with compost surrounding the root balls; compost was used everywhere. Earthworks were built to capture and concentrate the moisture.

---

[224] Ingham, Elaine. *Sustainable Design Masterclass*. 2016.
[225] Ibid.
[226] Ibid.

Dr. Elaine Ingham is always teaching and sharing her knowledge of soil and composting.

Despite it being a drought year with no rain and no irrigation used on the site, a small pond formed just from dewfall.[227] The native trees and plants that were transplanted partnered so effectively with the specifically tailored compost that it was a resounding success.

Dr. Elaine Ingham teaches online, speaks at conferences, has authored many books, is the author of the USDA's Soil Primer, consults with farmers, designs and oversees installations, and continues her experiments at Celebration Farm. Her work has radically changed the accepted scientific understanding of soil. As a consultant, she has designed a dewfall garden in Saudi Arabia, a commercial vineyard in the arid interior of Australia without irrigation, the George Bush presidential library grounds in Texas, and a carbon credit farm in South Africa where they sequester six tons of carbon per hectare a year, building 2-3 feet of soil annually.[228] The future is incredibly bright and encouraging if we can just apply Dr. Elaine Ingham's research and follow her example.

> **"Once you have the organisms in the soil, they will work for you -- as long as you don't kill them. Healthy soil, properly maintained, can reduce or eliminate the need for irrigation; the need for hoeing or weeding; the need for tilling; and can decrease water consumption up to 70% in some places, and increase yields**

---

[227] Ibid.
[228] Ingham, Elaine. *Lecture Notes, Work in Beauty workshop, Gallup, NM -- November 7, 2015*. 2015. 2015.

significantly, and increase nutrition in those foods. On an average 300 acre farm, we generally reduce costs for growers by $200,000 in the first year."
–Dr. Elaine Ingham, "Lecture notes, Work in Beauty workshop", 2015.

Rosemary "Rowe" Morrow

Permaculture pairing with humanitarian aid work is a natural partnership. Many well-known designers like Geoff Lawton and Darren J. Doherty do humanitarian aid work with permaculture, working with groups like the UN, World Bank, and more. Rosemary "Rowe" Morrow, born in Perth, Australia, was likely the first person to use permaculture in a post-war situation, but her model of coming in after or during conflict to help families grow food using permaculture became an enduring model that is replicated today by thousands of permaculturists across the globe.[229]

Initially teaching reading and writing informally to children in Lesotho, Africa on the border of apartheid-era South Africa, Rowe recognized the need for permaculture in areas of conflict which led to her serving in many areas of need—Vietnam, Bhutan, Ethiopia, Cambodia, Uganda, and more.[230] Her usual method is to train government officials or area leaders, so that they can then teach the people of the area using their own gardens as examples. For the people that Rowe serves, permaculture provides a way for them to grow more food in marginal spaces and in new, creative ways. Having more food in these areas makes an enormous difference in their quality of life, especially for children.[231]

Specifically in Cambodia, Rowe was instrumental in starting a program that was both successful and still endures to this day and continues to spread (in four provinces currently). In Rowe's own words:

**"Post-war food restoration with permaculture began in Cambodia when women extension workers in the Pursat Department of Women's Affairs (PDWA) learned permaculture. First, they spent six months developing their designs in their own gardens which they used as teaching sites. Then they went to villages, market places, houses and taught farmers. They taught them to design, then make gardens, next select and grow fruit trees and to re-use waste products more efficiently. They have gone further. Now there are small processing industries and nurseries. The economy and the environment have changed.
Next they went to another province, the 'white skirt' zone, no man's land, in Svey Rieng province. Here every living thing had been wiped out. Farmers were re-occupying a desert land.**

---

[229] Dawborn, Kerry and Smith, Caroline, editors. *Permaculture Pioneers: stories from the new frontier.* 2011. p. 152.
[230] Morrow, Rosemary. *Permaculture and the Forgotten. Teaching Permaculture in Places That Absolutely Need It. A Message of Hope with Rosemary Morrow.* (PVP068), 2015.
[231] Ibid.

The farmers learned permaculture in one year instead of the usual two or three. They were extraordinarily successful in restituting food supplies without damage to the environment. The PDWA has now taught in at least, four provinces.

And as farmers in the villages copy each other we don't know how many Cambodian farmers are permaculturists. But there are thousands and they are us.

The success of these projects, as in Vietnam and other countries is to embed the permaculture knowledge into local staff who then use it as part of their outreach. And the other factor is finding the local people or person with the imagination to see where permaculture can go. And it takes a little time - perhaps five years before 60% of the local population is practicing permaculture in one way or another.

Its success was that permaculture was absorbed through in-service training and became the extension curriculum for government departments for sustainability. This has also happened in one district of Uganda"

–Rosemary Morrow, 2016

Rowe continues today to teach teachers permaculture through her books, online courses, lectures, and classes. She is an enduring example of what People Care is in Permaculture.

Bhutan

Tucked between the two most populous countries in the world, India and China, Bhutan is a country of less than 700,000 people located in the Himalayas.[232] It is also the only carbon-negative country in the world. They sequester three times what they generate in CO_2.[233]

With a GDP of less than two billion dollars a year, Bhutan is categorized as less-developed country, but despite the categorization, Bhutan is managing to be more regenerative than all other countries especially considering that it is also the only carbon-neutral country.[234]

Despite Bhutan's small economy and self-imposed restraints, the government provides free healthcare, schooling, merit-based college education, electricity to rural farmers to prevent them burning wood, subsidies for clean-energy vehicles, and more with limited resources that are carefully managed. It is all part of their holistic plan:

---

[232] Tobgay, Tshering. *This Country Isn't Just Carbon Neutral - It's Carbon Negative.* 2016.
[233] Ibid.
[234] Ibid.

XXI

The Haa Valley in Western Bhutan

> **"We are thriving... balancing economic growth carefully with social development, environmental sustainability, and cultural preservation, all within the framework of good governance. We call this holistic approach to development 'Gross National Happiness,' or GNH. Back in the 1970s, our fourth king famously pronounced that for Bhutan, Gross National Happiness is more important than Gross National Product"**
>
> *—Tshering Tobgay, Prime Minister of Bhutan, TEDtalk 2016*

Bhutan is 72% forested. The Bhutan constitution stipulates that 60% must remain forested for all time.[235] The government is also paperless, and there is a focus on LED lights. Using the runoff from the Himalayan snow-melt, clean hydropower provides clean electricity that is sold to countries like India and China to offset their carbon usage. Because of this, Bhutan is also a world leader in carbon sinking and off-setting.

Despite all these amazing positives, climate change is affecting their country. The Himalayan glaciers are melting, as are glaciers worldwide. It is changing the ecology. Glacial flash flooding is now becoming common and increasingly dangerous. This has spurred government programs for action, support, and education. They've created biological

---
[235] Ibid.

corridors for large animals to move freely through the country which is increasingly important as habitats shrink and climate change forces habitat migrations.

Currently, Bhutan has almost reached its goal of raising enough funds to permanently set aside protected lands.[236] When it does, it will be firmly on the path to fulfilling its holistic goal of remaining carbon neutral forever. Let us hope that all countries everywhere in the world follow their example.

> **"Economic growth is important, but that economic growth must not come from undermining our unique culture or our pristine environment"**
> –Tshering Tobgay, Prime Minister of Bhutan, TEDtalk 2016

The Loess Plateau Water Rehabilitation Project reviewed by John D. Liu

The Loess Plateau in China has long been a place of plagued by erosion and desertification. It was once a fertile paradise and the birthplace of settled agriculture in China approximately 10,000 years ago.[237] The people from this area were the Han, who historically

---

[236] Ibid.
[237] Liu, John D. *Green Gold*. 2012.

were and currently are the largest ethnic group in China. As agriculture developed and bloomed, dynasties maintained their headquarters in this area, but as the fertility waned, it was abandoned. When agriculture become impossible, they turned to grazing with sheep and goats. The continuous grazing led to desertification which led to erosion which is the reason for the Yellow River's color and name.

The wind carries the silt in dust storms far into China, blocking the sky and making it difficult to see and breathe. Without vegetation, 95% of the water runs off, creating enormous gullies.[238] The flooding that occurs when the dikes break on the Yellow River is disastrous and has killed millions in single events as recently as 1931. It has been a persistent, real threat to

[238] Ibid.

Chinese daily life for millions for millennia which has earned it a second name, "China's Sorrow."

In under 10 years, the ecological damage which had accumulated over centuries has been dramatically reversed through the Loess Plateau restoration projects.[239] Documented by filmmaker John Dennis Liu, founder and director of the Environmental Education Media Project (EEMP), this project involved the local people, government officials, and the World Bank working together on a holistic goal of mutual benefit for each party and the bioregion.

The people were given long-term contracts on sections of land decided upon by the locals themselves, so they would be directly benefiting from the restorative work they did and involved in the allotment of the land itself. One critical component of the plan was the

[239] Ibid.

removal of the grazing animals from the overgrazed land. The other crucial component was a policy to stop all agricultural activities that involved disturbance of the soil on land with more than a 20 percent slope. In place of slope agriculture and terraces, perennials, shrubs, and trees were planted, which revegetated the landscape. The Loess Plateau lessons then became policy all over rural China—all slopes became no-go areas for ploughing, and the farmers received long-term subsidies for the apparent loss of immediate productivity.

The government paid for the goat and sheep feed during the transition which actually increased the protein in the animals' diets. The first work needed on the plateau was installing earthworks for water infiltration. Swale-like terraces were dug by hand with shovels, and then trees were planted, sometimes in bowls of packed earth. Since the rich loess silt only needs organic matter to be productive, there was an explosion of vegetative growth within only a few seasons of water retention.

A decade later, 500 million US dollars invested, and over 500,000km² restored, the project is heralded a success.[240] What began as one project with a budget of $150,000 US dollars with 15,600 km² of land to restore, gained an extension to grow to 35,000km², more funding, and led to Chinese laws being changed to allow for the rehabilitation practices to spread to other areas on the Loess Plateau ballooning the project impact to over 500,000km².[241][242] Today, the province is prospering, and its perennial model is being adopted all over the world to reverse the effects of desertification. The landscape is covered in green growth and is now producing perennial crops at such a rate that most citizens involved are earning more than they ever imagined possible. This has led to a locally shared exuberance and optimism about both the environment and the economy. The enthusiasm, the amazing rebound of growth, and the viral spread of regenerative practices are all similar in the way they spread and are all linked.

Whereas hunger was rampant in the 1960s, electricity was a rarity, and parents feared their children would only inherit a more degraded landscape, now hope and incomes are steadily on the rise. Being the solution to a problem that has plagued their region and people for thousands of years has fundamentally changed the people's outlook on life. The project's efforts alone restored 35,000 km² of land, the equivalent size of Belgium, yet at same time it spread virally, and they restored over 500,000km², the local economy, culture, and mental outlook.[243] The success of the project was facilitated by their conscious effort to balance the needs of the environment, the people, and the future. While their efforts continuously

---

[240] Ibid.
[241] The World Bank. *Loess Plateau Watershed Rehabilitation Project.* 2011.
[242] Liu, John D. *Email.* 2016.
[243] Ibid.

sequester carbon, the fruit trees, that were absent only a few years prior, are creating a new industry with the Loess Plateau in line to be a new, global supplier of tree fruit.

Reflecting on the Examples

In collecting together these examples of Permaculture in Action, it was interesting to find that each one includes experimentation and places consistent emphasis on adaptation. While some may have aspects that could be criticized, they all demonstrate successful examples of applied permaculture in varied scenarios, and they all share the same goal of becoming more regenerative in a holistic way. Often the most influential and successful examples are the ones that involved large communities and focused on linking social change to environmental change. More than growing food or teaching regenerative techniques, these examples demonstrate how People Care is central to Earth Care.

XXII. The Permaculture Lens

Permaculture is a way of seeing the world to solve problems holistically and ethically. It's not just how to garden, run a business, or design a farm; it is a lens that allows us to recognize the patterns and attributes of the natural world of which we are part of, inextricably.

When you study permaculture in-depth, it changes how you view everything. You see whole systems everywhere and whole systems within those systems. You start to identify and recognize the plants in your local area. You notice the sun path. You also start to include earth care and future care in your planning, thinking, problem solving, teaching, daily living, and relationships. Your peers will notice. Your family will notice. You will notice and reflect more than you ever have.

Once you begin to see yourself as part of a cycle, you begin to have more choices in your behavior both ecologically and socially. Permaculture helps us to recognize our place in the natural order but also the limitless possibilities. We see how we can help, how we can prosper, how we can connect beneficially, and how we can ensure a bright future for all. Despite all the damage we've done and continue to do collectively, we can be the most powerful creative forces on earth. We can set the stage for nature to restore the degraded and damaged ecosystems, and we can encourage and support that transition through our daily lives and work. We are not helpless; we are the only ones with the power to reverse the trend, and we can make an abundance and heal our communities at the same time.

Use this book, its references, the peer reviewers' courses, and your own research and experiences to lead you to a more effective regenerative patterns. There has never been a greater need for ethical, regenerative living at any point in human history than there is now.

You now carry with you the information and tools to redesign our world to be regenerative not destructive, to teach people to be producers not just consumers, and to live ethically not just lawfully. The future is in your hands. You are not alone; the most powerful forces on earth are working with you.

Go out and make it real in your own life.

Start small and go big.

Share your work. Go viral.

The best is yet to come.

The world's potential is beyond our wildest dreams.

MP

The Advanced Permaculture Student Online
an online course of regenerative solutions & career paths

We've Waited Long Enough
It's Time To Heal Our World & Ourselves!

Join Matt Powers & 55+ Experts in the Regenerative Spectrum of Permaculture
Learn How to Make Permaculture Your Lifestyle & Your Living

Permaculture Design - Holistic Management - Keyline Design - Advanced Soil Science - Large-Scale Land & Ocean Restoration - Social Permaculture - Mycology - Alternative Energy - Aquaculture - Permatecture - Gardening - Food Forestry - Ecological Landscaping - Plant Breeding - HM Grazing - ReGreening Deserts - Urban Permaculture - KNF - Bokashi/EM - Beekeeping - Conservation Hunting - Mead Making - Regenerative Business - Nonviolence - Life Planning - Probiotics - Earthworms & More!

Visit the
ThePermacultureStudent.com
for more information

Compost Teas • Compost Tea Brewers/Extractors • Packaged Biologicals • Humates • Mycorrhizae • Worm Castings
Kelp Solutions • Complete Organic Systems • Microbial Microorganisms • Education • Classes • More!

WHERE SCIENCE AND NATURE MEET

A complete circle of products, knowledge and instruction to help you **get the best from your grow**.

LIFE IN THE SOIL CLASS

Why is compaction so bad?
What prevents roots from growing deep?

These classes will show you the basic principles of the soil-food web-plant relationship as well as explaining fungal to bacterial biomass ratios. You will gain a fundamental understanding of the way soil biology drives plant nutrition; you will increase your understanding of how modern agriculture selects for disease and pests. Gain a complete understanding of soil health and healthy plants.

lifeinthesoilclasses.com

ANCIENT HUMATE

How can you organically assist plants with the uptake of micro-nutrients?

Ancient Humate is a powerful soil amendment that aids in micro-nutrient uptake. A concentrated liquid amendment for plants and soil. This is a carbon-based food for your plants that boosts compost tea to transform it's nutrients into a more plant-available form. Use Ancient Humate indoors, outdoors, in soil or containers and in hydroponic systems.

nature-technologies.com

THE MICROBE POSTER AND THE MICROBE MANUAL

How can you use sustainable methods to replicate mother nature?

This pair of resources offer extremely detailed scientific explanations of the photos on the microbe poster. Designed to help clarify what you see in your microscope, the microorganisms detailed are those typically found in soil, compost, worm castings and compost tea. Written by Elaine Ingham, PhD. and Carole Ann Rollins, PhD.

gardeningwithnature.net.com

Environment Celebration

environmentcelebration.com

nature-technologies.com
lifeinthesoilclasses.com
gardeningwithnature.net

NEW!

DISCOVER practical solutions beyond sustainability in the new *Permaculture Magazine, North America.*

Get **25% OFF** Digital Subscriptions at **PermacultureMag.org**
Use code **PMNA25**

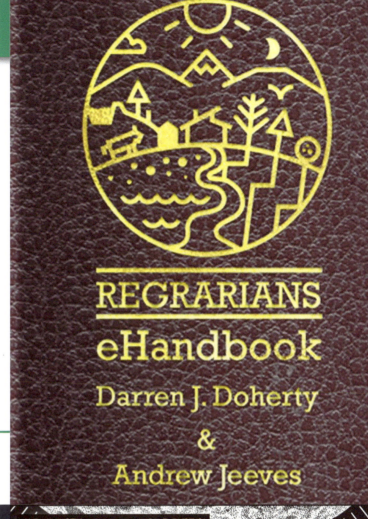

REGRARIANS
eHandbook
Darren J. Doherty
&
Andrew Jeeves

WEEKLY REGENERATIVE DESIGN CLASSES

EMPOWER YOUR FUTURE...

WWW.SUSTAINABLEDESIGNMASTERCLASS.COM

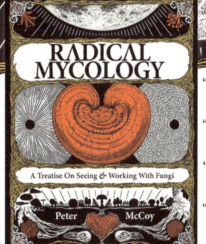

"An amazing compendium!"
—Sandor Katz, author of *The Art of Fermentation*

"A science for the people"
—Scott Kellogg, author of *Toolbox for Sustainable City Living*

"In-depth analysis...a must read"
—Tao Orion, author of *The War on Invasive Species*

"Passionate, entertaining, and very readable"
—Milkwood Permaculture

YOUR GUIDE TO THE FUTURE OF FUNGI

With over 650 pages of information on mushrooms and other fungi, *Radical Mycology* provides the skills needed for integrating fungi into any permaculture design.

Fungi in Systems – Easy & Effective Mushroom Cultivation
Medicinal Mushrooms – Mycoremediation – Fungal Ecology
Lichen & Mycorrhizae Cultivation – Cultural Perspectives
Mushroom Hunting – Fungal Ferments – 100 Species Profiles

SAMPLES AND FREE BOOK VIDEOS AT: **CHTHAEUS.COM**

Glossary

A

acidification - the process of becoming more acidic

advection - the movement of heat in air or liquid

aerate/aeration - to add in, mix in, or infuse with air; to oxygenate

aerobic - having oxygen readily available; having the quality of being aerated; having a dominant population of aerobic microbes

aggregate - a whole comprised of smaller elements

agroforestry - using perennial or "forest" in conjunction with agriculture. This can look like alley cropping or a food forest.

agronomist - an agricultural scientist

albedo - the ability to reflect sunlight: higher albedo means more light is reflected than lower albedo where more is absorbed

allelopathic - the ability of one plant to inhibit the growth or vigor of other plants

alley cropping - growing crops between alleys or rows of perennials, usually fruit or nut trees

anaerobic - lacking oxygen

aquifers - underground water reservoirs

aquifuge (see Keyway)

B

biochar - compost- or compost tea-infused charcoal

biocides - fungicides, pesticides, herbicides, and insecticides all kill the biology of the soil food web, so they collectively are life (bio) killing (-cide) substances

biodigester - an anaerobic fermentation chamber for organic matter (usually used for sewage) where methane gas released by the fermentation process can be used for heating purposes, used for cooking often

biodiversity - the number of different forms of life in a given system

biofuel - plant-based alcohol or oil that can be used to run a combustion engine

biointensive - a form of farming focused on high yields in a small space using hand tools and organic farming methods

biological farming - a form of farming that focuses on the soil biology and health.

biomes - a natural collection of biology found in a major habitat such as a forest or desert

bioregion - a geographically distinct region usually defined by a watershed; or specific ecosystem or small collection of ecosystems

bioremediate - to use biology to remove toxins or heal a landscape, soil, plant, or animal by natural, biological means

blackwater - sewage water, water with a fecal matter content

bokashi - a Japanese anaerobic fermentation method for organic matter

brittle - a term coined by Alan Savory to describe fragile ecosystems with little to no humidity and sparse precipitation.

BRIX Meter - a tool to measure sugar/starch levels of a liquid, usually plant juices.

C

carbonaceous - made of carbon

carbon farming - farming that sequesters carbon as one of the main goals

carcinogenic - able to cause cancer

chinampas - a growing method for wetlands where channels are cut for small boats and soil piled into islands for gardens

climate analogs - regions that share similar climatic factors

cob (cobb) - a mixture of earth, straw, water, and sometimes gravel or sand to make a natural building material like adobe (but that sometimes has manure in it as well)

compassion - care for the welfare of others

compost - a biological soil food web reaction that converts organic matter into humus, a sequestered form of carbon, infused with aerobic beneficial soil life with many of the nutrients that were present in original organic matter still available

compost tea - a liquid solution of aerobic soil life used to inoculate soils, plant roots, or plant surfaces

condensation - the natural action of atmospheric water distilling on surfaces caused by warm air cooling

conduction - the transference of heat through a solid

contour - a flat path along terrain, the lines in a topographic map

convection - the natural action of hot air rising and cold air falling

coppicing - cutting down to the ground (usually a tree or bush) to stimulate growth

cover crop - a planting that serves to protect the soil by covering it up and nourishing it with specifically effective nutrient-accumulating plants like legumes

cyanobacteria - photosynthesizing bacteria

D

decentralization - redistribution of responsibility, control or power; re-localization; to create a networked, lateral, cooperative system

desertification - the degradation of a landscape into a lifeless, barren desert

dikaryotic - having more than one nucleus

E

earthbagging - the building technique that uses bagged concrete or cob for quick construction

ecological - of or relating to the interactions between living organisms in an area they all share

ecosystems - complex living systems; an area shared by and comprised of a group of living things

ectomycorrhizal - root-associating fungi that do not penetrate, but usually encase plant cells within plant roots

enchytraeids - worm-like organisms that are smaller than earthworms and found in marine and terrestrial environments.

endomycorrhizal - root-associating fungi that penetrate plant cell walls

endophytic - symbiotic fungi found in all plant cells

endosymbionts - mutualistic living things that live within another organism

entropy - the process of losing energy over time; the idea that all systems over time are prone to chaos or disorder

epigenetics - the study of environmentally- and behaviorally-influenced genetic expression

ethics - guiding moral principles

eukaryote - cells with nuclei and organelles with membranes

eutrophication - excess nutrients in water that causes aquatic plant blooms that suffocate animal life

exudate - a substance exuded by plants, primarily sugars but with some proteins and carbohydrates

F

food forest - a forest of selected edible species

soil food web - the community of soil and soil-related organisms and their energy exchanges

freeboard - the distance between the top of the dam and the height of the dam water in an overflow event

G

gasification - conversion, of usually organic matter, into a gas

geothermal heating and cooling - a heating and cooling method that uses earth (usually under the building) to store and release hot or cold air

gley - an anaerobic fermentation process of plant material that makes a sticky, gel sealant that is used to seal ponds or any area needing to retain moisture - variable in efficacy.

GMO - Genetically Modified Organism

graywater - waste water that lacks fecal matter, heavy metals, or toxins

green manure - plants that are grown to fertilize the soil, usually legumes

H

Hadley cells - atmospheric phenomenon where hot air rising from equatorial regions falls approximately 30° away from the equator

homeschooling - home education of children by parents

Holacracy - a system of self-management that imitates the way ecosystems develop and behave

holistic (holism) - of the whole or in recognition of the whole; the idea that all things are part of a whole system and cannot be separated from their environment

holon - a whole system of interactions that contains smaller holons and is part of larger system of holons

hugelkultur - a soil-building technique that imitates deadfall in a forest by burying wood, can be used for gardening as well

humus - rich, dark-colored soil component comprised of thoroughly decomposed organic matter

hydraulic ram pump - a pump that uses the pressure of water to move that water

hyphae - the one-cell-thick fungal strands that make up mycelium

I

indeterminate (tomatoes) - tomatoes that continuously grow

inoculate - to coat, wet, or soak in mycorrhizal fungi or beneficial bacteria like nitrogen-fixing rhizobium

J

jet stream - the predominant, broad-scale, westerly atmospheric wind patterns

K

keyhole garden - a gardening arrangement that allows any one standing in the center of the raised bed to be within easy reach of the entire growing area

keyline - a contour line that extends off the keypoint

keyway - the impermeable core of the dam

keypoint - the point just after which the valley slopes change from places of erosion (convex) to deposition (concave).

L

land trust - a legal arrangement that takes land off the market for a specific purpose

legumes - a family of plants, many of which from symbiotic relationships with rhizobia bacteria to fix nitrogen form the atmosphere

M

microclimates - an area that has climatic conditions different from the larger context it is found within

monoculture - a cultivation of just one kind of plant

monokaryotic - having a single nucleus

mouldering - slow decomposition, characterized by neglect

mutualism - a relationship or interaction that benefits two or more organisms or parties

mycelium - the mycorrhizal body of hyphae

mycoremediation - purification or restoration of an environment or medium using fungi

mycorrhizal - rhizospheric fungi

N

natural farming - a farming method that imitates nature and uses the least amount of technology and inputs possible

non-brittle - Coined by Alan Savory to describe humid, resilient ecosystems with plenty of precipitation throughout the year.

nonviolent communication (NVC) - a communication method that focuses on recognizing universal needs to resolve conflicts

nuclei - small particles that attract moisture to form precipitation

nutrient - a vital substance for biological growth: minerals, lipids (fats), vitamins, proteins, carbohydrates, water, and air.

O

organic - plants grown without petrochemicals or synthetic chemicals; of or related to all living things

orographic effect - a climatic phenomenon that highlights the influence of altitude, land shape, and distance from the ocean

oxidized - to combine with oxygen—such as in burning or rusting

P

pasture cropping - the practice of raising crops like grains in pastures with native species

perennial - persisting for several years or longer; plants that live for 3 or more years

petrochemical - chemicals derived from fossil fuels

pH - a scale of measuring the hydrogen concentration in a liquid which determines how acidic or alkaline a substance is
photobiont - life dependent upon solar energy
photoperiodic - a characteristic in both plants and animals requiring specific durations of day or night hours to thrive
phthalates - endocrine-disrupting chemicals used to make plastics soft or malleable
phyla - above class yet below kingdom, a categorical term used in taxonomy
phytoremediation - healing or restoring an area or organism(s) with plants
plant guild - a beneficial polyculture
pollarding - the removal of the top or the branches of a tree to stimulate new growth
polyculture - a cultivation of 2 or more plants in the same area
principles - the foundational concepts or truths that form a system of belief or reasoning
pyrophytic - fire-tolerant

R

radiation - the action of moving outward from a center point - usually referring to energy
rain shadow - the visible wet and dry side to mountains and ridges formed when precipitation falls predominantly on the side facing the predominant winds - precipitation forms as moist air climbs in altitude
reed bed - a contained growing bed for reeds where graywater can be filtered by the plants
refractometer (see BRIX meter)
regenerative - restoring and improving beneficially and continuously without drawing down unsustainable and non-regenerative resources
respiration - the taking in of oxygen and exhalation of carbon dioxide for energy; breathing
restorative circles - a restorative justice practice
rewilding - restoration of a habitat to its wild, natural state, often involving reintroduction of species
Rhizobia - nitrogen-fixing root-associating bacteria that primarily partner with legume roots
rhizosphere - the root zone
riparian - the wetland areas around rivers and lakes
ripping - subsoil plowing
rocket mass heater - a masonry heat that burns cleanly with a j-tube rocket stove and runs its exhaust through a mass to conduct and radiate all the energy from the fire slowly over an extended time period
rocket stove - a j-tube stick fire stove that can burn at high enough temperatures to burn wood cleanly

S

salinity - salt level
self-reliance - the ability to provide for oneself
shadehouse - a growing area that is shaded

sheet mulching - a soil-building technique where paper and cardboard waste, mulch, and compost are combined
silviculture - forest management and cultivation
silvopasture - grazing animals in between or among orchard rows
sociocracy - a system of self-managing and self-organizing groups
solar pump - a tall black pipe that extends high above a building that heats up in the sun and draws off hot air or other gases from the building by convection
soluble - can be dissolved into water
spillway - a safe overflow for excess water in flood or high precipitation events
spillway pipe - overflow pipes that help keep bodies of water at a certain height in regular precipitation events
STUN method - Strategic Total Utter Neglect, a management technique term coined by Mark Shepard
sustainable - the ability to be repeated over time without drawing down or degrading non-renewable and non-sustainable resources
swales - tree system earthworks built on contour with flat, absorbent pathways, a soft lower berm, and spillways for overflow events
syntropic - the tendency for life to beget more life in ever greater complexities and stacking and interacting systems

T

thermophilic - heat loving; thriving in hot temperatures (106-252°F/ 41-122°C)
thermosiphon - a passive system using convection to draw air or liquid from one area to another
tree flagging - the displacement of the crown of a tree in relation to its trunk causes by strong prevailing winds
trickle pipe (see spillway pipe)
trompe - a passive system that creates compressed air using falling water
trophic - of the food web; related to nourishment

U

unschooling - self-directed education focuses on student-choice, often seen as a subset of homeschooling

V

vermicompost - worm castings made by worms digesting organic matter

WXYZ

walipini - a type of earth-sheltered greenhouse
water tables - the water saturation areas within the soil
wicking beds - a raised bed design where water is added below the beds and the water "wicks" up through the soil
willow water - a liquid for rooting plants made by boiling and steeping cuttings of new growth from a willow tree (any in the Salix genus) and allowing the rooting hormones naturally found in willow stems and branches to infuse the brew overnight.

Index

A
acidification 11, 38, 338
A-frame 154-155
albedo 27, 33, 243, 261
allelopathic 17
alley cropping 17, 18, 162, 219, 247, 340, 349, 358, 361
altitudinal effects 33
aquifuge *(see Keyway)* 61
arbuscular mycorrhizae fungi (AMF) 134, 135, 202
arthropod 85, 91-94, 98, 99, 129, 255
artificial reefs 272–274
aspect, section 178

B
backcut 64
banana circle 220
benching, section 157
berms or banks, section 157-158
biochar, section 221–222
biodigester 288, 295
biofuel, section 287–289
biointensive 203, 205, 344, 354, 357, 370, 373, 376
biological farming 337, 379, 380
bioluminescent 141, 286
Birch's Six Principles 7–8
bokashi 123–125
bokashi biochar 124
breeding, plant 214–215
Brittleness Scale 96–97
BRIX meter 130
broad climatic zones 21–22
Brooklyn Grange 305
BSFL (black soldier fly larvae) 240
bunyip level 156
Burlese Funnel 129

C
carbon cycle 34–39

carbon farming 337–338
casing a spring 57
chinampas 263–266
chop and drop 213–214
climate analog 21
cob 108, 163, 236, 308, 309, 311, 312
compassion 3, 12, 313, 323
compost *(see thermophilic, mouldering, or vermicompost)*
compost extract *(humic acid)* 122
compost tea, section 121
compost toilet 78–79
concreted or cemented soils 102
condensation 25–26
continental effects 33
convection 24, 27
conveyer belt of moisture 46–47
cooperatives/co-ops 329–330
coppicing 229–230
cover crop 117
crater garden 161
crop rotation 116-117, 204–206
crowdfunding 330
CSAs 330
currency 328

D
dam, common types 53–54
dam, building 57–65
decentralization 9, 168, 316
design ethics 2–3
dew 24–25
distillation 25
diversion banks and drains 70–71
dry farming, section 245–246

E
earthbagging 163–164
earth-sheltered greenhouses and homes 159–160
earthworm, section 93

edge cropping 17–18
edge effect 15–17
Edison iron-nickel battery 279–280
Edo period 303–304
energy audit 279
epigenetics 113
espalier 204–205
ethics (see design ethics)

F
farmshare farms 331
fibershed 332–333
floating gardens 261–262
fog 25
foliar spray 121–122
freeboard 64
frost 29
fruit walls 200–201
Fukuoka, Masanobu, section 367–370
functional design 170-171

G
gabion 54
gasification 287
geothermal heating and cooling 239
grafting, section 210–211
graywater 72–75, 225
graze/grazing, using holistic management 340-341
green manure 117
grow your own mushrooms 145–148
Guayaki 332
gullies 267–268

H
ha ha fence 160
hand tools 277–278
Hawaiian aquaculture 266
holacracy 322–323
holistic management 186–187, 340–341
Holmgren, David 2
holon 185
hugelkultur 107, 160–161

humanure 77, 79–80
hydrophobic soils 102

I
induce meandering 269
invisible pond 57
irrigation, section 208

J
Jar Soil Test 127–128
Jean Pain 120–121

K
keyhole garden 203
keyline 188-191
keyline scale of permanence 187
keyway 62–63
keypoint 53, 189, 190
Korean Natural Farming 125–126

L
land trust 319–320
landrace 110
landscape effects 33–34
latitudinal effects 34
layers of a forest 195-97
legumes, section - 108–109
levels and leveling 154–157
lichen 136
liquid culture 147–148
liquid inoculation jars 147–148

M
Malmö 304
micro-loans, micro-insurance, and micro-franchises 328–329
mouldering compost 79, 234
mudbrick 163
mulch plant 126
mycoremediation 109, 143, 144
mycorrhiza, section 133, 136

N
natural farming 338–339
natural farming, Korean 125–126
nematode, section 92
net and pan 162-163
nitrogen cycle, soil 87–90
nitrogen cycle, water 256

NonProfit Organizations (NPOs) 329–330
NonViolent Communication (NVC) 323–324
no-till 103–104
nursery, plant 214
nutrient accumulator 130

O
ocean repair 271-275
olla pot 248
one rock dam 268–269
orientation 176–178
orographic effect 23

P
papaya circle 220
passive systems 280–281
pasture cropping 200
Pelton wheel 284–285
Permaculture Skills Center 334
pH, soil section 97–98
Petra 54–56
pig pond 56
phytoremediation 109
pitfall trap 129
plant guides 197
plant guild, section 193
plant roots, section 94–95
pollarding 220
pollinators 206–208
potting soil, recipes 209
prime directive 3
protozoa, section 91
pruning 212-213

Q
qanat 244–245

R
rain shadow 23
rainwater catchment, calculating
rainwater tanks 60
rammed earth blocks 163
refractometer (see BRIX Meter)
restorative circles 324
rewilding 191–192
Rhizobia 86, 108, 110, 134

riparian 266–270
rocket mass heaters & rocket stoves 235–238
root cellar 240
rooting plants 210
runoff, calculating annual 60–64

S
sealing leaky dams 65
seed balls, section 198–199
seed saving 209
sewage 76–80
shadehouse 238–239
sheet mulching 127
silviculture 230–231
slope, how to calculate 179
slope, section 153–154
sociocracy 322–323
soil food web, section 83–85
soil structure, section 99–100
solar oven 281–282
solar radiation 26
soundscape 180–181
spillway 62–63
spore print 136–137
stocking, fish 257
stratification 215
STUN method 349
succession 101–102
sun angle, section 180–181
swales 67–70
swimming pool, natural 80–81, 263

T
terraces 159
thermal belt 28–29
thermophilic compost 118–120
thermosiphon 27–28
throw sow, section 198
transplanting 201
tree flagging 31
trickle pipe 63–64
trompe 291–292, 299–300
Tucson swales 158

U-V

vermicompost 120
vertical farming 272–273
Veta La Palma 264-265

W

walipini 159–160
water, purification of polluted 74–75
water level 154
water tanks 66
water wheel 282–285
watersheds, section 49
wheel pump 284
wicking bed 248
wind, section 30–31
windbreak 31–32

XY

Yeomans Plow 104–105

Z

zone planning and analysis 174–176
Zuni bowl 267–268

References

Please note that some references in this text come from emails and conversations with peer reviewers, authors, and other sources. If it is information easily found on Wikipedia or in a Google search, it is not included here.

A

- AgendaGotsch. *Films*. Documentary series with articles. Web. Accessed 2016. http://agendagotsch.com
- Alexandrina Council. *Environmental Health Fact Sheet: What is required with an application for a Reedbed (second-stage wastewater treatment) system?* Web. PDF. 2016. https://www.alexandrina.sa.gov.au/webdata/resources/files/Reedbeds.pdf
- Allen, Greg. *Toxic Chinese Drywall Creates A Housing Disaster.* NPR.org. 2009. Web. Accessed 2016. http://www.npr.org/templates/story/story.php?storyId=114182073
- aqysta.com *The Barsha Pump.* Web. Accessed 2016. http://www.aqysta.com
- Anderson, Jim, Beduhn, Rebecca, Current, Dean, Espeleta, Javier, Fissore, Cinzia, Gangeness, Bjorn, Harting, John, Hobbie, Sarah, Nater, Ed, and Reich, Peter. "Potential of Soils for Carbon Sequestration." *A Report to the Department of Natural Resources from the Minnesota Terrestrial Carbon Sequestration Initiative* (2008). The Potential for Terrestrial Carbon Sequestration in Minnesota. University of Minnesota, St. Paul, MN., Feb. 2008. Web. Oct. 2016. http://files.dnr.state.mn.us/aboutdnr/reports/carbon2008.pdf.
- ArcNews Online. *Willie Smits Shares Methodology for Sustainable Forests.* Winter 2009/2010. http://www.esri.com/news/arcnews/winter0910articles/willie-smits.html
- Asselin, Olivier. *The Permaculture Orchard: Beyond Organic.* Possible Media. Film. 2014.
- Aurovilleradio.org *Restorative Circles.* 2014. http://www.aurovilleradio.org/restorative-circles/
- Axe, Josh. *10 Turmeric Benefits: Superior to Medications?* DrAxe.com. web. 2016. Accessed 2016. https://draxe.com/turmeric-benefits/

B

- Baird, A., Baird, G., Hill, G., Hoeppner, E., Payne, M., Seymour, M. *Manual of Composting Toilet and Greywater Practice*. BC Ministry of Health, Health Protection Branch. February 2016 Draft for Consultation. Accessed 2016. http://www2.gov.bc.ca/assets/gov/environment/waste-management/sewage/composting_toilet_manual.pdf
- BBC.com. *First UK homes heated with 'poo power' gas from sewage.* BBC.com Oct. 1 2014. Accessed 2016. http://www.bbc.co.uk/news/uk-england-29443622
- Benson Agriculture and Food Institute. *Walipini Construction (The Underground Greenhouse).* Brigham Young University. Utah, 2002. Open Source Ecology. Web. n.d. Accessed 2016. http://opensourceecology.org/w/images/1/1c/Walipini.pdf
- *Blue Vinyl*. Biography/Documentary Film. 2002.
- Bradley, Kirsten. *Why Pasture Cropping is such a Big Deal.* Milkwood. 2010. Accessed 2016. https://www.milkwood.net/2010/12/07/why-pasture-cropping-is-such-a-big-deal/
- Brown, Azby. *The Edo Approach.* TEDxTokyo. 2010. Web. Accessed 2016. https://www.youtube.com/watch?v=1D7qc8nc2Ng
- Browne Trading Company. *Veta La Palma Seafood.* Web. Accessed 2016. https://www.brownetrading.com/veta-la-palma-seafood/
- Buck, John and Villines, Sharon. *We the People: Consenting to a Deeper Democracy.* Sociocracy.info, 2007.

C

- Catalyst. *Earth on Fire.* ABC. Aired June 2014. Accessed 2016. http://www.abc.net.au/catalyst/stories/4014144.htm
- Christensen, Ken. *Could a Mushroom Save the Honeybee?* NPR.org. Accessed 2016. http://www.npr.org/sections/thesalt/2015/10/09/446928755/could-a-mushroom-save-the-honeybee
- CompostGuy.com. *Bokashi - the fermentation of organic wastes.* The Compost Guy. 2012. Accessed 2016. http://www.compostguy.com/bokashi-resource-page/
- Cooper, Daniel. *Sweden debuts the world's first 'electric highway'.* engadget.com. June 2016. Accessed 2016. https://www.engadget.com/2016/06/24/sweden-electric-highway/
- Cornell University. *Marine microalgae, a new sustainable food and fuel source.* Eureka Alert. Web. Accessed 2016. https://www.eurekalert.org/pub_releases/2016-11/cu-mma112116.php
- Cotter, Tradd. *Organic Mushroom Farming and Mycoremediation.* Chelsea Green, 2014.

D

- Davis, Tony. *Tucson's Rain-catching Revolution.* High Country News. High Country News, 27 Apr. 2015. Web. 28 Oct. 2016. http://www.hcn.org/issues/47.7/tucsons-rain-catching-revolution
- Dawborn, Kerry and Smith, Caroline, editors. *Permaculture Pioneers: stories from the new frontier.* Collection of Essays. Meliodora Publishing, Victoria, Australia. 2011.
- Deppe, Carol. *Breed your own Vegetable Varieties.* Little, Brown, and Company. Boston, 1993.
- Deppe, Carol. *The Resilient Gardener: Food Production and Self-reliance in Uncertain times.* White River Junction, VT: Chelsea Green Pub., 2010.
- De Witte, Melissa. *Pesticide predicament for California's strawberry growers.* University of California Santa Cruz. Sept 28 2016. Accessed 2016. http://news.ucsc.edu/2016/09/guthman-strawberries.html
- Department of Environment and Primary Industries. *Trickle Flow Pipes for Farm Dams.* Agriculture Victoria. 2013. Web. Accessed 2016. http://agriculture.vic.gov.au/agriculture/farm-management/managing-dams/trickle-flow-pipes-for-farm-dams
- Design Coalition. *Measuring Sun Angles.* Web. n.d. Accessed 2016. http://designcoalition.org/kids/energyhouse/sunangles.htm

- Dervaes, Jordanne. *Email*. Message to Matt Powers. Aug 2016.
- *Determining Soil Texture by the Feel Method*. Baltimore Ecosystem Study. n.d. Accessed 2016. http://www.beslter.org/msp/institute.files/november/Determining%20Soil%20Texture%20by%20Feel%20Method.pdf
- Doherty, Darren J. and Jeeves, Andrew. *The Regrarians eHandbook*. Regrarians Limited, 2015.
- Doherty, Darren J. *Broad Acre Agroforestry Integration*. Permaculture Voices: PV3. March 2016. Conference.
- Drake, Nadia. *Will Humans Survive the Sixth Mass Extinction?* National Geographic. Web. June 23, 2015. Accessed 2016. http://news.nationalgeographic.com/2015/06/150623-sixth-extinction-kolbert-animals-conservation-science-world/
- Dumaresq, Charles. Cobalt Mining Legacy: Power to the Mines. 2009. Accessed 2018. http://www.cobaltmininglegacy.ca/power.php

E

- *Episode 8 Geoff Lawton on the future of Permaculture & Food Production, Children & Permaculture*. Permaculture Tonight. iTunes and Soundcloud Podcast. 2015.
- Evans, Ianto, Smith, Michael, G., and Smiley, Linda. *The Hand-Sculpted House: A Practical and Philosophical Guide to Building a Cob Cottage*. Chelsea Green. 2002.

F

- Farquhar, Brodie. *Wolf Reintroduction Changes Ecosystem*. YellowstonePark.com. 2016. National Park Trips Media. Web. Accesssed 2016. http://www.yellowstonepark.com/wolf-reintroduction-changes-ecosystem/
- fibershed.com. *About*. 2016. Accessed 2016. http://www.fibershed.com/about/
- Footer, Diego. *Permaculture Voices Interviews Willie Smits*. San Diego, 2014. Accessed 2016. https://vimeo.com/110548063
- Footer, Diego. *PV3: Permaculture Voices 3*. Conference. San Diego. 2016.
- Footer, Diego. *PV2: Permaculture Voices 2*. Conference. San Diego. 2015.
- Fortier, Jean-Martin. *The Market Gardener*. New Society Publishers, 2014.
- Fortier, Jean-Martin. *The Market Gardener with Jean-Martin Fortier, Six Figure Farming*. Living Web Farms. web. Youtube Playlist. Accessed 2016. https://www.youtube.com/playlist?list=PLCeA6DzL9P4uRadXW0_hj5Ct3EAqWH1zl
- Fortier, Jean-Martin. *Profitable small-scale farming. How design sets the stage for success*. Permaculture Voices: PV3. San Diego. March 2016.
- Freedman, Andrew. *Indonesia's peat fires make it the 4th-largest carbon emitter in the world*. mashable.com. web. 2015. http://mashable.com/2015/10/29/indonesia-peat-fires-largest-emitter/#fB5VYDcB.uqG
- Fu, Xiaowei, and Du, Qizhen. *Uptake of Di-(2-ethylhexyl) Phthalate of Vegetables from Plastic Film Greenhouses*. Journal of Agricultural and Food Chemistry. 2011, vol 59. Accessed 2016. pubs.acs.org/JAFC
- Fu, X.W., and Xia, H.L. *Uptake of di-(2-ethylhexyl)phthalate from plastic mulch film by vegetable plants*. AGRICOLA. United States Department of Agriculture, National Agriculture Library. 2009, vol. 26.
- Fukuoka, Masanobu. *Planting Seeds in the Desert*. Chelsea Green, Vermont. 2012.
- Fukuoka, Masanobu. *The One-Straw Revolution*. The New York Review of Books, 2009.

G

- Greenman, Eliza. *A New Fruit Culture!* Permaculture Voices: PV3. March 2016. Conference.
- Gronbeck, Christopher. Web. 2009. Accessed 2016. http://www.susdesign.com/sunangle/
- *Growing Power - A Model for Urban Agriculture*. Documentary. Oct 2010. Web. Youtube.com. Accessed 2016. https://www.youtube.com/watch?v=vs7BG4lH3m4
- GrowingPower.org. *Will Allen*. Web. 2014. Accessed 2016. http://www.growingpower.org/about/leadership/will-allen/
- Guevara- Stone, Leslie. *How a decaying Swedish city became an eco-friendly hub*. GreenBiz.com. 2014. Accessed 2016. https://www.greenbiz.com/blog/2014/10/01/how-decaying-industrial-city-became-eco-friendly-hub
- Guayaki.com *Sustainability*. web. 2016. Accessed 2016. http://guayaki.com/about/2231/Sustainability.html

H

- Harris, Nancy, Minnemeyer, Susan, Stolle, Fred, and Payne, Octavia Aris. *Indonesia's Fire Outbreaks Producing More Daily Emissions than Entire US Economy*. World Resources Institute. Web. 2015. Accessed 2016. http://www.wri.org/blog/2015/10/indonesia%E2%80%99s-fire-outbreaks-producing-more-daily-emissions-entire-us-economy
- Hemenway, Toby. *The Permaculture City*. Chelsea Green, 2015.
- Hickman, Leo. *Will the Brixton pound buy a brighter future?* The Guardian. Web. Sept 15 2009. Accessed 2016. https://www.theguardian.com/environment/2009/sep/16/will-brixton-pound-work
- Hollenhorst, John. *Are Wolves a "miracle" in Yellowstone? Science Seeks Answers*. DeseretNews.com. Deseret Digital Media, 26 Aug. 2016. Web. 29 Oct. 2016. http://www.deseretnews.com/article/865661057/Are-wolves-a-Miracle-in-Yellowstone-Science-seeks-answers.html?pg=all.
- Holmgren, David. *Permaculture: Principles and Pathways beyond Sustainability*. Hepburn, Victoria: Holmgren Design Services, 2002.
- Holzer Permaculture. *The Krameterhof*. web. 2016. Accessed 2016. http://www.holzerpermaculture.us/krameterhof.html
- Holzer, Sepp. *Desert or Paradise*. Chelsea Green, 2012.
- Holzer, Sepp. *Sepp Holzer's Permaculture: A Practical Guide to Small-Scale, Integrative Farming and Gardening*. Chelsea Green, 2011.
- HydrateLife.org. *Eco-Latrine of the future: Tiger Toilets*. Web. 2012. Accessed 2016. http://www.hydratelife.org/?p=539

- *Homegrown Revolution (Award winning short-film 2009)- The Urban Homestead, Dervaes*. The Urban Homestead. Youtube. 2011. Documentary. Accessed 2016. https://www.youtube.com/watch?v=7IbODJiEM5A

I

- Ingham, Elaine. *Biological Farming*. Permaculture Magazine North America. No. 3 Winter 2016. The Permaculture Bug, LLC. 2nd article in series.
- Ingham, Elaine. *Celebration Farm Tour*. Oroville, CA. 2016.
- Ingham, Elaine R., Moldenke, Andrew R., and Edwards, Clive A. *Soil Biology Primer*. Soil and Water Conservation Society in cooperation with the USDA Natural Resources Conservation Service. 2000.
- Ingham, Elaine. *Lecture Notes, Work in Beauty workshop, Gallup, NM -- November 7, 2015*. WorkInBeauty.org. 2015. Accessed 2016. http://bernalilloextension.nmsu.edu/mastercomposter/documents/2015-mc-project-mb.pdf
- Ingham, Elaine. *Email*. Message to Matt Powers. 22 March 2016. Email.
- Ingham, Elaine. *Restoring Your Soil Life, Increasing Yields, Lowering Costs*. Master Design Masterclass. Webinar. Sept 2016. http://www.sustainabledesignmasterclass.com
- *INHABIT: A Permaculture Perspective*. Director: Costa Boutsikaris. Producer: Emmett Brennan. Film. 2015.

J

- Japan for Sustainability Staff. *Japan's Sustainable Society in the Edo Period (1603-1867)*. Resilience.org. 2005. Web. Accessed 2016. http://www.resilience.org/stories/2005-04-05/japans-sustainable-society-edo-period-1603-1867
- Jenkins, Joseph. *The Humanure Handbook*. 2nd Ed. Chelsea Green. Vermont, 1999. Accessed 2016. https://humanurehandbook.com/downloads/H2.pdf
- Jones, Christine and Frisch, Tracy. "SOS: Save Our Soils." Acres U.S.A. Interview. March 2015. Vol 45, No. 3. Accessed 2016. http://www.amazingcarbon.com/PDF/Jones_ACRES_USA%20(March2015).pdf

K

- Kassam, Ashifa, Scammel, Rosie, Connolly, Kate, Orange, Richard, Willsher, Kim, and Ratcliffe, Rebecca. *Europe needs many more babies to avert a population disaster*. The Guardian. Aug 2015. web. Accessed 2016. https://www.theguardian.com/world/2015/aug/23/baby-crisis-europe-brink-depopulation-disaster
- Kjellman, Mikael. *Podride a practical and fun bicycle-car*. indigogo.com. web. 2016. Accessed 2016. https://www.indiegogo.com/projects/podride-a-practical-and-fun-bicycle-car-bike-bicycle#/
- Kourik, Robert. *Understanding Roots*. Metamorphic Press, 2015.

L

- Lal, Rattan. *Managing Soils and Ecosystems for Mitigating Anthropogenic Carbon Emissions and Advancing Global Food Security*. BioScience, Vol 60 no 9, Oct 2010. Web. Accessed 2016. http://tinread.usarb.md:8888/tinread/fulltext/lal/managing.pdf
- Lancaster, Brad. *Rainwater Harvesting for Drylands and Beyond*. Rain source Press, 2013. Distributed by Chelsea Green.
- Lawton, Geoff. *The Geoff Lawton Online Permaculture Design Course*. Permaculture Research Institute of Australia. New South Wales, Australia. 2014. http://www.geofflawtononline.com
- Lennox, James. *Phone Conversation with Matt Powers*. Operator at TransAlta's Ragged Chutes Hydro Station. September 2016.
- Lewis, Wayne and Lowenfels, Jeff. *Teaming with Microbes*. Timber Press, Inc., 2006.
- Liu, John D. *Email*. Group discussion with Rhamis Kent. 2016.
- Liu, John D. *Green Gold*. Environmental Education Media Project. Web. 2012. Accessed 2016. https://www.youtube.com/watch?v=YBLZmwlPa8A
- Liu, John D. *Lesson from the Loess Plateau*. Environmental Education Media Project. Web. 2012. Accessed 2016. https://www.youtube.com/watch?v=8QUSIJ80n50
- Living Web Farms. *The Market Gardener with Jean-Martin Fortier, Six Figure Farming*. Youtube Playlist. Web. 2015. Accessed 2016. https://www.youtube.com/playlist?list=PLCeA6DzL9P4uRadXW0_hj5Ct3EAqWH1zl

M

- Mainguy, Pierre. *Floating Gardens of SE Asia*. Email. Message to Matt Powers. April 28th 2016.
- Markham, Justin. *France to pave 1000km of roads with solar panels*. treehugger.com. web. January 2016. Accessed 2016. http://www.treehugger.com/solar-technology/france-pave-1000km-roads-solar-panels.html
- McCoy, Peter. *Radical Mycology*. Chthaeus Press, 2016.
- McCoy, Peter. *Radical Mycology Webinar 1: Seeing Fungi*. web. Youtube. 2016. https://www.youtube.com/watch?v=aB9JSky8x6k
- McCurry, Justin. *Japan's maglev train breaks world speed record with 600km/h test run*. The Guardian. April 2015. Web. Accessed 2016. https://www.theguardian.com/world/2015/apr/21/japans-maglev-train-notches-up-new-world-speed-record-in-test-run
- McFarland, Kathy. *The Brooklyn Grange*. Heirloom Gardener. Magazine. Spring 2013.
- Meisel, Ari. *A Resource Guide to Green Building*. Princeton Architectural Press, New York. 2010.
- Merton, Lisa, and Dater, Alan. *Taking Root: The Vision of Wangari Maathai*. Global Perspectives Collection. Film. 2009. http://itvs.org/films/taking-root
- Microponics. *Black Soldier Fly Larvae*. May 2009. Accessed 2016. https://www.microponics.net.au/diy-livestock-rations/black-soldier-fly-larvae/
- Middleton, Arthur. *Is the Wolf a Real American Hero?* NYTimes.com. The New York Times, 9 Mar. 2014. Web. Accessed 2016. http://www.nytimes.com/2014/03/10/opinion/is-the-wolf-a-real-american-hero.html?_r=1.
- Mollison, Bill. *Permaculture: A Designer's Manual*. Tagari Publications, Tasmania. 1989.
- Morrow, Rosemary. *Earth User's Guide to Teaching Permaculture*. Permanent Publications, 2014.
- Morrow, Rosemary. *Permaculture and the Forgotten. Teaching Permaculture in Places That Absolutely Need It. A Message of Hope*

- with Rosemary Morrow. (PVP068). Permaculture Voices. Podcast. http://www.permaculturevoices.com/permaculture-and-the-forgotten-teaching-permaculture-in-places-that-absolutey-need-it-a-message-of-hope-with-rosemary-morrow-pvp068/ Accessed 2015.
- Morrow, Rosemary. *Successful permaculture in Cambodia*. Email to Matt Powers. Aug 30 2016.
- *Mushroom Companion Plants*. The Medicine Garden. Accessed 2016. http://www.medicinegarden.co.za/about/companion-planting/mushroom-companion-plants/
- *Mycorrhizae-Compatible Plants*. Fungi.com. Accessed 2016. http://www.fungi.com/plant-list.html

N

- Natural Capital LLC. *Natural Capital Plant Database*. Web Database. 2014. Accessed 2016. http://www.permacultureplantdata.com/
- Nelson, K.D. *Design and Construction of Small Earth Dams*. Inkata Press. Melbourne, 1991. Accessed 2016 with nominal fee. https://soilandhealth.org
- Nguyen, Tuan C. *Can an Algae-Powered Lamp Quench Our Thirst For Energy?* Smithsonian.com. 2013. Web. Accessed 2016. http://www.smithsonianmag.com/innovation/can-an-algae-powered-lamp-quench-our-thirst-for-energy-3509307/?no-ist

O

- OAEC.org. *Compost Toilet Research Project*. Occidental Arts and Ecology Center. 2016. Accessed 2016. https://oaec.org/our-work/projects-and-partnerships/compost-toilet-project/
- Oettershagen, Philipp - Representative of AtlantikSolar. *Re: Photo Request for Alternative Energy chapter in the first permaculture high school textbook*. Email to Matt powers. 2016.
- Olaizola, M., T. Bridges, S. Flores, L. Griswold, J. Morency, and T. Nakamura. *Microalgal Removal of CO_2 from Flue Gases: CO_2 Capture from a Coal Combustor* (2004): n. pag. National Energy Technology Laboratory. Physical Sciences Inc., Andover MA, 2004. Web. 2016.. https://www.netl.doe.gov/publications/proceedings/04/carbon-seq/123.pdf

P

- Pain, Ida and Jean. *The Methods of Jean Pean or "Another Kind of Garden"*. Ancienne Imprimerie, 1972. archive.org. Accessed 2016. https://archive.org/details/Another_Kind_of_Garden-The_Methods_of_Jean_Pain
- Permaculture Research Institute. *OVER 200 FOOD PLANTS ON JUST A TINY 1/10TH ACRE OF COLD CLIMATE URBAN LAND*. Web. PRI AU. 2014. Accessed 2016. http://permaculturenews.org/2014/01/18/perennial-abundance-200-food-plants-on-1-10th-acre-cold-climate-urban-land/
- Phytotrade Africa. *The Baobab Tree and its Fruit*. web. Posted 2012. Accessed 2016. http://phytotrade.com/products/baobab/
- Pickerell, John. *Oceans Found to Absorb Half of All Man-Made Carbon Dioxide*. National Geographic News. online. 2004. Accessed 2016. http://news.nationalgeographic.com/news/2004/07/0715_040715_oceancarbon.html
- Pimm, S.L., Russell, G.J., Gittleman, J.L., and Brooks, T.M. *The Future of Biodiversity*, Science 269: 347-350 (1995). http://www.rachel.org/files/document/The_Future_of_Biodiversity.pdf
- Pineault, Jonathan, and Fortier, Jean-Martin. *Permaculture Meets Market Gardening*. Permaculture Voices: PV3. San Diego. March 2016.
- Pittet, Jennifer. *A Farmer Turns Wasteland into Rainforest*. Farm Radio International. 2000. Accessed 2016. http://www.farmradio.org/radio-resource-packs/package-55-agroforestry-for-the-small-farmer/a-farmer-turns-wasteland-into-rainforest/
- Plakias, Anastasia Cole. *Re: Can I get a picture of the grange to feature in my book?* Email to Matt Powers. Oct. 14 2016.
- *Polyfaces*. Directors: Lisa Heenan and Isaebella Doherty. Producer: Lisa Heenan. Film. 2015.
- *Ponds: Planning, Design, Construction*. Natural Resource Conservation Service, United States Department of Agriculture. Agriculture Handbook 590. 1997. Web. Accessed 2016. http://soiltesting.tamu.edu/publications/USDAPONDS.pdf
- Powers, Matthew. *Where Did the Water Go in Central Valley California? PowerTalk*. youtube.com. Sept 18 2014. Accessed 2016. https://www.youtube.com/watch?v=OpxYnvNZHDQ
- PRI Zaytuna Farm, NSW, Australia. PermacultureGlobal.org. web. 2011. Accessed 2016. https://permacultureglobal.org/projects/3-pri-zaytuna-farm-nsw-australia

Q-R

- *RAHT RACER: Cycling vehicle - pedal as fast as a car*. Youtube.com. 2015. web. Accessed 2016. https://www.youtube.com/watch?v=W4ZGTDUO0DQ
- RestorativeCircles.org. *Restorative Circles*. 2014. web. Accessed 2016. https://www.restorativecircles.org
- Ripple, William J., Estes, James A., Beschta, Robert L., Wilmers, Christopher C., Ritchie, Euan G., Hebblewhite, Mark, Berger, Joel, Elmhagen, Bodil, Letnic, Mike, Nelson, Michael P., Schmitz, Oswald J., Smith, Douglas W., Wallach, Arian D., Wirsing, Aaron J. *Status and Ecological Effects of the World's Largest Carnivores*. Science. 10 Jan 2014. American Association for the Advancement of Science. Web. Accessed 2016. http://science.sciencemag.org/content/343/6167/1241484
- Robertson, Brian. *Holacracy: The New Management System for a Rapidly Changing World*. Holacracy One LLC, 2015. Macmillan Audio, 2015.
- Rodale Press. *One Straw Revolution - by Masanobu Fukuoka*. Youtube. 2014. Accessed 2016. https://www.youtube.com/watch?v=8atbgaiekZI
- Rosenberg, Marshall. *Nonviolent Communication Training Course Marshall Rosenberg CNVC org*. CNVC.org. Posted on Youtube.com. Aug 24 2014. Accessed 2015. https://www.youtube.com/watch?v=O4tUVqsjQ2I
- Rosgen, David. *Dave Rosgen PhD River Hydrologist on Buffalo Bayou & Braes Bayou*. Youtube.com. 2012. Accessed 2016. https://www.youtube.com/watch?v=Jz625ybka8U
- Rotheroe, Dom. *The Coconut Revolution*. Stampede Films. 2001. Hosted youtube.com. Accessed 2016. https://www.youtube.com/watch?v=mGUBuuUEC0s

S

- Savitz, Jackie. *Save the Oceans, Feed the World!* ted.com. TEDxMidAtlantic, 2013. Accessed 2016. http://www.ted.com/talks/jackie_savitz_save_the_oceans_feed_the_world
- Savory, Alan. *Holistic Management: A New Framework for Decision Making*. Island Press, 1999.
- Sawada, Kozue, and Koki, Toyota. *Effects of the Application of Digestates from Wet and Dry Anaerobic Fermentation to Japanese Paddy and Upland Soils on Short-Term Nitrification*. Microbes and Environments. The Japanese Society of Microbial Ecology (JSME)/ The Japanese Society of Soil Microbiology (JSSM), 30 Mar. 2015. Web. 28 Oct. 2016. https://www.ncbi.nlm.nih.gov/pmc/articles/PMC4356462/
- *Sepp Holzer The Agro Rebel*. Documentary. Youtube. web. Accessed 2016. https://www.youtube.com/watch?v=Ekub958v7Ks
- Shepard, Mark. *Farming. It's Damn Hard*. An interview with Mark Shepard. (PVP091). Permaculture Voices. iTunes and Soundcloud Podcast. Accessed 2016. https://soundcloud.com/permaculturevoices/mark-shepard-restoration-agriculture-pvp091
- Shepard, Mark. *Restoration Agriculture: Real-World Permaculture for Farmers*. Acres U.S.A., 2013.
- Sheil, Douglas, and Murdiyarso, Daniel. *How Forests Attract Rain: An Examination of a New Hypothesis*. American Institute of Biological Sciences. Oxford Journal: BioScience. Vol. 50. Apr. 2009. Accessed 2016. http://bioscience.oxfordjournals.org/content/59/4/341.full
- Schultz, Grant. *Permaculture 2.0, Designing a Profitable Broadacre Perennial Farm with Grant Schultz. (PVP034)*. Permaculture Voices. iTunes and Soundcloud Podcast. 2015. Accessed 2016. https://soundcloud.com/permaculturevoices/pvp034-02212014
- Simcox, Joseph. *Episode 33 Joseph Simcox - The Botanical Explorer & the Gardens Across America Project*. Permaculture Tonight. Podcast (iTunes & Soundcloud). Interview. 2016. https://soundcloud.com/permaculturetonight/episode-33-joseph-simcox-the-botanical-explorer-the-gardens-across-america-project
- Smits, Willie. *How to Restore a Rainforest*. TED talk. Accessed 2016. https://www.ted.com/talks/willie_smits_restores_a_rainforest?language=en
- Sobkowiak, Stefan. *History - Miracle Farm*. Web. 2016. Accessed 2016. http://miracle.farm/en/history/
- Sobkowiak, Stefan. *Miracle Farm*. Facebook Message to Matt Powers. 2016.
- Spackman, Neal. *10 Keys for Greening Any Desert*. Sustainable Design Masterclass, LLC. 2016. Webinar. Accessed 2016. http://www.sustainabledesignmasterclass.com
- Spackman, Neal. *Facing Fear and Stepping into the Unknown - The Al Baydha Project*. Permaculture Voices: PV2. San Diego. March 2015.
- Spackman, Neal. *Email*. Message to Matt Powers. Ongoing 2015-2016.
- Spadaccini, Michael. *The Basics of Business Structure*. Entrepreneur.com. March 9 2009. Web. Accessed 2016. https://www.entrepreneur.com/article/200516
- Staiger, Christiane. *Comfrey: A Clinical Overview*. US National Library of Medicine, National Institutes of Health. Feb 2012. Web. Accessed 2016. https://www.ncbi.nlm.nih.gov/pmc/articles/PMC3491633/
- Stamets, Paul. *How Mushrooms Can Clean Up Radioactive Contamination - An 8 Step Plan*. Permaculture Magazine UK. Web. 2011. Accessed 2016. http://www.permaculture.co.uk/articles/how-mushrooms-can-clean-radioactive-contamination-8-step-plan
- Stamets, Paul. Assisted by Yao, Dusty Wu. *MycoMedicinals: An Informational Treatise on Mushrooms*. 3rd Ed. Paul Stamets, 2002. MycoMedia Productions, Fungi Perfecti LLC.
- Stone, Curtis. *The Urban Farmer: Growing Food for Profit on Leased and Borrowed Land*. New Society Publishers, 2015.
- Stone, Curtis. *Re: Case Study on Green City Acres*. Email to Matt Powers. 2016.
- Stone, Curtis. *Tool review: Quick Greens Harvester*. Urban Farmer Curtis Stone. Web. March 2014. Youtube. Accessed 2016. https://www.youtube.com/watch?v=8Axy37RytoA
- Stone, Curtis. *$75,000 on 1/3 acre. Profitable Urban Farm Tour. Green City Acres*. Urban Farmer Curtis Stone. Web. Sept 2015. Youtube. Accessed 2016. https://www.youtube.com/watch?v=adW3GCQGHug
- Stone, Curtis. *$80,000 on Half An Acre Farming Vegetables - Profitable Mini-Farming with Curtis Stone*. Permaculture Voices: PV1. Web. San Diego, 2014. Accessed 2016. https://www.youtube.com/watch?v=1MNhtcagNO0
- Stone, Nathan. *Renovating Leaky Ponds*. Southern Regional Aquaculture Center. 1999. Accessed 2016. http://aqua.ucdavis.edu/DatabaseRoot/pdf/105FS.PDF

T

- Teutsch, Betsy. *100 under $100: One Hundred Tools for Empowering Global Women*. She Writes Press. China, 2015.
- ThePermacultureOrchard.com. *The Farm*. 2016. Web. Accessed 2016. http://www.permacultureorchard.com/the-farm/
- The Corporation of the Town of Cobalt. *Ragged Chutes: A Modern Wonder*. 2016. Web. Accessed 2016. https://cobalt.ca/ragged-chutes/
- The New Heroes. *PBS New Heroes Ep3 03 Muhammad Yunus Microcredit Bangladesh*. PBS. web. Posted on Youtube 2012. Accessed 2016. https://www.youtube.com/watch?v=4NGU5gkI6-Y
- The Pennsylvania State University. *Inoculation of Legumes for Maximum Nitrogen Fixation*. Penn State College of Agricultural Sciences. 2016. Accessed 2016. http://extension.psu.edu/plants/crops/forages/successful-forage-establishment/inoculation-of-legumes-for-maximum-nitrogen-fixation
- *The Salatin Semester*. DVD Course & Book set. Verge Permaculture/Acres USA. 2016. http://salatinsemester.com
- The Urban Homestead. *By the Numbers*. UrbanHomestead.org. web. 2016. Accessed 2016. http://urbanhomestead.org/about/by-the-numbers/
- The World Bank. *Loess Plateau Watershed Rehabilitation Project*. 2011. Accessed 2016. http://projects.worldbank.org/P003540/loess-plateau-watershed-rehabilitation-project?lang=en
- Tobgay, Tshering. *This Country Isn't Just Carbon Neutral - It's Carbon Negative*. TED.com 2016. Accessed 2016. http://

- www.ted.com/talks/ tshering_tobgay_this_country_isn_t_just_carbon_neutral_it_s_carbon_negative
- Toensmeier, Eric. *Paradise Lot*. Chelsea Green, 2013.
- Toensmeier, Eric. *The Carbon Farming Solution*. Chelsea Green, 2016.

U

- US Census Bureau. *Commuting in the United States: 2009*. US Department of Commerce, Economics and Statistics Administration. Sept 2011. Accessed 2016. https://www.census.gov/prod/2011pubs/acs-15.pdf

V-W

- Water Powered "Air Compressor and Water Pump". The "Trompe Hammer", Trompe and Water Ram. Account: MrTeslonian. web. 2015. youtube.com. Accessed 2016. https://www.youtube.com/watch?v=xv1lQA-tnwo
- Watters, Ethan. *DNA is not Destiny*. Discover Magazine. Nov 22 2006. Web. Accessed 2015. http://discovermagazine.com/2006/nov/cover
- Weiss, Zachary. *Elemental Ecology*. Permaculture Voices: PV3. Conference. March 2016.
- Weiss, Zachary. *Email*. Email to Matt Powers. Oct 8 2016.
- Weller, Chris. *There's a new path to Harvard and it's not in a classroom*. BusinessInsider.com. Sept 2015. Accessed 2016. http://www.businessinsider.com/homeschooling-is-the-new-path-to-harvard-2015-9
- Wheaton, Paul. *Can Pigs Build Ponds?* MakeItMissoula.com. web. 2012. Accessed 2016. http://www.makeitmissoula.com/2012/07/paul-wheaton-can-pigs-build-ponds/
- Wisner, Erica and Ernie. *The Rocket Mass Heater Builder's Guide*. New Society Publishers, 2016.
- Wisner, Ernie. *Email*. Feb 13 2016.
- Wood, A.D., and Richardson, E.V. Design of Small Storage and Erosion Control Dams. Department of Civil Engineering. Colorado State University. 1975. Accessed 2016. https://dspace.library.colostate.edu/bitstream/handle/10217/52669/CER_Wood.pdf?sequence=1
- *Woody Agriculture – Breeding Trees, Restoring a Piece of America's Past and Establishing a Piece of Our Agricultural Future with Phil Rutter – Part 1 of 2 (PVP057)*. Permaculture Voices. iTunes and Soundcloud Podcast. Posted 2015. Accessed 2015. http://www.permaculturevoices.com/woody-agriculture-breeding-trees-restoring-a-piece-of-americas-past-and-establishing-a-piece-of-our-agricultural-future-with-phil-rutter-part-1-pvp057/
- Worrich, Anja , Stryhanyuk, Hryhoriy , Musat, Niculina , König, Sara, Banitz, Thomas , Centler, Florian , Frank, Karin , Thullner, Martin, Harms, Hauke, Richnow, Hans-Hermann, Miltner, Anja , Kästner, Matthias, & Wick, Lukas Y. *Mycelium-mediated transfer of water and nutrients stimulates bacterial activity in dry and oligotrophic environments*. nature.com Published 2017. Accessed 2018. https://www.nature.com/articles/ncomms15472
- Wright, Sara F; Nichols, Kristine A. *Glomalin: Hiding place for a third of the world's stored soil carbon*. Agricultural Research; Washington (Sep 2002). http://search.proquest.com/openview/32cd9540e48f8e2ace82786043736c1c/1?pq-origsite=gscholar&cbl=42132
- Wuerthner, George. *Climate Change and Livestock Grazing*. Counterpunch.org. Counter Punch, 04 Oct. 2015. Web. 28 Oct. 2016. http://www.counterpunch.org/2015/02/06/climate-change-and-livestock-grazing/#_edn3

X-Y

- Yeomans, Allan J. *Priority One: Together We Can Beat Global Warming*. Keyline Publishing Co., Australia. 2005.
- Yeomans, PA. *The Challenge of Landscape*. Keyline Publishing Pty. Limited. 1958. Soil and Health Library. Web. n.d. Accessed 2016. **http://soilandhealth.org/wp-content/uploads/01aglibrary/010126yeomansII/010126ch4.html**

Z

- Zeedyk, Bill. *2013 Quivira Conference, Bill Zeedyk*. Youtube.com. Web. Accessed 2016. https://www.youtube.com/watch?v=V3d85D4xlbA
- Zeedyk, Bill. *Understanding Slope Wetlands*. Slides from Quivira 2014 Conference. Zeedyk Ecological Consulting, LLC. Web. Accessed 2016. http://quiviracoalition.org/images/pdfs/5908-Zeedyk_1%2520and%25202.pdf

Editors/Peer Reviewers

Gabrielle Harris, Lead Editor
Gabrielle has studied with John D. Liu in China, received her PDC at Zaytuna Farms, and received a specialized PDC in Aid and Development. She is a writer, editor, journalist, and designer. Linkedin.com/in/gabrielle-harris-0855bb6

Neal Spackman
Neal is the project manager and primary designer at the Al Baydha Project in Saudi Arabia and co-founder of Sustainable Design Masterclass LLC. Neal teaches internationally, blogs, makes videos, writes, and consults. SustainableDesignMasterclass.com

Peter McCoy
Peter is the author of *Radical Mycology* and the co-founder of the grassroots movement and organization by the same name. He also blogs, makes videos, speaks at events, runs workshops, teaches, and even hosts annual Radical Mycology convergences. RadicalMycology.com

Elaine Ingham, Phd.
Dr. Elaine Ingham is a soil scientist, researcher, consultant, educator, speaker, and president of Soil Foodweb Inc. She has served as chief scientists at the Rhodale Institute and taught at Oregon State University. SoilFoodWeb.com

Javan K. Bernakevitch
Javan is a land designer, consultant, writer, educator and speaker with All Points Land Design and a trainer and coach with All Points Life Design in North America and internationally. Allpointsland.ca Allpointslife.com Permaculturebc.com

Danial Lawton
Danial Lawton is a second-generation permaculturist, consultant, educator, entrepreneur, philanthropist, and founder of Permaculture Tools. Permaculturetools.com.au

Darren J. Doherty
Darren is an experienced regenerative landscape, farm, ranch, and business designer, educator, author, and consultant. He is founder and director of Regrarians Ltd. Regrarians.org

Peer Reviewers

Hannah Apricot Eckberg
Hannah has studied permaculture for over 20 years, traveling to many sites around the world. She is a writer, gardener, conversationalist, editor, and co-founder of *Permaculture Magazine, North America*. PermacultureMag.org

Cassie Langstraat
Cassie is a writer, gardener, philosopher, feminist, editor, and co-founder of *Permaculture Magazine, North America*. PermacultureMag.org

Troy Martz
Troy Martz is a technologist, blogger, permaculturist, and alternative energy hacker, living with his beautiful family in tropical paradise. OffGridPro.com

Ernie and Erica Wisner
Ernie and Erica are educators, researchers, authors of *The Rocket Mass Heater Builder's Guide*, and innovators focused on rocket mass heaters and natural building techniques like cob. ErnieAndErica.info

Jean-Martin Fortier
JM is a market gardener, biointensive farmer, and author of *The Market Gardener* in Quebec, Canada. FermeQuatreTemps.com LaGrelinette.com

Stuart Muir Wilson
Stuart is a designer, a teacher, and the grandson of permaculture co-founder, Bill Mollison, and has been immersed in permaculture his entire life and focused specifically on permatecture and the Third ethic. DesignForHumanity.org.au

Stefan Sobkowiak
Stefan is a teacher and farmer in Quebec, Canada. His farm, the Miracle Farm, has been featured in the documentary *The Permaculture Orchard*. PermacultureOrchard.com

Larry Korn
Student of Masanobu Fukuoka, Larry served as editor for *The One-Straw Revolution* and *Sowing Seeds in the Desert* for Fukuoka. Larry is also a teacher and published writer. Onestrawrevolution.net

Geoff Lawton
World-renowned permaculture educator, farmer, designer, and humanitarian aid worker from Australia who co-taught with Bill Mollison. GeoffLawtonOnline.com

Joel Salatin
Joel is a regenerative US farmer, author, speaker, and visionary that has been featured in dozens of documentary films. PolyfaceFarms.com

Curtis Stone
Curtis is an author, teacher, and urban farmer in British Columbia, Canada, using just a third of an acre (just over 1000m^2) to run a robust, thriving business that can be replicated anywhere. TheUrbanFarmer.co

John D. Liu
Filmmaker and director of the Environmental Education Media Project, John has documented large-scale land restoration primarily in China but many other parts of the world as well. Eempc.org

Willie Smits, Phd.
Founder of Masarang and the Borneo Orangutan Survival Foundation, Dr. Smits is restoring the rainforests and biodiversity in parts of Indonesia while trying to preserve the culture and local economies.

Rob Avis
Rob is an teacher, designer, consultant, and co-founder of Verge Permaculture with his wife Michelle in Calgary, Canada. VergePermaculture.ca

Rhamis Kent
Rhamis serves as a co-director of the Permaculture Research Institute and works internationally as a PRI-certified instructor and consultant. PermacultureGlobal.org/users/51

Grant Schultz
Grant is a farmer, teacher, consultant, and innovator who established Versaland farm in Iowa and an online nursery business. Versaland.com

Zach Weiss
Zach is a designer, teacher, consultant, and certified as a designer by Sepp Holzer. ElementalEcosystems.com

Erik Ohlsen
Erik is co-founder and director at the Permaculture Skills Center in Sebastopol, CA. He is an author, educator, entrepreneur, and innovator. PermacultureSkillsCenter.org

Cuauhtemoc Villa
Cuauhtemoc has taught in classrooms all over Sonoma county, California, representing the Sonoma Biochar Initiative. He is an educator and consultant.

Teacher's Rationale

This book is firmly rooted in inquiry-based science that engages students in improving their scientific literacy to make better choices in their daily lives. Permaculture blends many different scientific practices and increases scientific fluency as it encourages synthesis, application, and critical thinking (the highest three tiers of cognition according to *Bloom's taxonomy*). Permaculture uses this blend of disciplines to better explain and interact with the natural world which is the role of science first and foremost. It uses hands-on problem-solving with science to connect all learning modalities with the skills, understanding, and context for that skill or process. It is both collegiate and accessible to advanced high school students.

In this book molecular biology, environmental science, physics, meteorology, life science, social science, physical science, agricultural science, chemistry, earth science, geology, and the history of science are all blended in a format designed to invite the reader or student to become scientists in their own lives. To experiment, test, measure, reflect, theorize, and apply science in their daily lives to make their lives better. The ability to see the connections between sciences and use different sciences interchangeably and in combination is an incredibly important skill.

This book easily qualifies as one that fulfills standards throughout both the National Science Education Standards (NSES) and Next Generation Science Standards (NGSS), but it does not easily fit into just one scientific genre. It involves history, mathematics, social studies, government, health, art, architecture, landscaping, wood working, metal working, reading, and writing. This book embodies the potential for a science class that introduces all the sciences in a balanced, true-to-life format that favors all learning modalities, making science even easier to love and apply. It is the ultimate science elective for those who have struggled with traditional science classes, and judging by the national statistics, it is much needed. Permaculture can be the catalyst for re-engaging millions of students in science class throughout the United States and the world.

For school gardens, this is how you can justify having a garden and greenhouse. This is how they can be leveraged into productive and profitable additions to the school as well as educational components of nearly every classroom.

About the Author

Matt Powers is a teacher, author, videographer, seed saver, plant breeder, researcher, gardener, and family guy. Matt and his family lived in the foothills of the Sierra Nevada mountains in the Central Valley California almost a decade - before that Matt and Adriana lived in NYC where Matt was a professional musician.

Other books by Matt Powers

The Permaculture Student 2 The Workbook
The Regenerative Career Guide
The Permaculture Student 1
The Permaculture Student Workbook
The Magic Beans
Permaculture for School Gardens
The Forgotten Food Forest
and Translations

Courses with Matt Powers

The Advanced Permaculture Student Online
The Permaculture Student Online K-12
The Permaculture Student Online - for Adult Learners
Permaculture Gardening with Matt Powers
Best Practices for Regenerative Entrepreneurs & Educators

CPSIA information can be obtained
at www.ICGtesting.com
Printed in the USA
LVHW071931210620
658653LV00008B/276